# THE HIMALAYAN ENVIRONMENT ISSUES AND CHALLENGES

Eds.
Puran Ch. Pande
Ravindra K. Pande
Rajnish Pande

1998

DAYA PUBLISHING HOUSE
DELHI-110035

ISBN 81–7035–187–1


PURAN CH. PANDE ( b. 1942- )
RAVINDRA K. PANDE (b. 1957- )
RAJNISH PANDE (b. 1962- )



*Published by* : **DAYA PUBLISHING HOUSE**
1123/74, Deva Ram Park
Tri Nagar, Delhi-110035
Phone : 7103999
Fax: (011) 7198902

*Laser Composed by* : **DELITE PRINTERS**
Delhi-110035,
Ph: 7182135

*Printed at* : **CHAWLA OFFSET PRINTERS**
Delhi

**PRINTED IN INDIA**

# Preface

In the Himalaya the deteriorating quality of environment and the consequent imbalances in ecosystem has assumed huge dimensions and is becoming a matter of grave concern. The resource potentials are hampered by 'ecological handicap'. We have adequate testimony of these issues and challenges and the same have been discussed in this volume. Though, micro-regionally varying in their nature and importance, socio-political and geo-economic problems are not alone in hampering eco-system of Himalaya, specially in the rugged and fragile inner-zones—the deteriorating environmental qualities are enhanced by repugnant ecological fettles. It would, however, be wrong to regard ecological factors as the sole cause of the environmental menace. In gestalic perspective, the main ecological problems in Himalaya, must, in practice, include excessive landuse, exorbitant rate of erosion of fine soil, increased evaporation of even the lower lying soil water, further impoverishment of naturally poor, soil growing settlement pressure on higher slopes and altitudes bringing steep slopes under cultivation, decreasing share of able-bodied male population in agricultural activities, unnecessarily increasing number of workers per unit of land, uneconomic cattle pressure, small and sub-economic size of holdings, over grazing, deforestation and so on. Some of these problems are discussed in this volume.

The present volume focuses on the unity amidst diversity of the information base in the Himalayan environment. However, the papers incorporated in the volume represent an extremely modest selection of articles covering some vital issues related to the environment of this young and fragile mountain. The editors are well aware of the gaps that have not been covered herein. This volume is brought to fore and need for intensified efforts in the sustainable development of this region about whom even the government share the common belief that the full potential of the resources have yet to be tapped and ecologically monitored.

The editors are deeply indebted to a number of distinguished people who have so unselfishly assisted in the preparation of the manuscript. These include Prof. G.C. Pande, Professor O.P. Singh, Dr. (Smt.) Neeta Pande and Dr. H.S. Bhakuni. We thank all the contributors to this volume for a timely response. We express our gratitude to several authors and publishers whose whole parts of the figures, tables, etc. are included herein.

P.C. Pande
R.K. Pande
Nainital Rajnish Pande

# Contents

# List of Contributors

Anantharaman, M.S.; Department of Geology, D.S.B. College, Dehradun- 248001.

Bawa, R.; Department of Forest Biology and Tree Improvement, Dr. Y.S. Parmar University of Horticulture & Forestry, Nauni (Solan)-173230.

Bhandari, N.S.; Department of Chemistry Kumaon University, Almora Campus, Almora-263601.

Bisht, B.S.; Pt. G.B. Pant Institute of Himalayan Environment & Development, Kosi (Katarmal).

Dangwal, R.C.; Department of Commerce, H.N.B. Garhwal University Campus, Tehri (Garhwal).

Datt, Devi ;Department of Geography, H.N.B. Garhwal University, Srinagar (Garhwal).

Ghildyal, B.P., International Rice Research Institute, Liaison Office-India, 11, Jor Bagh, New Delhi-110003.

Joshi, B.P.; Department of Medicine, C.S.A. University of Agriculture and Technology, Mathura Campus, Mathura-281002.

Joshi, Bhagwati; Department of Geography, Kumaon University, Naini Tal- 263002.

Joshi, G.C.; Department of Botany, Kumaon University, Alomra Campus, Almora- 263601.

Joshi, Jyoti, Department of Geography, Kumaon University Almora Campus, Almora- 263601.

Joshi, K.N. Department of Economics, H.N.B. Garhwal University, Srinagar (Garhwal).

Joshi, Pramila; Department of Botany, Kumaon University Almora Campus, Almora- 263601.

Kuniyal, J.C. Scientist A, G.B. Pant Institute of Himalayan Environment & Development, Dhopur, Kullu, Himachal Pradesh.

Lodhiyal, L.S.; Department of Forestry, Kumaon University, Naini Tal-263002.

Mehta, I.S.; Department of Botany, Kumaon University Almora Campus, Almora - 263601.

Negi, K.S.; Faculty of Commerce, H.N.B. Garhwal University, Srinagar (Garhwal).

Padoshi, A.R., Head of the Department of Economics, Goa University, Bambolin, Goa.

Pande, D.C.; Department of Economics, Kumaon University, Naini Tal- 263002.

Pande, D.C. ; Department of Geography, Kumaon University, Naini Tal - 263002.

Pande, H.C.; Department of Botany, Kumaon University Almora Campus, Almora-263601.

Pande, Neeta; Department of Botany, Kumaon University, Naini Tal- 263002.

Pande, P.C.; Department of Economics, Kumaon University, Naini Tal-263002.

Pande, P.C., Department of Botany, Kumaon University, Alomra Campus, Almora- 263601.

Pande, R.K.; Department of Geography, Kumaon University, Naini Tal-263002.

Pande, Rajnish; Department of Economics, Kumaon University Almora Campus, Almora-263601.

Pande. Vineeta; 261, Hari Sadan, Cheenakhan Lines, Tallital, Naini Tal-263002.

Pande, Y.C.; Department of Geography, Kumaon University Almora Campus, Almora- 263601.

Pant, G.B.; Department of Economics, Kumaon University, Naini Tal-263002.

Sarkar, A. K.; Department of Commerce, H.N.B. Garhwal University Campus-Tehri (Garhwal).

Sehgal, R.K. ; Wadia Institute of Himalayan Geology, Dehradun-248801.

Sehgal, R.N.; Department of Forest Biology and Tree Improvement, Dr. Y.S. Parmar University of Horticulture & Forestry, Nauni (Solan)-173230.

Singh, A.K.; Department of Geography, Kumaon Unversity Almora Campus, Almora-263601.

Singh, O.P.; Department of Geography, Kumaon University, Naini Tal-263002.

Singh, Raj Laxmi; Department of Geography, Kumaon University, Naini Tal- 263002.

Singh, R.P.; Department of Forestry, Kumaon University, Naini Tal- 263002.

Tewari, P.C.; Department of Geography, Kumaon University, Naini Tal- 263002.

Upadhyay, H.C., Department of Sociology, Kumaon University, Naini Tal-263002.

Valdiya, K.S.; Advance Scientific Research Centre, Jakkur Camps, P.O. Bangalore - 560064.

## Chapter 1

# Uttarkashi Earthquake of October 20, 1991 in Garhwal Himalaya : A Warning

*K.S. Valdiya*

As the Uttarkashi-Tehri region is Garhwal Himalaya slipped on one or the other of the active faults, it was thrown into violent tremors in the wee hours of October 20,1991. This earthquake... which violently shook a large part of north-central India...killed over 1000 people, wrecked more than 28,220 homes and damaged another 20,658 houses. The zones of damage and destruction follow belts of active faults, which are many. This is quite understandable, for this narrow belt fault-ridden, densely populated Lesser Himalayan terrain is sliding under the ever-snowy Great Himalaya on what the geologists call the Main Central Thrust (MCT). It was noticed that south of the lower boundary of the MCT zone, the ground was heaved up and down by east-west surface waves, causing extensive destruction and damage to old structures built on terraces made of soft, unconsolidated sediments deposited geologically recently by the rivers on their banks. North of the MCT boundary, in contrast, the north-south surface waves wrecked houses constructed on very loose aggregates of debris deposited by ancient landslides and mountain torrents on steep slopes.

Everywhere in the quake-ravaged belts, there is a perceptible change in the spring discharges...some springs have gone dry or have diminished discharges, while others are emitting more water. Mountain-tops have cracked and crumbled in the severely rocked segments that slipped. However, the disastrous impacts of the changes in the circulation of groundwater and mass-movements on ruptured slopes will be evident only after the rains. The hazards of landslides and ground-subsidence would be much more severe and distressing in couple of years to come.

The perceptive seismologists believe that not all the stored strain has been relaxed by the present earthquake of magnitude M 6.6 and that the region is likely to experience a trauma of another earthquake of much larger magnitude (of the order of M 8 or above) in not a distant future. It will therefore be prudent to prepare for coping with the hazards of future earthquake of much higher magnitude.

The controversial Tehri Dam is located in this seismicity-prone region close to the active Shrinagar Thrust, which has registered a few pulses of uplift in the geologically recent past. Its tremendous height (260.5 m) is bound to induce earthquakes, when water is impounded to high levels above 100 m.

**Much Faulted Terrain**

As already pointed out the Lesser Himalayan terrain is sliding, of course episodically, under the formidable Great Himalayan domain. The plane of sliding is known as the Main Central Thrust (MCT) (Fig. 1) Inclined 15-20° and 30-40° northwards, the MCT zone is made up of succession of severely sheared and considerably crushed rocks, split up by a large number of thrust planes into many tectonic slabs. Repeated movements on these faults have not only weakened the rocks considerably, but are progressively lifting up the Great Himalaya at the rate of 0.7 to 1.1 mm per year. The continuing tectonic tumult at depth is mainfest in periodic shaking, trembling and twitching of the fault-ridden terrain as evident from great many smaller earthquakes of magnitude 3 or less recorded in the Uttarkashi region (Fig. 1). Many of the faults, from which earthquakes originate, are active and have registered vertical as well as horizontal movement in the geologically recent times.

It appears that on October 20 at the depth of about 12 km, there was sudden slipping on one or more of these active faults immediately beneath the MCT zone boundary. Rocks of other faults zones (Fig.1) in the proximity may have snapped, generating violent shock waves.

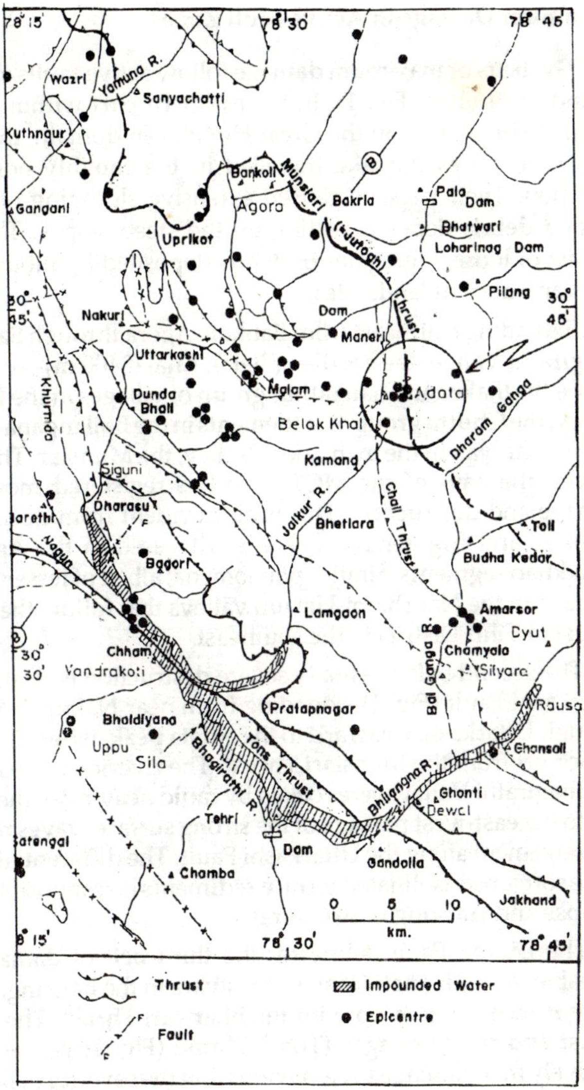


Fig.1: Distribution of epicentres of minor earthquakes of magnitude M3 or less in the much-faulted terrain of the Uttarkashi-Tihri districts in Garhwal. Note the relationship of microseismicity with the faults and thrusts.

## Maximum Damage in Active Fault Zone

The belts of maximum damage follow active faults and active boundary thrusts (Fig. 1). In the northern part within the main Central Thrust zone in the Great Himalayan domain, the surface waves of the earthquake travelled in the roughly north-south direction. Their passage caused extensive slumping of the fan-shaped debris deposits resting on the steep slopes. These fans consist of loose aggregates of debris deposited by mountain torrents and ancient landslides.

Extending southeast from Bankoli-Agora through Sainj in the Bhagirathi Valley, the saddle of Belak Khal (3400 m)-Adata (3424 m) the Budhakedar-Thatikathur group of villages in the Balganga Valley, the Ghuttu group of settlements in the Bhilangana Valley to Triyuginarayan in the mandakini Valley, the Munsiari Thrust that defines the base of the MCT zone has registered most severe damage and destruction. This is also evident from deep ground ruptures and frightening rockfalls witnessed in the Agora and Belakkhan segments. Similar phenomena, albeit of lesser severity, are seen in the Nailchami-Helaun valleys that follow the Batwari Thrust in Tihri District to the southeast.

The second belt of equally severe destruction is related to the Uttarkashi Fault (Fig. 1) extending from near Nakuri in the west through Uttarkashi eastward to the Adata peak, where it seems to branch off from the Munsiari Thrust. The riverine terraces lining the Bhagirathi Valley were rocked by rapid heaving of the ground due to the east-west passage of the strong surface waves related to the movement along the Uttarkashi Fault. The differential settling of the soft, unconsolidated terrace sediments is responsible for the collapse and cracking of structures.

The Dunda Fault delineates the third belt of damage. It is traceable through Thati Dhanari to Silyara in the Balganga Valley. There it seems to join up with the Bhatwari Thrust. The Berinag Thrust and the Shrinagar (Tons) Thrust (Fig. 1) registered but relatively mild damages. It is obvious that the severity of the earthquake shocks diminished progressively southwards.

## High Mortality

Even though a moderate earthquake of M 6.6., the Uttarkashi event caused death of over 1000 people. This happened because the inmates got trapped inside the collapsing masonry houses, as a result of the doors getting jammed or stuck within their distorted door frames. This distortion occurred owing to heaving and swaying of those buildings which were oriented in the travel directions of the ground waves——north -south in the MCTzone of the Great Himalaya and east-west in the Lesser Himalayan terrain.

**Why Earthquakes in Uttarkashi Region**

A careful look at Fig. 4 would demonstrate that the Himalaya mountain is being pressed by the northward drifting Indian Subcontinent. The Aravali Ridge of the Indian subcontinent is strongly and persistently pressing the Himalaya-rather prodding the mountain like a colossal battering ram (Fig. 2). Therefore, considerable stresses are accumulating in the body frame of the Himalaya in front of the prodding ridge. The earthquake ravaged Uttarkashi region lies directly in the line of the Aravali Ridge. The last major earthquake of M 7.5 occurred 164 years ago in 1828; and not all the accumulated stresses (of the last 164 years) was released by the October 20 earthquake of M 6.6. It is implied that a significant amount of the stresses is still left there, waiting to be released by a future earthquake, whose magnitude has to be M 8 or higher. The future earthquake will cause much severe damage and the extent of destruction and damage would stretch far south beyond the Shrinagar Thrust.

The Shrinagar Thrust, it may be mentioned, is an active fault. This means that it has registered pulses of movements including uplift in the geologically recent past-possibly also in the last eleven thousand years or less. This is quite evident from a number of geomorphological features in the stretch where the Bhagirathi River flows south of the Shrinagar Thrust after traversing it. There are six levels of river- made terraces within this stretch compared to the normal three elsewhere. The old Bhagirathi, which usually flows in its wide mature valley with gentle slopes, shows incised meandering and canyon course south of the Shrinagar Thrust. The north-flowing streams have extremely steep gradient and tumble down into the Bhagirathi. There are huge fans of debris avalanches, triggered presumably by earthquake related the uplift of the southern massif along the Shrinagar Thrust.

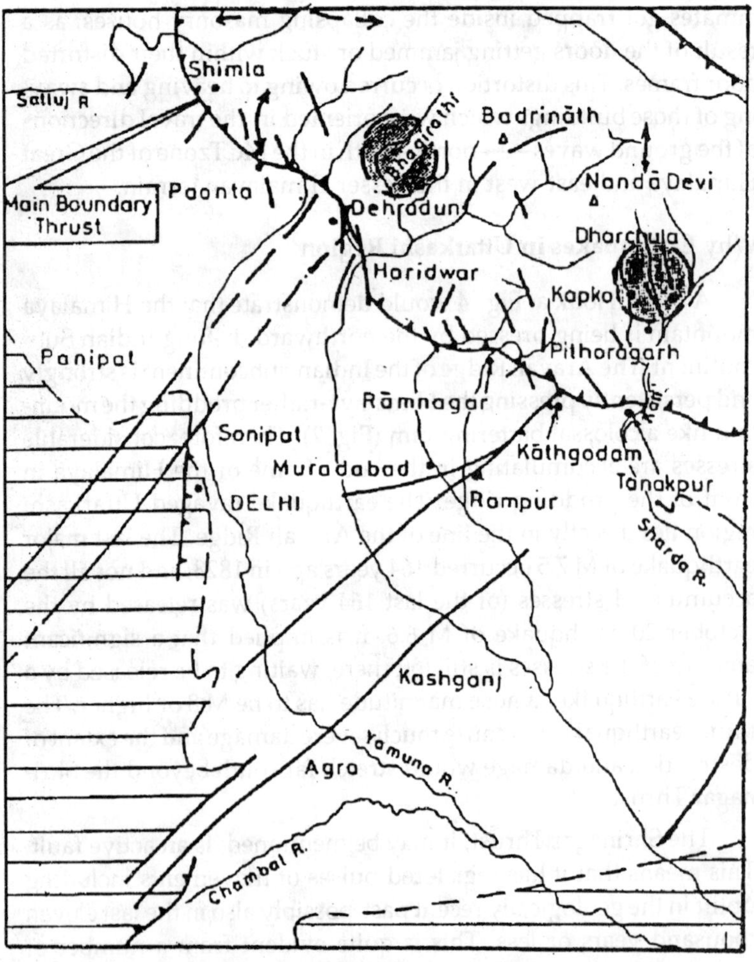


Fig. 2 : The Aravali Ridge is prodding the Himalayan Ranges like a huge battering ram. Consequently, very strong stresses are generated in front of the promontary, such as the quake-ravaged, Uttarkashi-Tihri area. Another candidate for further earthquake of larger magnitude is Kapkot-Dharchula are in northeastern Kumaun—directly in line of the active though hidden, Muradabad Fault.

The controversial 260.5 m high Tehri Dam will come up on a location about 4 km south of the active Shrinagar Thrust. One cannot help being very apprehensive about the safety of all kinds of structures (dam, buildings and bridges) and the wellbeing of the people in the proximity of this active fault in the region likely to be rocked by major earthquake in future. It would not be fair to compel the people of Tehri district to bear the entire ecological and social cost of the Rs. 3008-crore project. Even if there is in recurrence of natural earthquake of higher magnitude, the induced seismicity after full impoundment of water behind the dam will make the lives of people very hazardous and miserable. There must be people in the government of the people for the people who would care for the 70,000 people living in the valleys of the Bhagirathi and Bhilangana, whose fate will be doomed when their villages are inundated.

# Chapter 2

# A Geomorphological Appraisal of Landslides in Garhwal and Kumaun Himalaya (U.P.)

*M.S. Anantharaman*
*R.K.Sehgal*

Landslides and allied phenomena in Garhwal and Kumaun Himalaya are manacing to the lives of the inhabitants of these areas directly or indirectly by the way of disruption in communication loss of agricultural lands, loss of life, property, cattle etc. Landslides and rockslides are so common here that hardly any year passes without a few catastrophies.

Catastrophic landslides and rock debris along with the subsidence of hill slopes around this region now had become a matter of anxiety because of the growing development in this area. The magnitude of the damage caused every year by the sudden and unexpected landslides is very large and is becoming costlier to tackle the problems associated with these which become aggrevated during and soon after heavy rainfalls. To a layman deforestation and disturbing the existing hill slopes through construction of roads etc. appear to be two of the main causes which might be accentuating the problem but infact there are many other causes which are responsible for landslides. It has been observed that each landslide is peculiar by itself and therefore, it is most essential to understand the mechanism of the initiation and propagation together with the causes responsible for these in order to evolve a rational solution.

### Causes of Landslides

As stated above each landslide is peculiar by itself. Neither all landslides have the same cause nor all the causes are needed for one landslide. Some landslides occur due to causes depending upon the local conditions, climate, geomorphology and geology.

Some of the most common causes for landslides are discussed here :

### Slope Failure

The slope as a factor of gravity has resulted into many prominent landslides of Garhwal and Kumaun Himalaya. The hill side slope or valley side fails die to certain geological, climate and biotic factors like highly folded and fractured nature of rocks, extremes conditions of temperature, destruction of forests etc. The landslides of Rudraprayag-Tilwara-Mayali area of Garhwal Himalaya are the result of such a cause along with the action of water (Negi, Pandey and Sinha, 1982).

### Slump Type of Sliding

The slump type of sliding takes place due to saturation of debris material and on account of toe erosion by the tributaries, thereby increasing the shear stresses and thus decreasing the shear strength due to the development of pore pressure caused by saturation. On super-saturation and movement of the debris material, the soil structure of the clay matrix gets disturbed and the upper part of the slide gets transformed into a landslide. This phenomena is well exhibited by the Hilang—Dungri—Waresi landslides near Joshimath, Chamoli district where the landslide has occupied an area of about 13 Sq. Km. and Nandprayag landslide on Hardwar-Badrinath Road. (Krishnaswamy and Jain, 1975).

### Case History of Chamoli Landslide

In 1972 a group of six villages namely (i) Hilang- Dungri—Waresi (ii) Nandprayag (iii) Munyati — Gandasu (iv) Thirpak (v) Rangoti — Golim and (vi) Manasu were affected by landslides. Out of these the Hilang — Dungri — Waresi area was very badly affected and it was considered economically impractisable to retain the habitat there. It covered an area of about 13 Sq. Km. (Fig. 1 & 2).

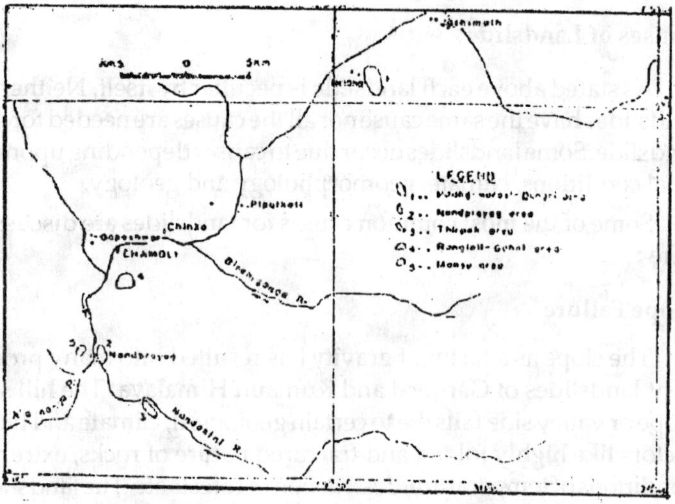


**Fig 1. :** Index Plan of Slide Affected Areas of Chamoli District (Modified after Krishnaswamy and Jain, 1975)

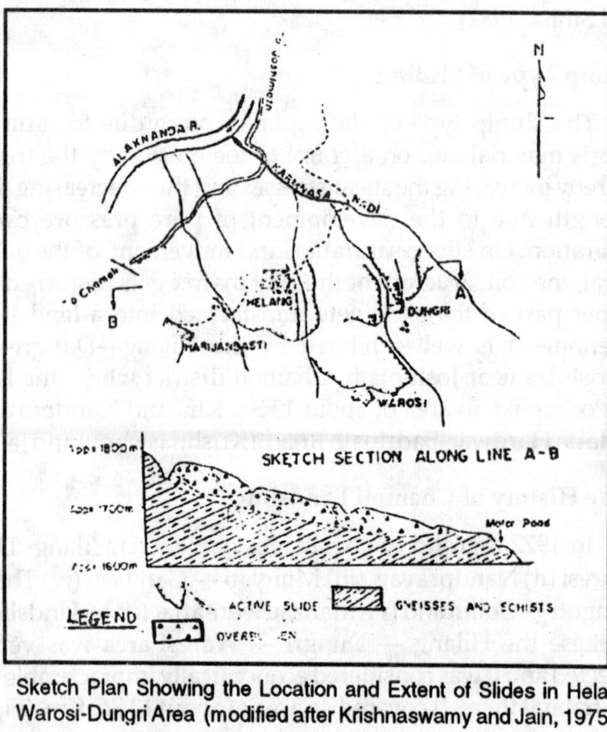


**Fig 2. :** Sketch Plan Showing the Location and Extent of Slides in Helang—Warosi-Dungri Area (modified after Krishnaswamy and Jain, 1975)

## Case History of Nandprayag Landslide

This landslide is located around 216 km. post on the Hardwar-Badrinath road and is classified as a slump slide. It covered an area which is 200m. long and 160 m. wide. The slide zone is located in an embayment of a tributory of the river of Alaknanda and is bounded on two sides by small drainages.

The reason for these landslides is that the talcose rock unit, on weathering and saturation losses cohesion and tends to flow out. The outward flow of the saturated sericitic talcose schist removes the toe support of the overlying jointed gneisses also and move down, thus enlarging the zone of the slide year after year (Fig. 3).

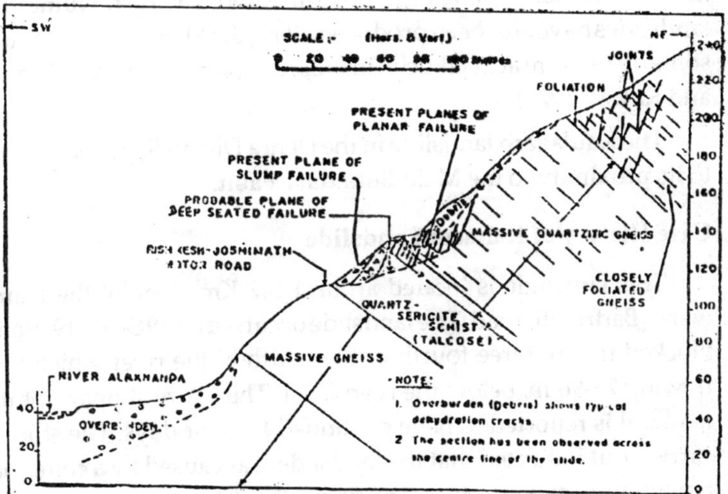


Fig. 3.: Geological section Across the Nandprayag Landslide (modified after Krishnaswamy and Jain, 1975)

## Severe Crushing

In the Nalota-Kalagarh landslides on Dehra Dun-Mussoorie Road, the rocks of existing slide area might have been crushed severly and disturbed due to tectonic activities. The rain water gets an easy access into the pores and thus removal of the fine particles from the weathered portion. This leads to the rock falls from the steep slope and accumulation of the debris at the lower level (Prasad and Mathur, 1976).

**Extensive Toe Erosion**

The extensive toe erosion by the river at the point of the bend in its course followed by gradual failure of the overlying material had caused the Kaliasor landslide located around 147 Km. stone of the Hardwar—Badrinath road at the Kaliasor village (Krishnaswamy and Jain, 1975).

Geological thrusts (disturbance zones) also contribute to the occurrence of landslides. The Kaliasor landslide is due to its being in the disturbance zone of the North Almora Thrust. Numerous landslides have also been produced along the Main Central Thrust, some of them are at Belakuchi, in the area north of Ghansali (Prasad and Rawat, 1979).

The Kaulagarh landslide in the Dehra Dun valley is partly due to its proximity to the Main Boundary Fault.

**Case History of Kaliasor Landslide**

This landslide is located around 147 Km stone of the Hardwar—Badrinath road. The landslide occurred on 19 Sept., 1969 and blocked nearly three fourth of the width of the river which was flowing 75-90 m. below the road level. The active slippage of the material is reported to have continued for four days. The studies carried out indicated that the landslide was caused by a combination of three major factors. The first is the extensive toe erosion by river at the point of bend in its course followed by gradual failure of the overlying material. The second factor was the heavy precipitation in the area during the failure, which saturated the sheared and the closely jointed rock mass. The third factor was probably the inherent insatisfactory of closely jointed and highly sheared rock-mass itself (Fig. 4).

**Tectonic Activity**

The close connection between the causes of the tectonic activity

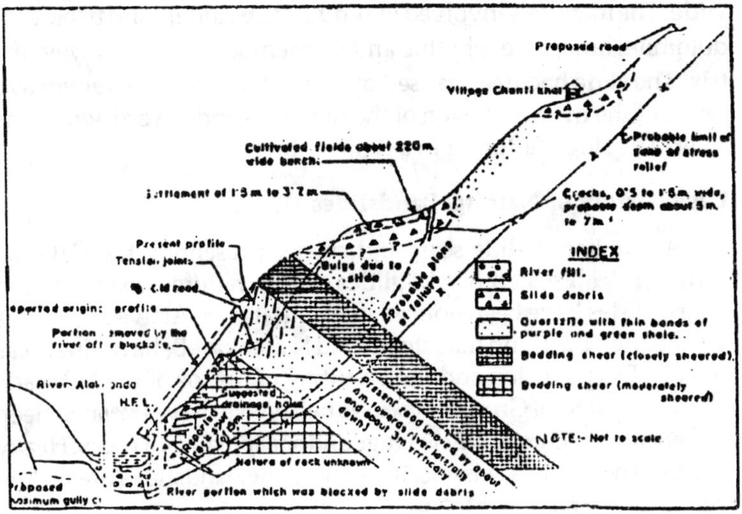


**Fig. 4 :** Sketch Geological section of the slide at km. 147, Kaliasor, Hardwar-Badrinath Road. (modified after Krishnaswamy and Jain, 1975)

of individual massifs, the complicated folding, nappe structure and the quantity and duration of precipitation are combined causes which measure the degree and extent of landslides in Garhwal-Kumaun Himalaya (Jan Kalvoda, 1971).

Pandey (1974) assumed the movement along the shear zones and development of convex slopes leads landslides in Nainital region. During heavy rains erosion along the shear planes, under-cuttings at the base of the hills, influenced by gravity causes major landslides.

The nature of the joint patterns in the area also cause landslides. For example in Nainital area joints are common. Joints are dipping (45°-90°) in the direction (NW-SW, E-W, N-S, NW-SE). Water circulating along the joint plane during or after heavy rains helps in loosening the blocks and thus causes landslide (Pandey 1974).

**Case History of Sial Landslide**

This slide is located at 249 Km. stone on the Hardwar-Badrinath road. It covers an area which is 180 m. long and 300 m.

wide. The materials involved in sliding are boulders and pebbles of dolomite, limestone, phyllite and slate embedded in a clayey matrix. The slide had been caused by reduction in shear strength as a result of the oversaturation of the materials and toe erosion by the river. The slope failure is in a typical slip circle.

### Case History of Nainital Landslides

A number of slide scars and cracks, presenting translational slides or deep seated blocks glides, slump failures, rock falls and debris slides have been noticed in Nainital area. Active slide areas are the Sher-Ka-Dand hills slopes and Kailankan-Balia ravine areas (Fig. 5). Deep erosion and undercutting of the toe of the hill made up of jointed Krol Group of rocks is believed to have initiated deep seated movements along cleavage and shear planes of rocks. Heavy rainfall and being a seismically active zone, landslides are prominent in Nainital (Fig. 5).

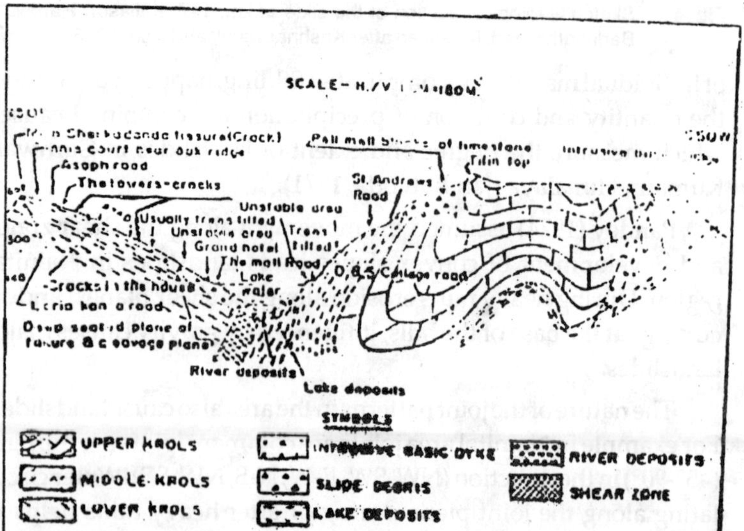


Fig. 5. : Geological Section Along Sher Ka Danda to Land S end Via Grand Hotel, Nainital. (modified after Krishnaswamy and Jain, 1975)

### Construction Activities

The construction of the roads and highways on the steep slope of mountains produces small slides. For example 30 km. north of Srinagar (Garhwal) the slope varies from 40° to 60° and the rock strata are almost perpendicular, due to this a 700 m long stretch of mountain road is continually being overwhelmed by sliding materials of limy quartizite and slates.

Excessive and arbitary destruction of the vegetative cover of deforestation is also one of the major cause for Landslides. Due to deforestation soil of the area becomes loose, fine materials is eroded, thus pore spaces are developed. During or after rains water perculate in these pores and causes mudflows. Some of the landslides in Mussoorie region are the result (Negi and Anantharaman, 1980).

**Heavy Rainfall**

Heavy rain fall is also one of the cause for landslides in Mussoorie. Landslides occur more frequently in Garhwal-Kumaun region during heavy rainfall or monsoon. The unprotected rain soaked soil often with clayey, talcose, quartzitic material becomes highly viscous and slips under the gravity action, to lower levels.

**Case History of Mussoorie Landslides**

Along the Mussoorie -Dehra Dun road near Rajpur, Nalota and Kalagarh landslips are occurring. Also on Rajpur- Mussoorie road near Kulukhet recently a landslip occurred. Due to this the traffic was blocked for a week. The Kalagarh landslip is along an escarpment at the head of the Kalagarh stream, associated with formation of gullys over the cliff and the deposition of enormous talus fan in the front, also the rocks of the area are fractured slates and phyllites which during the rainy season become loosened and tumble down.

Earthquake shocks increase shearing stress along a potential surface of failure, whereas the shearing resistance remains largely unchanged. This in turn reduces the factor of slope which causes landslide. For example Nainital is a seismically active area. Also it is gathered that the earlier major landslide were preceded by earthquake (Krishnaswamy and Jain, 1975).

The mechanical disintegration of rocks by extremes of temperature and erratic and intense local precipitation on and around vulnerable slopes are also responsible for landslides and rock falls. Many small landslides in Badrinath and Kedarnath are due to this.

### Case History of Belakuchi Landslide

This slide near 253 Km. stone on the Hardwar- Badrinath road was caused by toe erosion and movement along the foliation plane of a thick talc schist band of the Garhwal groups of rocks. It was an old slide and was reactivated in July 1970 floods. As a result of this slide, the village Belakuchi was completely wiped off.

Landslides also occur when the base of the talus exists beside the river valley and it is eroded by the undercutting action of river water during the flood period as was the case in the Alaknanda Landslides of 1970 (Nityanand and Prasad, 1972).

Sometimes in an area water table start rising due to some reason. As a result of this pore pressure increases promotingly and thereby causes landslides. In Alaknanda Valley the water table is steep and slopes towards the valley. The small landslides often causes the rise of water table because of the blockage of spring orifice and due to this heavy flood result. This has a reciprocative action as due to rise of water table, pore pressure increases promotingly and thereby the landslide (Ravindra and Negi, 1979).

### Case History of Alaknanda Landslide

Due to the fact that landslides located on the bank of the river Alaknanda and its tributaries a flood of very high intensity occurred on 20th July, 1970, causing heavy damage to life and property. The landslides, rockslips and debris slide in Alaknanda valley of Garhwal Himalaya occur on account of the multiplying results of geographical factors like mass-wasting, soils-erosion, soil creeping, torrentional run off, slope failure and severe infiltration as a consequence of unscientific farming quarrying and blasting for road and dam construction, overgrazing, overfollowing, over-ploughing and deforestation. It has been observed that these geographical factors accelerate their functions where slates and phyllites have been faulted, crushed, fractured and powdered due to severe effect of the thrust zone of the Garhwal Himalaya, trending perpendicular to the Alaknanda from NW-SE (Fig. 6).

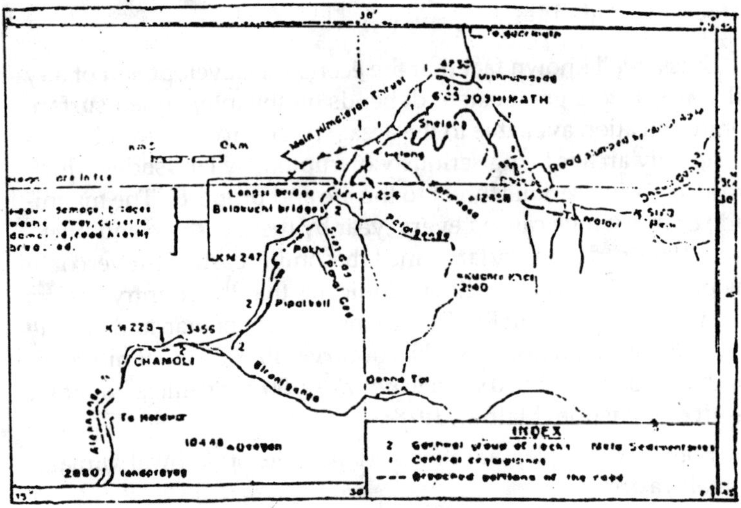


**Fig. 6 :** Plan Showing Affected portions of the roads during July, 1970 Floods, Alaknanda Valley. (modified after Krishnaswamy and Jain, 1975)

Some times water may enter into the joint planes, fractures, fissures, of the rock and due to prevailing very low temperature may be transformed into ice. Water when solidifies into iee increases in volume and this may cause the displacement of the blocks of the rocks. The ice may also act as a lubricant. Many of the small landslides near Badrinath, Kedarnath and Reni (Joshimath) have been initiated by this process.

Water-falls existing adjacent to the road create shatters in rock formations due to vibration which may result in rock falling.

Cultivation of crops like paddy and cardemom on the slopes is risky because these crops need standing water. The standing water may seep inside the ground and supersaturate the whole hill side which ultimately may result into small landslides.

The reactivation of fossil landslides are responsible for tiggering new landslides around Chamoli (Kumar and Shome, 1970).

Some of the Garhwal- Kumaun Himalayan glaciers are receding each year. Receding glaciers leaves behind huge unvegetated tracts composed of unconsolidated debris which are favourable factors for accelerating erosion and hence landslide.

## Effects of Landslides

It is a well known fact that the economic development of any hilly area is to a great extent depends on the interrupted surface communication available to the area. The surface communication in the hilly area is being seriously disrupted, by landslides which are also posing serious threat to the life and property. The magnitude of the damge caused every year by these sudden and unexpected landslides is very large and is becoming costlier to overcome the problems. A rough estimate indicates that the country suffers an annual loss of about Rs 35 crores on account of man and vehicle hours. 50,000 man hours on the country's 10,000 km. of hill road, Rs. 2 per hour and 5,000 vehicle hours Rs. 50 per vehicle hour as a result of landslide (Mathur, 1978).

Some of the unpleasant effects of landslides of Garhwal Kumaun Himalaya are :

Landslides disrupt road communication resulting in hold up of traffic, the duration depending upon the magnitude of slide, which act as a hurdle in the economic development of the area. For example recently landslide near Kulukhet (Mussoorie) caused the blockage of road for a week. Similarly on 22-23 August 1985 Kumaun was cut off from rest of the country due to landslides occurring over the highways.

Landslides cause direct and indirect loss to the public and to the administration. Direct loss consists of loss of life and injuries, damage to forest wealth, loss of property, assests and expenditure on restoration measures apart from relief measures. Indirect loss consist of additional cost of transportation, rise in price of essential commodities, scarcity conditions, loss or revenue and loss of time in all activities in all walk of life.

Villages, buildings, roads, power houses etc. located above or below the slides in the hills are in danger of being damaged or destroyed.

Landslides effect vital sectors of the national development like highways and communication routes, irrigation and power project sites, rural and urban settlement and man's ecological environment.

Landslides along with the other geomorphological setup are responsible for the evaluation of lakes around Nainital. Evidences indicate an ancient river meander in the cresent shaped Nainital - Khurpa Tal Valley which has been truncated from the original course due to subcrecent movements along the Manora thrust and the valley within the meander portion has been blocked due to huge landslides from the valley slopes leading to the formation of Khurpa Tal, Saria Tal, Sukha Tal and Nainital. It has been confirmed by exploratory drilling on the rim of Nainital lake (Hakku, Srivastava and Jaitle, 1974).

As a result of landslide on the hill side, the valuable agricultural land in Joshimath, Chamoli district is being washed away. In ordinary parlance one may not consider this to be a serious economic loss, comparable for example to a length of road or a part of a town washed away but for a poor cultivator whose fields have disappeared, it means the loss of livilihood for himself and his family.

Sudden floods in Garhwal-Kumaun Himalaya which have resulted in severe catastrophies are caused by landslides. For example - (i) A flood of very high intensity occurred in the river Alaknanda and its eastern tributary on 20th July, 1970. (ii) In 1978, massive landslides blocked the confluence of the river Bhaghirathi and the Kanauldia bed destroying over 1000 hectares of valuable forests. As a result, the famous hamlet of Gangnani near Uttarkashi was wiped out of existence. The village was covered under 8 mts of silt, debris, stone and boulders.

Landslides and rock cum snow avalanches pose problems of safe operation of communication routes, intake structures of hydel facilities and often safety of valuable monuments and temples like those of Badrinath and Kedarnath.

Sometimes due to landslides some of the important geological formations are either washed out of get hidden by the soil debris. Important fossils are eroded and important structures are destroyed.

The areas where water table is steeply dipping towards the valley, the landslides in such area are often causing the rise of the water table because of the blockage of spring orifice which ultimately results in floods.

An unmeasurable effect is the psychological setbacks which include misery and suffering of the people affected. The delays, longer routes and extra travel time, additional expenditure, uncertainty about the time of reaching destination and schedules and likely long wait on roads causes psychological effects on human beings.

Landslides along with avalanches are a constant threat to the mountaineers, trekkers and tourists visiting high areas.

**Preventive Measures of Landslides**

Landslides are problems that if not attended grow with time. It is therefore, important that they should be tackled at the earliest opportunity. Moreover, every landslide is an individual problem, despite common initiating principles. The solution to each problem has to be determined for each individual site, though the solution may lie in a combination of a few of the well established solutions.

Some of the general preventive measures for landslides are :

The upstream river should be diverted away from the slide area and underground drainage should be adopted for small channels and streamlets. Adequate facilities for the disposal of rainwater are to be done by making channels in the slope and at the toe of the slope.

Unloading of the slope above the plane of slum failures and if this is not possible a good lateral support should be built.

High power explosions should not be used for road and dam constructions.

Cribs made of timber, concrete and filled with coarse rocks should be used as retaining walls.

Dressing of slope in accordance with the angle of rest of the constituent materials should be done. This is done by cutting the slope backwards and thus reducing the gradient. Concrete gascade are made on and along the slope to dissipate the energy and reduce the velocity of the swift water.

Stout embankments or breast wall back packed with boulders/gravels should be constructed at the toe of the slopes and on the upper side of the roads. Heavy toe buttress of complicated earth is made for lateral support of the slope.

Deforestation, overlapping and overgrazing and bare-fallowing should be checked (Saxena, Anantharaman and Pandey, 1979).

Afforestation should be done on large scale. Moreover, horiculture should be given more importance than agriculture in hilly regions (Anantharaman, Saxena and Pandey, 1982). Large scale cultivation of *Agave sislana, Vetiver* and *Khus* should be done along the slopes (of hills prone to landslides) as these are very good soil binders.

The dry and perennial steam flowing accross the roads should be allowed to drain through the cement pipes or iron pipes.

Rock bolting is one of the easymethod that can be adopted in field to prevent rockslides. In this method, holes are drilled at suitable intervals in such a manner that the hole passes through two or more chunks. Then drill rods or bars of appropriate diameter are inserted in these holes. This will prevent rocks from sliding. The bars, underpinning and roof bolts are put to prevent rocks falls in sedimentary rocks which are plan of fissuration and igneous rocks with one or two plane of fissuration.

The strength of the moisted soil should be increased, for example, by injecting the cement or hydrated lime slurry into the soil mass under pressure forces the grout through paths of least resistances moving along fractures, bedding planes, sand lenses seams, roots and fractures created by jetting slurry. Grout can be lime, cement, sodium silicate solution, bitumen emulsion etc.

The use of Geo-textile along the slopes can prevent the soil erosion long the slopes. Geo-textile is a mesh of jute spread on slopes with grasses and other soil binding plants.

The landslides can be predicted by the study of aerial photographs, landsat imagres and rainfall data. The stereoscopic study of the aerial photos give a 3-D birds view of the entire landscape. Many fractures, not visible from ground view are apparent. The appreciation is much better than from the study of topo-maps. Probable nature of the rock types, the joint pattern, slope distribution, drainage characteristics, flood effects, landuse vegetation patterns are all deciphed from the aerial photos and landsat imagres.

Since Himalayan terrains are composed of different rocks varying in structural competency clothed upon by varying surface mantle and superimposed by the drainage network. Moreover, it is sculptured by the glacial, periglacial and fluvial agencies resulting in varied distribution of surficial deposits, therefore, the knowledge of geology and geomorphic processes will help in understanding the intricacies of landslides.

On existing roads, a detailed geological log and strip geomorphic map would greatly help in understanding the problems of landslides.

Unscientific farming on valley slopes should be avoided.

Cooperation among the geologists and engineers can check over the problems of landslide easily in an area.

The concerned state should establish landslide protection and control agency for initiating projects for detailed study of slides and their economic rehabilitation.

The department of forest, agriculture, roads, mines, quarries etc. who derive benefit from the development of hill areas should year mark certain portion of their budgets for investing in studies of slope stability, slope protection wall and their observation.

Determination of geohydrological environments of selected areas and data on watershed management should be taken into consideration.

**Acknowledgements**

Grateful thanks due to Dr. V.C.Thakur, Director, Wadia Institute of Himalayan Geology, Dehra Dun and Dr. B.B. Raizada, Principal, D.B.S. College, Dehra Dun for providing necessary facilities. Thanks are also due to Dr. A.C.Nanda, Scientist, Wadia Institute of Himalayan Geology, Dehra Dun and Dr. P.B. Saxena of Geography Department, D.B.S. Colleges, Dehra Dun for their constructive remarks in the preparation of this paper.

## References


Anantharaman, M.S., P.B. Saxena, and B.K.Pandey, 1982. A Study of Soil Morphogenesis, its Depletion and Conservational Measures in the Dehra Dun Valley (Garhwal Himalaya). *Him. Geol.*, Vo. 12, : 225-235.

Bloom, A.L., 1978. *Geomorphology - A Systematic Analysis of Late Cenozic Landforms.* Prentice- Hall, Inc., Englewood Cliffs, New Jersey : 1-510.

Chansarkar, R. A., 1975. Geologic and Geomorphic factors in Landslide Investigations. *Pro. Indian Soc. Engg. Geol. : 66-69.*

Hukku, B.M., A.K.Srivastava, and G.N.Jaitle, 1974. Evolution of Lakes around Nainital and the Problem of Hillside Instability. *Him. Geol.*, Vol. IV (1), : 516-534.

Kalvoda, J. 1972. Geomorphological studies in the Himalaya with special reference to the Landslides and Allied Phenomena. *Him. Geol.*, Vol. 11 : 301-316.

Krishnaswamy. V.S. and M.S. Jain, 1975: A review of some of the Major Landslides in Northern and North-western Himalaya. *Prov. Indian Soc. Engg. Geol :* 3-40.

Narayanan, E.V., 1975. Hurdles on Himalayan Highways. *Proc. Indian Soc. Engg. Geol* : 51-53.

Negi, S.S. 1982 (a). Causes and Effects of Deteriorating Himalayan Environment: A Preliminary report. *Jour. Env. Sc.* : 47-50.

Negi., S.S., 1982 (b). *Environment Problems in the Himalayas*, Ind. Book Co., Dehra Dun: 1-187.

Negi. S. S., & M.S.Anantharaman, 1980. Landslides—The Tragedy of Garhwal Himalayas. *Him. Time.* Vol. XXXII (19): 2-4.

Negi, S.S., B.K.Pandey, and A.K.Sinha, 1982. Slope studies in Rudraprayag.- Tilwara- Mayali area of Garhwal Himalayas with special reference to the slope stability, *Him. Geol.*, Vol. XII : 330-344.

Pal, D. and S.S. Merh, 1975. Geomorphic Evolution of the Nainital area. Kumaun Himalaya. *Him. Geol.*, Vol. V : 327-337.

Pande, J.C., 1974. Tectonic Interpretation of the Geology of Nainital Area. *Him. Geol.* ,Vol- IV (1): 532-546.

Prasad, C. and H.N. Mathur, 1976. Study of Landslips Along the Main Boundary Fault Near Rajpur, Dehradun. *Him. Geol.* Vol. VI: 517-525.

Prasad, C. and G.S. Rawat, 1979. Bhagirathi Flash Floods. A Geomorphological Appraisal. *Him. Geol.*, Vol. IX (2): 734-743.

Ravindra R. and R.S.Negi, 1979. Observations on Alaknanda Tragedy. *Him. Geol.* Vol. 1X (2) : 724-733.

Saxena. P.B., M.S.Anatharaman, and B.K. Pandey, 1979. A Resource Appraisal of Springs of Krols in Dehra Dun Valley (Garhwal Himalaya). *Him. Geol.*, Vol. 1X (II): 845-853.

Saxena, P.B., P.C.Bahukhandi, and B.K.Pandey, 1979. A Study of Landsliding and its Land Depletion in the Alaknanda valley of the Garhwal Himalaya. *Him. Geol.* Vol. IX (2): 716-723.

Sharma, H.S., 1981. *Perspectives in Geomorphology, Recent Trends.* Concept Publishing Co. New Delhi, Vol. I-IV: 1-432.

Subramaniam, V. and R.A. Dalawi, 1980. Some Aspects of Stream Erosion in the Himalaya. *Him. Geol.*, Vol VIII (2) : 822-834.

Valdiya, K.S., 1980. Geology of Kumaun Lesser Himalaya. W.I.H.G. Publication: 1-291.

# Chapter 3

# Anthropogenic and Technogenic Landforms and their Effect on Human Life in a Lesser Himalayan Drainage Basin

*Jyoti Joshi*

Man has profoundly altered the earth's land surfaces by deforestation, crop cultivation and urbanization (Strahler 1977). Man's activities have been responsible for the production of highly distinctive landforms of both erosional and depositional origin (Bradshaw, et. al., 1978). Gulkind (1953) saw man and nature as twin agents in shaping the face of the earth. Bull (1975) considered man's impact on the environment so profound that one must consider man an independent variable in many geomorphic systems. Regarding the importance of human activities in geography, Dylik (1957) has stated that "the earth surface whose diversification constitutes the object of geomorphological studies, is the meeting point of the lithosphere, hydrosphere and atmosphere. Furthermore, it is closely related to the biosphere... Landform surfaces and their elements are ... also surfaces of economic exploitation".

The last decade had witnessed great concern over the extent of man's impact on environment. In physical geography, this concern has been expressed by studies in which the earth surface has been changed directly and indirectly by man, and in geomorphology it has been registered by an increased interest in anthropogeomorphology (Gregory and Walling, 1973). Gregory and Walling (1973) have further stated that the major changes in the drainage basin are those expressed in modifications of the basin character, particularly its landuse cover; in changes of river courses arising from dam and reservoir construction, conservation practices and irrigation schemes; and in changes of water and sediment distribution by human use of drainage basin water derived from rivers and ground water sources.

In the Himalayan region, traditional agricultural practices, high cost of inputs, unrestricted use of grazing lands, increased deforestation activities, poor socio-economic conditions and increased defence and development activities (road construction, hydroelectric power generation, mining for minerals) have caused an alarming situation as far as the balance in ecosystem is concerned (Gupta, 1977).

Deforestation produces increased immediate runoff of summer rains, with soil erosion, gullying and landsliding and, conversely reduced river flow during the dry season (Ives, 1985). Tewari (1984) has reported that within the Indian Central Himalaya only 28 per cent land is covered by forest and only 4.4 per cent is covered by forest having more than 60 per cent crown cover. Deforestation in the Himalayan region has led towards several environmental disasters.

Road construction in the Himalayan region has led to several environmental disasters and is responsible for the development of several landforms, i.e., debris, landslides and debris chutes etc. Many of the roads constructed in the Himalayan region have been unimaginably planned and their construction is very irrationally executed, leading to destabilization of hillsides and production of colossal amount of debris. After the roads are built, they are recurrently ravaged by landslides and rockfalls during rains. Recently, Valdiya (1985) has reported that the network of roads in the Himalayan region is approximately 44,000 km. long. Each year every kilometer of these roads produces 550 $m^3$ of debris. If, on an average, 550 $m^3$/km/year of debris is generated on the already constructed roads, the 20.5 km road network in the Ninglat basin must be producing about 11275 $m^3$ of debris every year in addition to the removal of 2650 million tonnes debris at the rate of approximately 60,000 $m^3$/km of road when constructed.

The interaction of processes of mineral exploitation with the environment leads to ecological imbalance. Case studies of Mussooriee hills, Jhiroli Magnesite, Dehradun phosphorite and Sikkim base metals by Misra (1985) leads towards the conclusion that a planned and scientific apprasial of mineral resources and of the requirements of environmental preservation should be carried out.

## The Study Area

The study area viz., the Ninglat drainage basin lies between 29° 30′40″ and 29° 27′30″N Latitudes and 79° 25′ and 79° 32′30″ E longitudes in the Lesser Himalayan terrain (Fig. 1). Administratively, the study area lies in district Nainital (UP). The basin is approached by motorable roads from reach to mouth by Haldwani-Almora road constructed along Ninglat stream (Fig. 1). An unmetalled road in the south-eastern part crosses from the upper part of Kilbury Hillspur. Before the construction of the Haldwani-Nainital-Almora motor road, this area was connected by 'pacca' tracks and briddle paths. The settlements far from the motor road are still connected by these 'pacca' tracks and briddle paths. Other common means of transportation in the area are unmetalled roads and foot paths. Geologicallly the basin constitutes a part of the intensively folded and faulted Nainital massif of the Krol tectonic unit. The Nainital Hills in which the basin lies, are experiencing serious environmental hazards related to the instability of hillslopes, manifest in massive landslides and deep erosion (Valdiya, 1988).

## Anthropogenic Landforms in the Ninglat Drainage Basin

Various types of human activities in the Ninglat basin have generated a new landscape, geomorphological known as anthropogenic landscape. The major activities by which the anthropogentic landscape (which is a assemblage of landforms caused by human activities) within the study area is caused, are deforestation and agricultural activities.

### Deforestation

Out of the total study area about 25 per cent is deforested. It includes the barren hills (2 per cent) and agricultural land (23 per cent). Due to deforestation, the Ninglat drainage basin is suffering from ecological imbalance in a number of ways. The natural water

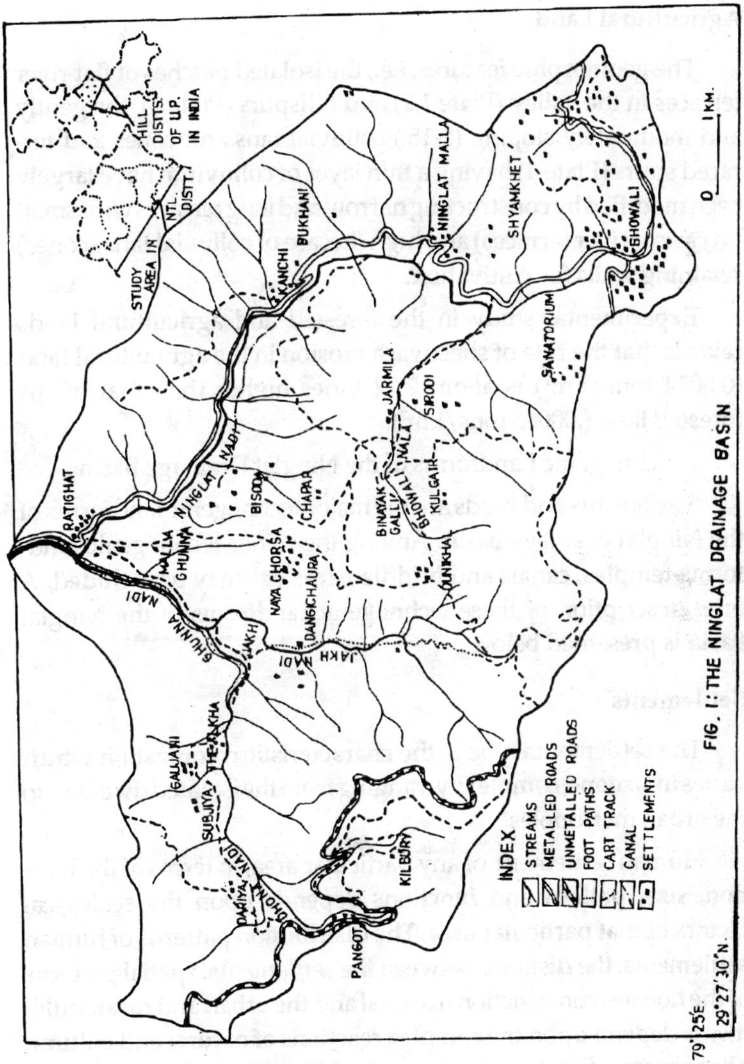


FIG. 1: THE NINGLAT DRAINAGE BASIN

sources in the hills are drying up. Studies within the micro catchments of the Ninglat drainage basin reveal that in 1986-87 the rate of denudation was about 5.7 times higher (4.36 mm/year) in the deforested land than the forested land (0.76 mm/year) (Shastri and Rawat 1992).

## Agricultural Land

The geomorphic features, i.e., the isolated patches of flat river terraces in the valley (Plate 1,2) and hillspurs (Plate 3), the gently and moderately sloping (5-15°) colluvial fans and cones and terraced spurs (Plate 4) having a thin layer of colluvium have largely been modified by constructing narrow and long terraces with small (in case of river terraces) and high (in case of colluvial fans/cones) retaining walls for cultivation.

Experimental study in the forested and agricultural lands reveals that the rate of sheetwash erosion in the agricultural land (0.0014 tons/km$^2$) is about 2.22 times higher than that of the forested land (.00063 tons/km$^2$).

Technogenic Landforms in the Ninglat Drainage Basin

Settlements and roads are the major technogenic landforms of the Ninglat drainage basin. Among the minor technogenic landforms-temples, canals and briddle paths etc. may be included. A brief description of these technogenic landforms of the Ninglat basin is presented below :

## Settlements

The settlements type is the characteristic colonization which varies in size and complexity, ranging from the isolated dwelling to the urban metropoles.

Human settlement of any particular area, in terms of the location, size, pattern and functions depends upon the ecological factors of that particular area. The distribution patterns of human settlements, the distance between the settlements, spatial patterns of the houses, construction material and the urban and rural settlements depend upon the complex reactions of natural and cultural phenomena of that particular region. These include the physical background, population and its structure, stage of technology, the level of cultural contact with other regions and ecological successions of that particular region (Kaushic, 1959).

The settlements of the Ninglat drainage basin may be grouped into three patterns. These are isolated dwellings, compact and linear pattern.

Plate 1 : An Isolated patch of flat river terrace in the valley near Talla Ninglat
Plate 2 : An Isolated patch of flat river terrace in the valley near Kainchi

Plate 3 : Flat terraces in hillspur near Thelkha village

Plate 4 : Moderately shoping colluvial fans modified by construction of terraces for cultivation near Ninglat

Defining isolated dwellings in Kumaun, Joshi et. al. (1983) have stated that the disemination of settlements in this region is largely a result of varying topographical conditions where the slope and the different elevations are usually the principal determinants of site for settlements. The conditions of slope always determines the availability of arable land. The next important fact is water, which through an uneven distribution and limited supply, greatly restricts tendencies of agglomeration. The predominance of slope over level areas has largely conditioned the fertility, depth, extent of surface and the moisture retaining capacity of the soil. Consequently, areas of fertile soil are limited and scattered in different parts. Thus, the paucity of arable land, combined with the difficulties of inaccessibility, often causes settlements to scatter. Moreover, the irregular distribution of resources in the village, such as arable land and water etc., and extreme fragmentation of holding necessitates the cultivator to live close to his fields. Such dispersed settlements are characteristics of those upland villages where irrigation is not practiced. The settlements may have a common or independent water supply but the houses are scattered throughout the length and breadth of the village and each family had its holdings near the house. Most of the cultivators have occupancy rights in the soil, hence they are free to build their dwellings anywhere on their holdings.

In the Ninglat drainage basin the isolated settlement pattern is discernible particularly in Ghunna, Purana Chorsa, Jakh (Plate 5), Budlakot, Padli, Bisoda and Dariya villages due to the collective affect of the above factors.

The compact settlement pattern is confined in the urban center where due to the availability of various civic amenities the settlement growth is rapid and gives rise to compact settlement. This settlement pattern is discernible particularly at Bhowali (Plate 7.) Arrangement of settlement along Bhowali-Almora road shows the linear settlement (Plate 6).

**Settlement Density :** Settlement density expresses number of settlements per square kilometre. Survey of Indian topographic sheets were used to obtain the settlement density of the Ninglat drainage basin.

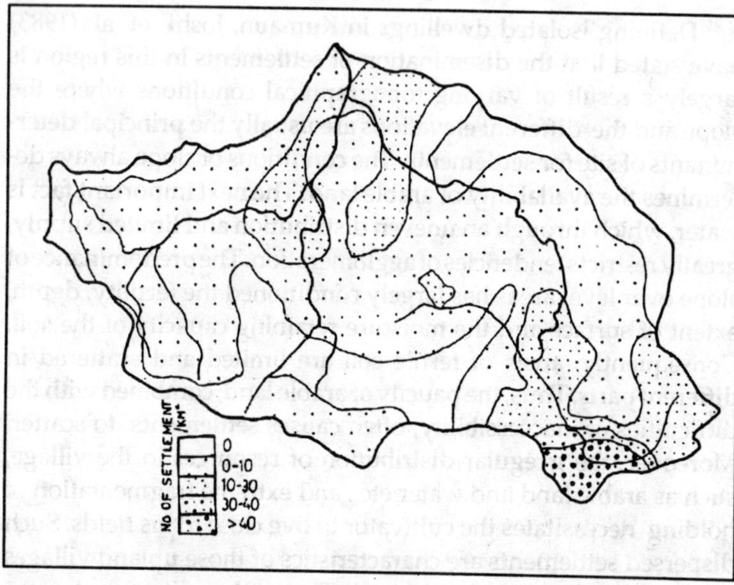


**Fig. 2 : Settlement Density Map of the Ninglat Drainage Basin Based on S.O.I. Toposheet $53\frac{0}{7}$ and $53\frac{0}{11}$**

A small part, i.e., 0.78 per cent (0.5 sq km) which lies in the southeastern part of the drainage basin falls under this region. In this area the settlement density is high (more than 40 settlements sq. km).

A negligible part of the drainage basin, i.e., 0.39 per cent (0.25 sq km) falls under moderate density where the average settlement density varies in between 30 and 40 settlements sq. km. This area includes the outer boundaries of Bhowali urban area and Ninglat village (Plate 7).

About one third of the drainage basin area falls under low density. In this region the settlement density ranges in between 10 and 30 settlements/ sq. km. Ratighat. Purana Chorsa, Budlakot, Gaugani, Thelkha, Jakh, Naya Chorsa, Padli, Bisoda, Ninglat, Kainchi, Selwakanja and Dadar villages falls under this group of low settlement density (Plate 6).

A large part of the drainage basin aera, i.e., 42.8 per cent (275 sq. km.) of the total area falls under this region. In this region, the settlement density ranges in between 1 and 10 settlements / sq. km.

Plate 5 : Isolated settlement pattern near Jakh village
Plate 6 : Linear settlement pattern along the Bhowali-Almora Road at Ratighat

Plate 7 : Compact settlement pattern at Bhowall town

Gairkhan, Bukhani, Dangichaura, Subjiya, Chaurani, Dunikhal, Bhowaligaon, Kulgair, Binayak and Galli villages fall under this group of very low settlement denstiy.

About one fifth area of the Ninglat drainage basin (i.e., 20.23 per cent (13 sq. km.) is forested either by dense or scattered forest cover which is devoid of settlements.

## Road Construction

The total length of the metalled (13 km) and unmetalled (7.5 km) roads in the Ninglat drainage basin is 20.5 km. The metalled road runs along Ninglat stream from reach (Bhowali) to mouth (Ratighat) and the unmetalled road from Nainital to Kilbury forest area.

Landslides are major hazards along these roads. Geomorphological study of one of the roads of the Ninglat drainage basin i.e., Kilbury road (Haigh, et. al., 1988) reveals that the road is cut into rocks of the Blaini-Infra Krol and lower Krol formations. The lower Krol are mainly slates with some thinly bedded marls and limestones. They grade downwards into often black, carbonaceous pyritous slates interbedded with siltstones of the Infra Krol. Beneath are distinctive purple slates, siltstones, sandstones and conglomerates of the Blaini Formation (Pangoat member). The dip of these strata is 34° (SD 15) and bearing of the dip averages 263°. The trend of the road is NW/SE and the bearing of roadcut is 70°. The road runs through reserve forest of pine, oak and rhododendron. Trees cover 57 per cent of the hillside above the road. Downslope, damage from roadbed debris, cleared off the downslope road edge had reduced to the tree cover to 36 per cent. About 39 per cent of this roadbed is affected by rockfall (Table 2). There are about 24 debris chutes per kilometer. These data provide a clear insight into the downslope impact of the road generated debris. Downslope tree and ground cover is a fraction of that upslope of the road. A very high percentage of the road is affeeted by undermining of the roadbed. These circumstances are both related to the presence of extensive loose materials created below the road by the clearance of landslide debris (Haigh et. al. 1988).

**Table 1 :** Distribution of area under different settlement density groups in the Ninglat drainage basin, Kumaun Lesser Himalayas

| *Settlement density* | *Area (m km²)* | *Area (in %)* | *Remark* |
|---|---|---|---|
| 0 | 13.00 | 20.23 | — |
| 0-10 | 27.50 | 42.80 | Very Low |
| 10-30 | 23.00 | 35.80 | Low |
| 30-40 | 0.25 | 0.39 | Medium |
| 40 | 0.50 | 0.78 | High |
| Total | 64.25 | 100.00 | — |

## Others

Apart from major technogenic landforms, there are numerous other minor man made features which constitute a considerable part of the technogenic landforms of the Ninglat drainage basin *i.e.,* temples, bridges, foot paths and pack tracks (60 km long) etc. The distribution of these minor tecchnogenic landforms in the Ninglat drainag basin has been presented in Fig. 1.

**Table 2 :** Morphometric and environmental charateristics of the Kilbury Road in the Ninglat drainage basin (Source-Haigh, Rawat and Bartariya, 1980)

| *Code* | *Variables* | *Occurrence (Units)* | *SD* |
|---|---|---|---|
| RFL | Rockfall | 39% | 23.5 |
| SLP | Slumping | 38% | 18.1 |
| DL | Debris Chutes | 4.8 per 200 M | 7.7 |
| UM | Undermining | 69% | 22.3 |
| WD | Retaining Wall (down) | 50% | 23.1 |
| WV | Retaining Wall (up) | 6% | 14.3 |
| RH | Roadcut height | 7.7 metres | 3.8 |
| RA | Roadcut angle | 74% | 6.4 |
| RW | Road width | 6.2 metres | 1.3 |
| SV | Hillslope (Up) | 35° | 9.3 |
| SD | Hillslope (Down) | 46° | 10.2 |
| TV | Tree cover (down slope) | 57° | 18.5 |
| TD | Tree cover (down slope) | 36° | 20.1 |
| VV | Vegetation cover (downslope) | 35° | 17.1 |
| RG | Regolith Deph | 1.1 metres | 1.0 |
| DI | Rock dip | 34° | 14.6 |

## Discussion and Suggestions

Study of anthropogenic and technogenic landforms in the study area suggests that anthropogenic landforms of the Ninglat drainage basin dominates all the physical processes of the drainage basin. The area is geologically sensitive and tectonically active. Therefore, the increased phese of human activities leads towards environmental hazards in the study area which can be dangerous for the inhabitants of the area. For the restoration and rehabilitation of the badly degraded basin, there is an immediate need to initiate Integrated Watershed Management Programme (IWMP).

The strategy of IWMP envisages ecological regeneration and productive approach for optimum utilization of land, water and human resources in scientific way. Under this programme, the main thrust should be on integrating various developmental activities having a bearing an ecological balance and development of potential resources. Under this programme, efforts should be directed towards insuring optimum utilization of land, water and forest resources on the one hand, and man and animal resources on the other.

**Acknowledgements**

The author is thankful to CSIR (Government of India) for financial assistance.

## REFERENCES

Bloom A.L., 1969. *The Surface of the Earth,* New Jersey : 152.

Bradshaw, M.J., 1978. *The Earth's Changing Surface.*

A.J. Appott, and A.P. Gelsthorpe, Edinburgh : 115.

Bull, W.H., 1975. Allometric Changes of Landforms, *Geol,. Soc. of Amer. Bull.*, Vol. 86, No. 2.

Demangeon. A., 1962. The Origin and causes of settlement types. *Reading in Cultural Geography,* Chicago : 516.

Dylik, J., 1975. Dynamic geomorphology, its nature and method, *Bull. De. La. Soc. Des. Sciences Ed. Des. Letters De Lodz Casee III, VIII* (12) : 1-42.

Gregory , K.J. and D.E. Walling, 1973. *Drainage Basin : Form and Process : A Geomorphological Approach,* Arnold, London: 400-407.

Gupta, R.K., 1977. Watershed - the right approach to eco-system management in the Himalayas, *Dialogue,* 76/77 : 142-154.

Gupta, R.K. and Y.K. Arora. 1985. Technology for the rejuvenation of degraded lands in the Himalaya, In *Environmental Regeneration in Himalaya : Concept and Strategies* (ed.) J.S. Singh, CHEA and Gyanodaya Prakashan, Nainital : 169-186.

Haigh, M.J., J.S. Rawat, and S. Bartariya, 1988. Environmental correlation of landslide frequency along new highways in the Himalaya : Preliminary results. *CATENA,* B-3302: 39-553.

Ives, J.D., 1985. Himalayan regeneration : an overview what are the problems and how can they be tackled? In Environmental Regeneration in Himalaya: Concepts and Strategies (ed.) J.S.Singh. CHEA and Gyanodaya Prakashan, Nainital : 12-38.

Joshi, S.C., D.R. Joshi, and D.D. Dani, 1983. *Kumaun Himalaya : A Geographical Perspective on Resource Development Gyanodaya Prakashan, Nainital* : 324.

Kaushic, S.D., 1959. Habitations in Dehradun Valley. *Geog. Review of India,* 21, No. 2: 4-6.

Kaushic, S. D., 1988. Human Geography. Rastogi Pub. Meerut : 458-503.

Misra, R. C., 1985. Environmental problems created by exploration of mineral resources in Himalaya. In *Environmental Regeneration in Himalaya : Concepts and Strategies* (ed.) J.S. Singh. CHEA and Gyanodaya Prakashan, Nainital : 155-168.

Strahler, A.H. and A.N. Strahler, 1975. *Geography and Man's Environment.* John Wiley and Sons, New York : 214-221.

Valdiya, K.S., 1985. Accelerated erosion and landslide prone zones in the Central Himalayan region. In *Environmental Regeneration in Himalaya : Concepts and Strategies.* (ed.) J.S. Singh. CHEA and Gyanodaya Prakashan, Nainital : 12-38.

Valdiya, K.S., 1988. *Geology and Natural Environment of Nainital Hills-Kumaun Himalaya. Gyanodaya Prakashan, Nainital* : 52-80.

# Chapter 4

# Pteridology in Kumaun Himalaya : Existing State of Art

*P.C. Pande*
*H.C. Pande*

The study of Pteridophytes, which occupy a unique position between non-seedbearing and seedbearing plants, is extremely fascinating from the angle of their phylogenetical and morphological characters.

On the conservative estimate, approximately 346 species of ferns and fern-allies occurs in Kumaun Himalaya, which is roughly 35 per cent of the Pteridophytic flora of India.

Vascular cryptogams are very abundant in this region, probably, due to altitude variations ranging from 500 to 7000 m, differences in the composition of the dominating forest vegetation, a large number of small and big rivers and watersheds, and the geographical factors. Pteridophytes, of Kumaun, inspite of the above facts, have failed to draw, attention of the past as well as present pteridologists.

The reason of taxonomic disinterest on these plants is most probably, due to their distribution in inaccessible areas, which lack transportation and communication facilities. Information too, on these places of Botanical interest is scanty.

This paper is an attempt to compile information on the pteridophytic flora, of this region (Fig.1).

Fig. 1

## Nainital and Udham Singh Nagar District

(A) **Nainital and Udham Singh Nagar Plains (Kumaun Plain)** : The Kumaun plains constituting about 10 per cent of the total area are still unexplored. The main places of interest are Haldwani, Kashipur, Ramnagar, portions of Jim Corbett park, Khatima and Tanakpur. Though this region is not rich in pteridophytic flora, still a few low altitude vascular cryptogams may be present in Tanakpur, Khatima and portions of Jim Corbett Park. The rare plants reported from this area are:

*Helminthostachys zeylanica* (Tanakpur : Joshi & Pande, 1987), *Psilotum nudum* (Jim Corbett Park : Pant, 1986), *Lygodium circinatum* (Tanakpur : Dhir, 1980).

(B) **Nainital Hills** : Nainital has attracted the attention of plant collector since 1846. Their interest, however, remains concentrated only at Nainital and its surroundings, viz., Naina Peak, Lariakanta, Land's End, Kilbary, Bhowali, Dhobighat, Bhujiaghat, Jeolikote etc. The other fern rich areas like Naukuchia Tal, Padampuri, Ramgarh, Mukteshwar and Paharpani are still unexplored and promise great potential. A total 102 fern species have been reported from Nainital hills (Loyal and Verma, 1960; Bir and Trikha, 1974; Dhir and Rani, 1979; Verma and Khullar, 1980; Pangtey *et. al.*, 1986; Pangtey and Samant, 1987; Pangtey *et. al.*, 1986 Pangtey and Rawat, 1987). This number could be possibly enhanced through detailed methodical exploration in explored and unexplored areas.

The following rare and interesting ferns and fern-allies have been reported from Nainital hills.

*Anogramma leptophylla* (Nainital : Verma and Khullar, 1980), *Arthromeris repandula* (Below Nainital : Trotter, 1891 vide Satija and Bir, 1985), *Athyrium duthie* (Nainital : Hope, 1861, Dhir, 1980), *Botrychium ternatum* (Lariakanata: Verma and Khullar, 1980), *Colysis pothifolia* (Jeolikot below Nainital : Pangtey et. al., 1986), *Didymoglossum insigene* (Dhobikhud : Verma and Khullar, 1980). *Drynaria tibetica* (Cheena Peak : Kullar, 1984), *Equisetum debile* (Dogaon enroute Nainital : Pande and Pangtey, 1987), *Equisetum hymale* (Dogaon enroute Nainital : Saxena, 1976), *Nephrolepis falcata* (Below Nainital : Hope, 1899-1904), *Ophioglossum petiolatum* (Khurpatal : Verma and Khullar, 1980), *Paradavallodes membranulosum* (Patwadanagar near Nainital : Pangtey and Punetha, 1987). *Phymatopteris stracheyi* (Nainital : Verma and Khullar, 1980), *Pronephrium nudatum* (Bhujiaghat : Pangtey *et. al.*, 1986).

**Almora District**

Almora lies between 29°30′ to 30°20′N latitudes and 79°2′ to 80°20[1] E longitudes and is located in the central part of Kumaun

Himalaya. The district had a total area of 5,397.72 sq. km, which ranges from 500-8000 m in altitude having subtropical to alpine regions.

Exploration in Almora is limited to a few places like Almora proper, Binsar, Bageshwar, Gwaldam, Jageshwar, Kapkot, Dhakuri, Khati, Dwali and Pindari. A total of 181 fern species have been reported from the district (Clarke, 1980; Beddome, 1882, 1892; Hope 1899-1904; Duthie, 1906; Pande, 1973; Pangtey *et. al.*, 1982; Awasthi and Pande, 1984; Pande *et. al.*, 1984; Pangtey *et. al.*, 1985; Pande and Pande, 1987; Pande, 1989).

Many important places described below viz. Bhatkot, Pandukholi, Ghatti and Machhor, Aerdeo, Penag, Sundardhunga and Kaphani glaciers still remain unexplored. Many more interesting plants can be collected from these fern rich areas.

The following rare and interesting plants are collected from the Almora District.

*Actinopteris radiata* (Ranikhet ; Pande, 1973; Kumaun: Clarke, 1880), *Adiantum pedatum* (Dwali: Pande, 1989: Dhir, 1980), *Botrychium virginianum* (Khati : Dhir, 1980), *Ceratopteris thalictroides* (Atpaisia near Berinag : Pande and Basera, 1988), *Dryoathrium boryanum* (Dhaulchena : Pande, 1989), *Equisetum arvense* (Pindari: Clarke, 1880; Pande, 1989), *Lindsaea cultrata* (Bharari : Dhir, 1980), *Loxogramma lanceolate* (Khati -Dwali Road : Dhir, 1980), *Microsorium zippellii* (Jageshwar : Dhir, 1980), *Notholaena marnatae* Pindari gorges : Strachey and Winterbottom, 1846-49: Duthie, 1986), *Onychium fragile* (Jageshwar : Pande and Bir, 1985), *O. japonicum* (Almora: Pande and Basera, 1989), *Ophioglossum gramineum* (Ranikhet · Pangtey *et al.*, 1985), *Osmunda regalis* (Khati : Pande, 1989), *Pallaea hastata* (Below Almora : Duthie 1906), *Phymatopteris stewartii* (Binsar : Duthie, 1906), *Phymatopteris stracheyi* (Khati : Dhir, 1980), *Psilotum nudum* (Almora : Pande, 1985), *Selaginella bryopteris* (Ranikhet : Pande, 1973), *S. ciliaris* (Almora : Pande and Pande, 1989), *Stegnogramma pozoi* (Jageshwar : Pande *et. al.*, 1984; Duthie, 1906), *Vandenboschia pyxidifera* (Pindari gorges : Hope, 1900), *Woodsia andersonii* (Sundardhunga glacier : Beddome, 1883), *W. lanosa* (Pindar valleys : Duthie, 1906).

**Pithoragarh District**

District Pithoragarh (Latitude 29°32′ N to 30°47′N, longitude 79°45′E to 81°21′E) bordering Tibet and Nepal, is a treasure house of attractions for tourists, botanists and naturalists. The district is also priveledged being a pathway for the great Kailash Mansarover pilgrimge. On the other hand the holy rivers like Kali, Sarju, Ramganga (eastern) and Gori flow here. Chalthi, Dharchula, Ghat and Thal etc. are its valleys. Most of the district is dominated by dense and virgin forests which are the natural abodes of ferns and fern-allies. After the British Botanist (Strachey and Winterbottom, Beddome, Clarke, Hope, Duthie and his collectors), who collected the fern from Gori and Kali valleys, and Didihat-Askot area only, the following ferns rich places, viz. Nainipatal, Chandak, Thalkedar, Dhwaj, Thal, Sukhidhank, Champawat, Berinag, Chaukori, Munsiyari, Nachani, Lohaghat were explored by Niranjan and Ray, (1978-79) ; Kaur and Punetha, (1984); Rawat and Aswal, (1984); Kaur and Punetha (1985); Punetha. (1985 a, b); Punetha *et. al.*, (1985); Pangtey *et. al.*, (1985); Pande and Kandpal. (1986); Pangtey and Punetha (1987); Pande and Basera (1988); Pande and Dashila (1988) in recent times. A total 121 fern and fern-allies are reported from the district (Cf. Punetha and Kaur, 1987). This number can be enhanced by exploration in a number of explored and unexplored areas of the district. The following places are rich in ferns and fern-allies need further exploration:

(1) Thal area

(2) Didihat-Askot area,

(3) Devidhura-Champawat area,

(4) Gori valley (Dharchula to Garbyang),

(5) Kali valley (Munsiyari locality upto Milam).

A good number of East Himalayan ferns have their last limit in this region and are not represented in the North-West Himalayas. Some interesting and rare fern and fern-allies reported from this district are as follows :

*Angiopteris evecta* (Dhwaj and Thalkedar : Pangtey ad Punetha, 1987), *Arthromeris lungtanensis* (Pithroagrah : Kaur and Punetha, 1984), *Asplenium, cheilosrum* (Lilam and Bogdwar : Pangtey and

Punetha, 1987), *Asplenium sarelii* (Thalkedar, Didihart : Punetha and Kaur, 1987), *A. nidus* (Ramganga valley : Duthie, 1906; Pangtey and Punetha, 1987), *Athyrium duthie* (Darma and Byans valley: Beddome, 1892). *A. hohenackerianum* (Gangolihat and Pithoragarh : Niranjan and Roy, 1978-79), *A. kumaonicum* (Lohaghat : Punetha, 1985), *Blechnum orientale* (Nolra : Pangtey *et. al.*, 1985), *Botrychium lunaria* (Milam : Pangtey and Punetha, 1987), *B. ternatum* (Didihat: Pande and Kandpal, 1986), *Ctenopteris truncata* (Gini, Sossa and Kalamuni : Duthie, 1884, 1906), *Cyathea spinulosa* (Didihat : Rawat and Aswal, 1984), *Dicranopteris linearis* var. *hirta* (Didihat : Kaur and Punetha, 1984), *Diplazium heterophlebium* ( Munsiyari : Punetha and Kaur, 1987), *Elaphoglossum stelligerum* (Above Thal : Pangtey *et. al.*, 1985), *Gymnopteris delavayi* Garbyang : Ghosh & Ghosh, 1986), *Metalhelypteris gracilescens* (Pithoragarh : Pangtey and Punetha, 1987), *Microlepia hirta* (Ramganga valley: Strachey, 1848, vide Dhir, 1980), *Notholaena marantae* (Kali valley : Strachey and Winterbottim, 1846-49; Duthie 1906 vide Dhir, 1980), *Pneumatopteris truncata* (Pithoragrah : Pangtey and Punetha, 1987), *Polypodioides atkinsonii* (Chandak, Champawat and Thalkedar : Punetha and Kaur, 1985), *Polystichum duthie* (Lebong Pass : Duthie, 1884 and 1886), *Pseudocyclosorus ochthodes* (Chanpatti : Pande and Kandpal, 1986), *Pyrrosia subfurfuracea* (Chandak : Punetha and Kaur, 1985). *Tectaria dubia* (Thal : Pangtey and Punetha, 1987), *Thelypteris palustris* (Deochula: Pande and Kandpal, 1986), *Vandenboschia radicans* (Didiat; Pande and Kandpal, 1986), *Vittaria himlayensis* (Kalamuni : Punetha and Kholia, 1988), *Woodsia alpine* (Lidder valley : Duthie, 1893).

Some rare and interesting species reported from Kumaun Himalaya without definite localities are as follows :

1. *Archniodes aristata* (Kumaon : Hope, 1903),
2. *Bolbities virens* (Kumaion : Duthie, 1906),
3. *Christella kumaunicum* (Kumaun : Holttum, 1976),
4. *Pronephrium repandum* (British Kumaon : Clarke, 1880; Hope, 1903; Duthie, 1906 ; *Kullar* et. al., 1983).

**Conclusions**

It can thus be inferred that there exists a big vacuum in our knowledge of the pteridophytic flora of this extremely rich region.

Many remote valleys and unaccessible hills have yet to be explored and additional data can be gathered. Besides, there is considerable need for intensive study, for understanding of ecological, phyto-geographical and evolutionary processes better. Currently, exploration of pteridophytic flora are being undertaken by Dr. P.C.Pande, Kumaun University, Almora and Dr. N. Punetha, Govt. P.G. College, Pithoragarh, so that the plant resources of this area could be better utilized to the advantage of the people and the country.

## General Bibliography

The authors while engaged in the study of Pteridophytes observed that a considerable amount of work had been done in Kumaun by British Pteridologists 1846-1918. However, a general slackness of research upon Pteridophytes was observed during the second quarter of the twentieth century. Thereafter, a good deal of work on Pteridophytic flora of Kumanun Himalaya was done, 1959, onwards, by several workers. The resuslts and findings of these workers have been published in various scientific and technical journals, bulletins, reports, from time to time, which are summarized as follows.

### Taxonomy Systematics and Morphology

Alston, A.H.G., 1945. An enumeration of the Indian species of *Selaginella Prioc nat Inst Sci India* 11 (3) : 211-235.

Baker, J.C., 1887. Handbook of Fern-Allies. George Bell & Sons, London.

Beddome, R.H., 1892. A Handbook to the ferns of British India, Ceylon and Malayan Peninsula with supplement. Thacker Spink & Co., Calcutta.

Bir, S. S., 1962. Taxonomic account of the Indian species of genus *Vittaria*. *Res Bull* (NS) *Pan Uni* 13 (1-11) : 15-24.

Bir, S.S., 1962. Taxonomic account of Indian members of family Aspleniaceae, *Bull Bot Surv Ind.* 4 : 1-16.

Bir, S.S. & Kamla Devi, 1968. Taxonomic revision of Polypodiaceous genera of India-II *Phymatodes* Presl. *Bull Bot Surv India* 10 : 196-219.

Bir, S. S. & C.K. Trikha 1968. Taxonomic revision of the Polypodiaceous genera of India-I *Microsorium* Link *Bull Bot Surv India* 10 : 133-143.

Bir, S. S. & C.K. Trikha, 1968. Taxonomic revision of Polypodiaceous genera of India-III *Pleopeltis* Humboldt and Bonpl. *Amer Fern J.* 85 119-125.

Bir, S.S. & C.K. Trikha, 1969. Taxonomic reviosn of Polypodiaceous genera of India-IV *Polypodium lineare* complex and allied species, *Bull Bot Surv India* 11 : 260-276.

Bir, S.S. & C.K.Trikha 1973, Taxonomy of Indian species of genus *Cystoptens* Bernh. *Nova Nedwigia* 24 : 1-21.

Bir, S.S. & C.K. Trikha 1974. Taxonomic revision of Polypodiaceous genera of India-VI *Lepisorus excavatus* group. *Amer Fern* J. 64 : 49-63

Bir, S.S., C.K. Trikha & S.M. Vasudeva, 1964. Taxonomic revision of the Polypodiaceous genera of India -V *Polypodium* Linn and *Goniophlebium* (BL) Presl *New Botanist* (New Delhi) 1 (3 &4) : 142-159.

Ching, R. C., Lin Youxing & Wu Sukung 1983. A Taxonomic revision on *Lepisorus clathratus* (Clarke) Ching complex in the Sino-Himalayan region. *Acta Botanica Yunnanica* 5 (1) : 1-23.

Ching, R. C. & Reichstein, 1980. *Asplenium nesii* Christ (Aspleniaceae : Pteridophyta) Studies in *Asplenium* for Flora Iranica 2. *Candollea* 36 : 195-202.

Chowdhury, N.P. 1937. Notes on some Indian species of *Lycopodium* with remarks on the distribution of the genus in India. *Trans National Inst Sci India* 1(9) : 187-226.

Chowdhury, N.P., 1959. Observation on the structure and ecology of a xerophytic *Selaginella* from India *S. bryopteris* (L.) Baker. *Jap Jour Bot Tokyo* 17 : 101-119.

Clarke, C.B., 1880. A review of ferns of Northern India *Trans Linn Soc. Ser 2 (Bot)* 1 : 425-619.

Copeland, E.B., 1947. Genera Filicum. The Genera of Ferns Waltham Annales Cryptogamici et. Phytopathologici No. 5. Chronica Botanica Walthem Mass. USA.

Dhir, K.K., 1980. Ferns of North-western Himalayas *Bibliotheca Pteridologia 1* : 1-158.

Dhir, K.K. & K. Rani. 1977. Ferns of Nainital *Proc 64 Indian Sci Congr* (Part -III). 79.

Dhir, K.K. & K. Rani. 1979. The genus *Athyrium* Roth in Nainital Hills. *J. Bomb Nat Hist Soc 76* : 49-58.

Dixit, R.D., 1981. The genus *Vittaria* Sm in India *J. Econ Taxo Bot 2* : 209-22.

Dixit R. D., 1984. On the occurrence of *Lycopodium clavatum* and *L. japonicum* Thunb in India. *Indian J. For 7* (1) : 72-74.

Dixit, R.D., 1984. A Census Indian Pteridophytes. Flora of India Series IV, B.S. I. Howrah.

Dixit R. D., 1988. Lycopodiaceae of India. Bishen Singh Mahendra Pal Singh Dehra Dun.

Dixit, R. D. 1989. Taxonomy of the Indian *Selaginella*, P. Beauv. Internat. Symp. Present and Future Perspectives Pterid Univ. Rajasthan Jaipur Abst. Vol. 25-26.

Dixit, R. D. & Anjali Das, 1977. Notes on the species of *Coniogramme* Fee in India *J. Indian bot Soc 56*. 254-258.

Dixit, R. D. & Anjali. Das, 1979. The genus *Conigramme* Fee in *India Proc Indian Acad Sci* 88 (B) II 4 : 253-268.

Duthie, J.F., 1906. Catalogue of the Plants of Kumaon and the Adjacent Portions of Garhwal and Tibet (Catalogue originally prepared by Strachey in 1952 in collections made by Strachey and Winterbottom between the years 1846-1849). Lovell Reeve & Co. Ltd. London.

Fraser-Jenking, C. R., 1984. An Introduction to the Fern genera of the Indian subcontinent. *Bull Brit Mus Nat Hist Bot Sci 12* : 37-73.

Fraser-Jenkins, C. R., 1986. A classification of the genus *Dryopteris* (Pteridophyta Dryopteridaceas), *Bull Brit Mus Nat Hist Bot Sci. 14* : 183-218.

Fraser-Jenkins C. R. & S.P. Khullar, 1982. *Athyrium attenuatum-* a confused Sino-Himalayan Fern. In Aspect of Plant Sciences (ed.) S. S. Bir. Today and Tomorrow Pub. New Delhi 6 : 94-102.

Fraser-Jenkins, C. R. & S.P. Khullar, 1985. The nomenclature of some confused Himalayan species of *Polystichum* Roth Indian Fern J. 2. : 1-16.

Fraser-Jenkins, C., R. & T. Reichstein, 1982. *Asplenium aitchinsonii* Fraser-Jenkins & Reichstein spec nova (Aspleniaceae : Pteridophyta) *Candollea* 37 : 339-347.

Ghose, S. R. & R.K. Ghose, 1985. *Gymnopteris delavayi* (Bak. ) A new fern record for India *J. Bomb Nat Hist Soc. 82* (1) : 243.

Holttum, R. E., 1972. Studides in the family Thelypteridaceae IV The genus *Pronephrium* Presl *Blumea* 20 : 105-126.

Holttum, R.E., 1976 b. The genus *Christella* Leveille. Sect *Christella* Studies in the family Thelypteridaceae XI *Kew Bulletin 31:* 293-339.

Holttum, R.E. & J.W. Grimes 1979. The genus *Pseudocyclosorus* Ching Thelypteridaceae *Kew Bull* 34 : 499-316.

Hope, C. W., 1899-1904. The ferns of North Western India including Afghanistan, the Trans-Indus protected States and Kashmir *J. Bomb nat Hist Soc* 12 : 315-325, 527-538, 621-633, 1899; 13: 25-36, 236-251, 1900; 443-461, 657-671, 1901; 14 : 119-127, 252-266, 458-480; 1902; 720-740, 1903; 15 : 67-111, 415-429, 1904.

Joshi, G.C. & P.C. Pande, 1987. *Helminthostachys zeylanica* (L.) Hook (Ophioglossaceaae) A new record for Western Himalaya *J. Bomb nat Hist Soc 84* (3) : 724-724.

Kaur S. & N. Punetha, 1984. *Arthromeris lungtauensis* Ching - a new record for western Himalya *J. Bomb nat Hist Soc.* 81 : 727-739.

Kaur S. &. N. Punetha, 1984. A new variety of *Dicranopteris linearis* (Burm. f.) Underw. From Indian *J. Indian Bot Soc. 63 :* 34-36.

Kaur S. &. N. Punetha, 1985. Some new records of Ferns from Kumaun Himalaya *New Botanist* 12 (2-4) : 166-168.

Khullar, S. P., 1976. Some lesser known ferns from the W. Himalaya- 1 *Cheilanthes anceps* var *Brevifrons Amer Fern J. 66 :* 21-24.

Kullar S. P., 1979. Morphological studies of some Himalayan *Polystichums* In Lower Plants of the Indian Subcontinent -II *ex Beiheft Nova Hedwigia 63 :* 41-47.

Kullar S. P., 1984. The ferns of Western Himalaya - a new additions, corrections and annotations *Indian Fern J.* 1 (1-2) : 89-95.

Khullar, S. P. 1988. The ferns of W. Himalaya. An enumeration (as in the illustrated Fern Flora of W. Himalaya) *J. Himalayan Research & Develop.* 7 : 59- 67.

Khullar, S. P., 1989. The genus *Asplenium* in the W. Himalaya. Internat Symp. Present and Future Perspective Pterid. Uni Rajasthan Jaipur. Abst. vol. 69-70.

Khullar S. P. & S.C. Gupta, 1979. On the occurrence of *Polystichum aculeatum* In the Himalayas in Recent Reserches in Plant Science (ed.) S. S. Bir. Kalyani Publishers. New Delhi : 66-70.

Khullar, S. P. & S.C Gupta 1980. Systematics of the genus *Polystichum* from the Himalayas *Nova Hedwigia 32 :* 415-44.

Khullar, S. P., N. Punetha & B.S., Kholia, 1988. The species of the genus *Dryopteris* in the Pithoragarh district of Kumaun (W. Himalaya) *Indian Fern J.5*: 127-135.

Khullar, S. P. & S.S. Sharma, 1979. Systematics sof the Himalayan species of the genus *Onychium*. In Aspects of Plant Sciences (ed) S.S. Bir. Today and Tomorrow's Printers and Publishers, New Delhi 3 : 63-88.

Khullar, S. P. S. S., Sharma, 1986. *Asplenium capillipes Makino* (Aspleniaceae) A sino Japanese fern in the Western Himalaya *J. Bomb nat Hist Soc 83* (2) : 463-466.

Khullar, S. P., S. S. Sharma & S. Paramjit, 1983. A little known fern from the Himalaya- *Asplenium nesii* Christ and nomenclature of *A exiguum* Bedd. (Aspleniaceae ) *J Bomb nat Hist Soc 80*: 262-265.

Khullar, S. P. & S. Verma 1990. The genus *Lepisorus* in the Western Himalaya . In Frontiers in Botanical Res. A national Symp., Abst. Vol : 70-71.

Kramer, K. U., 1972. The Lindsaeoid ferns of the old world VI-Continental Asia, Japan and Taiwan *The Gardern Bulletins Singapore 26* (1) : 1-48.

Loyal, D. S., Vijay Kumar, 1983. Cytotaxonomy of 3 *Selaginella* species with a note on some chromosome number and typical leaf vasculature. *Res Bull* (Sci) *Punjab Uni 34* (3 & 4): 139-147.

Loyal, D. S. & Vijay Kumar, 1984. Cytotaxonomic observations on three *Selaginella* species from North-Western Himalayas *Indian Fern J . 1 :* 59-62.

Loyal, D. S. & S. C. Verma, 1960. Ferns of Nainital. *J. Bomb nat Hist Soc 57* : 479-490.

Nayar, B.K., 1961. Ferns of India I *Adiantum Bull Nat Bot Garden* Lucknow No 52.

Nayar, B.K., 1961. Ferns of India II *Drynaria* and *Pseudodrynarea. Bull Nat Bot Garden* Lucknow No. 56.

Nayar, B.K., 1962. Ferns of India III *Microsorium Bull Nat Bot Garden* Lucknow No 58.

Nayar, B.K., 1964. Ferns of India VII *Egenolfia. Bull Nat Bot Garden* Lucknow No. 100.

Nayar, B.K., 1965. Ferns of India XV *Pyrrosia* Mirbel. *Bull nat Bot Gardens* Lucknow No. 117.

Pande, H. C. & P. C. Pande, 1990. An enumeration of Pteridophytes of Shyamtal hills of Kumaun Himalaya. *Vegetos* 3 (2) : 132-134.

Pande, P. C., 1973. Pteridophytic flora of Ranikhet. *Indian for 99 :* 49-52.

Pande, P. C., 1985. *Psilotum nudum* (L) Griseb from Kumaun Himalaya rediscovered *J. Indian Bot Soc.* 64 : 214 - 215.

Pande, P. C., 1985. *Peranema cytheoides* D Don from Kumaun Himalaya. *J. Indian Bot Soc. 64 : 277.*

Pande, P. C., 1988. A contribution to the aquatic of Mashy plants of Almora district. In : Perspectives in Aquatic Biology, (ed.)R. D. Khulbe, New Delhi 407-17.

Pande, P.C., 1989. Contribution to the Fern Flora of Almora district *Geobios New Report* 8 (2) : 103-110.

Pande, P.C., 1990. An annotated check list of the fern-allies of Kumaun Himalaya. *Indian Fern. Journ.* 7 : 58-65.

Pande, P. C., 1990. A Census of Kumaun Ferns (north-Western Himalaya). *Indian Fern. J.* 7 : 140-195.

Pande, P.C. B.D. Andola & G.S. Rawat, 1984. An enumeration of ferns of Jageshwar area in Almora district. *Himal Res & Dev* 3 (2) : 39-40.

Pande, P.C,. & P.S.Basera, 1988. Pteridophytic flora of Berinag hills (Western Himalaya) *Indian Fern J.* 5 : 150-161.

Pande, P. C. & P.S. Basera, 1989. *Onychium Japonicum* (Thurb). Kze. (Cryptogrammaceae) a rare species recollected from Kumaun. *Actas. Botanica. Indica* 17 : 93-94.

Pande, P.C.& P.S., Basera, 1989. *Coniogramme falcata* (Don) Salom (Hemionifidaceae)- A new record for North-western Himalayas. *Acta Botanica Indica* 17 : 233.

Pande, P. C. & S. S. Bir, 1985. Occurrnce of *Onychium fragile* Verma et Khullar from Kumaun Himalaya. *J. Bomb Nat Hist Soc.* 82 (2) : 441

Pande, P. C. & R. S. Dashila, 1988. Ferns of Kunapokhari, Nachani and Munsyari hills of Kumaun Himalaya. *Indian Fern J.* 5 : 78-88.

Pande, P. C. & M.M. Kandpal, 1986. *Thelypteris palustris (Salisb)* Schott New record fro U.P. Hills. *J. Bomb nat Hist Soc 82* : 474.

Pande, P. C. & M.M. Kandpal, 1986. Pteridophytic flora of Didihat (W. Himalaya) *Acta Botanica Indica 14* (Suppl); 115-122.

Pande, P. C. and H. C. Pande, 1989. *Crytomium hookerianum* (Presl). C. Chr. (Aspidiaceae) a new record for Western Himalayas. *Jour. Bom. Nat.* Hist. Soc. 86 (3) : 479.

Pande, P. C. & H. C. Pande, 1989. The *Polystichum* of Kuman Himalaya (W. Himalaya) Internat Symp Present and Future Perspectives Pterid Uni. Rajasthan Jaipur Abst. Vol. pp. 71.

Pande P. C. & H.C. Pande, 1990. Systematics of the genus *Dryopteris* from Almora district of Kumaun Himalaya. In Frontiers in Botanical Res. a nat. symp. Abst. Vol. P. 71.

Pande, P. C. & H.C. Pande, 1990. Systematic of the genus *Selaginella* from Kumaun Himalaya. *Indian Fern J.* 7 : 5-17.

Pande, P. C. & H. C. Pande, 1990. Pteridophytic Hora of Mukteshwar and Ramgarh area (Kumaun Himalaya). *Vegetos* 3 (1) : 56-61.

Pande, P. C. & H.C. Pande, 1990. *Elaphoglossum marginatium,* (Wall ex. Fee). T. Moore, A new record for Western Himalaya *Res J. Pl. Eviron* 6 (1) : 101-102.

Pande, P. C. & H. C. Pande. 1991, Rediscovery of a rare fern (*Archniodes aristata* (Frost. f. Tindale) from Kumaun Himalaya. *Jour Bomb. Nat. Hist. Soc.* (In press).

Pande, P. C. & H.C. Pande, 1991. *Stenochlaena plaustines* (Burm). Bedd.- new record for Western Himalaya. *Vegitos* (In press).

Pande, P. C. & H.C. Pande 1991 The *Polystichums* of Kumaun Himalaya (W. Himalaya) In. Aspects of Plant Sciences (eds.) T. N. Bharadwaja). Today & Tomorrow's Printers and Publishers, New Delhi, 13 : (In press).

Pande, P.C. & T.D. Pande, 1987. An enumeration of ferns of Binsar (W Himalaya) *Acta Botanica Indica 15* : 101-103.

Pande, P.C. & Y. P. S. Pangtey, 1987. New records of fern-allies from Kumaun Himalaya, *Indian J. For 10* : 151-152.

Pangtey Y.P.S., B.S. Kalakoti & G.S. Rawat, 1985. Hitherto unrecorded species of *Ophioglossum* Linn (Ophioglossaclae) fern Kumaun Himalaya Indian. J *For* 8 : 331-332.

Pangtey Y.P.S., B.S. Kalakoti, G.S. Rawat & P.C. Pande, 1982, Observation on the fern flora of Pindari area. *Himal Res & Dev* 1 : 156-160.

Pangtey Y.P.S. & N. Punetha, 1987. Pteridophytic Flora of Kumaun Himalaya an Updated list. In Western Himalaya (eds). Y.P.S. Pangtey & S. C. Joshi. Gyanodaya Prakashan, Nainital *1* : 390-412.

Pangtey Y.P.S., G.S.Rawat & S.S. Samant, 1986. *Colysis pothifolia* (Ham ex D Don) H Ito (polypodiaceae) from Nainital- New Record for North Western Himalaya. *J. Bomb Nat Hist Soc* 82 : 472-473.

Pangtey Y.P.S., G.S. Rawat & S. S. Samant, 1986. Addition to the Pteridophytic Flora of Nainital. *J. Bomb. Nat. Hist. Soc. 83* : 683-684.

Pangtey Y.P.S. & S. S. Samant, 1987. On the recollections of rare ferns from Nainital. *J. Econ Tax Bot 9* : 485-488.

Pangtey, Y.P.S. & S. S. Samant, 1987. Additions to the Pteridophytic Fora of Nainital *J. Econ. Tax. Bot. 11* : 5-8.

Pangtey, Y. P. S. & S.S. Samant, 1987. Miscelaneous notes 38. Additions to the pteridophytic flora of Kumaun and Nainital (Western Himalaya) *J. Bomb. Nat. His. Soc.* 84. 486-488.

Pangtey, Y. P. S. & S. S. Samant, 1988. *Asplenium bullatum* Wall ex. Mett. (Aspleniaceae). A new record for North. Western Himalaya from Kumaun Hills. *Jour. Bomb. Nat. Hish. Soc.* 85 : 237-238.

Pangtey, Y. P.S. & S. S. Samant, 1989. Ferns of Pindar & Sarju valley (Kumaun Himalaya). Nat. Symp. Plant. Syst. Adv. & Trends. Dept. Botany, Panjab. Univ. Chandigarh, Abst. Vol., 22.

Pangtey, Y. P. S., S. S. Samant & S. S. Bankoti, 1988. *Mattenccia Orientalis* (Hook) Trev. (Onocleaceae) a new record for Kumaun Himalaya. *Indian J. Forestry 11* : 170-171.

Pangtey, Y. P. S., S.S.Samant & P.C.Pande, 1987. A list of species and genera of ferns not included in J. F. Duthie's Catalogue of Plants of Kumaun. *J. Econ Tax Bot* 9 : 476-480.

Pangtey, Y. P. S., S. S. Samant & G. S. Rawat, 1985. *Elaphoglossum stelligerum* in the North West Himalaya. *J. Econ Tax Bot 6* : 490-492.

Pangtey, Y.P.S., S. S. Samant & G.S. Rawat, 1985. A note on the occurrence of *Blechnum orientale* Linn in Kumaun Hills. *J. Econ Tax. Bot 6* : 490-492.

Pangtey, Y.P.S., S. S. Samant and Sandeep Verma, 1990. A new natural hybrid in the genus *Pteris* (Pteridaceae :Pteridophyta) from the Kumaun Himalaya. *Fern. Gaz.* 13 (6). 357-360.

Panigrahi, G. & R. D. Dixit, 1966. Studies in the systematics of Indian *Selaginella* III. *Proc nat Acad Sci India* (B) *36* : 102-108 .

Panigrahi, G. &. R. D. Dixit, 1967. Studies in the systematics of Indian *Selaginella II.* J. *Indian bot Soc.* 46 : 221-223.

Panigrahi, G. & R. D. Dixit, 1967. Studies in the Pteridophytes II the family Osmundaceae in India. *J. Indian Bot Soc. 48.* 96-101.

Panigrahi, G. & R. D. Dixit, 1969. Studies in the systematics of Indian *Selaginella* I. *Proc nat Inst Sci India 34* : 191-209.

Panigrathi, G. & R. D.Dixit, 1969. Studies in the Indian Pteridophytes IV The family Ophioglossaceae in India *Proc nat Inst Sci Indian Series B* 35 : 230-266.

Pani P. C., 1986. Flora of Corbett National Park Series 4 B. S. I. Howrah : 186.

Punetha N. 1985. Taxonomic observation of some *Athyrioide* ferns From Pithoragarh district of Kumaun Himalaya (Western Himalaya) *Indian Fern J. 2* : 22-31

Punetha N., 1985. Taxonomic observations on some species of *Pteris* from Pithoragarh district of Kumaun Himalaya (Western Himalaya) *Indian Fern* J. 2 : 65-72.

Puntetha N. &. S. Kaur, 1987. Pteridophytic flora of Pithoragarh district of Kumaun (West Himalayas) *J. Econ Tax Bot. 9* : 269-286.

Punetha, N. & S. Kaur, 1990. Taxonomic studies in the family Gleicheniaceae *Indian. Fern* J. 7 : 86-93.

Punetha, N. & S. Kaur. 1990. Sporophyte morphology and Phylogeny of the family Gleicheniaceae. *Canadian J. Bot.* (in press).

Punetha N &. B. S. Kholia, 1988. A new natural hybrid of *Pteris* from the Indo-Himalayas Nat Conf Pterid NBRI Abst Vol. 17.

Punetha N. & B. S. Kholia, 1988. *Vittaria himalayensis* Ching A new record for Western Himalayas. *Curr sci.* 57 (6) : 346-348.

Punetha N. & B. S. Kholia 1988. Taxonomic distinction of some species of *Dryopteris* Adanson from Pithoragarh district of Kumaun (North Western Himalayas) Proc 75th Indian Sci Cong Part III Abstract 111.

Punetha, N & B. S. Kholia, 1988., On the occurrence of *Dryopteris conjugata* Ching in Western Himalaya. *Indian J. Forestry* 11 (4) : 340-341.

Punetha, N & B. S. Kholia, 1989. *Polystichum makinoi*(Tag) Tag A new record for India. *Curr. Sci.* 58 (6) : 317-318.

Punetha, N. & B. S. Kholia 1989. Additions to the Pteridophytic Flora of Pithoragarh district of Kumaon (West Himalaya). New Botanist 16 (1-4) : 115-126.

Punetha, N & B. S. Kholia, 1990. On the occurrence of *Cornopteris atroviridis* W.C. *crenulatoserrulata* in Indo-Himalaya In : Frontiers in Botanical Research, a national Symp. Chandigarh, Atst. Vol. 76.

Punetha, N. & B. S. Kohlia, 1990. *Lepisorus pseudoclathratus* Ching & Wu. a new record for India. *Indian Fern* J. 7 : 40-42.

Punetha N., Manjusha Tewari & S. Kaur, 1985. Taxonomic studies on the families Loxogrammaceae and Polypodiaceae of Pithoragarh district of Kumaun (North Western Himalaya) *Indian Fern J*, 2 : 73-89.

Rao, T. A., 1959. Report on a botanical tour to Milam Glacier. *Bull Bot Surv Ind. 1* : 97-120.

Rao, T. A., 1960. A botanical tour to Pindari Glacier and Kumaun hills *Bull Bot Surv Ind.* 2 : 61-94.

Rao, T. A. 1964. Observation on the vegetation of Eastern Kumaun bordering the Nepal Frontier *Bull Bot Surv Ind. 6* : 47-57.

Rasback Helga & Tadeus Reichstein, 1990. The Chromosome number of *Annogramma leptophylla* (Adiantaceae : Petridophyta). from Europe. *Feru Gaz.* 13 (6).

Rawat, G. S., B. S. Aswal 1984. On the occurrence of a tree fern *Cyathea spinosa* Wall ex Hook (Cyatheaceae) in Kumaun Himalaya. *Himal Res & Dev.* 3 (2) : 44.

Satija, C. K. & S. S. Bir, 1985. Polypodiaceous Fern of India : Aspects of Plant Sciences VIII. Today & Tomorrow's Printers and Publishers, New Delhi.

Satija, C. K., S. S. Bir A. K. Bhardwaj, 1983. Taxonomic revision of the Polypodiaceous genera of India *Pyrrosia* Mirbel. *Bull Bot Surv India* 25 : 62-89.

Saxena D. K., 1976. Notes on the occurrence of *Equisetum hymale* L. in India. *Curr Sci 45* (18) : 679.

Singh T. C. N., 1931. A study of ferns from Nainital District *Proc 18th Indian Science Congr* (Nagpur) : 268-269.

Sleep Anne & T. Reichstein, 1984. Typification of *Asplenium Varians* Wall ex Hook & Grev (Aspleniaceae, Pteridophyta); Studies in Asplenium for Flora Iranica 5. *Candollea* 39 : 675-691.

Strachey, R. 1882. Catalogue of Plants of Kumaon and Adjacent Portion of Garhwal and Tibet. In E. t. Atkinson's Gazetteer of North-West Provinces, Cosmo Publications, New Delhi.

Vasudeva, S. M & S. Chhibber, 1989. Taxonomic studies on some species of Pteris Linn in India. *Indian Fern J.* 6 : 204-206.

Verma, S. C. & S. P. Khullar. 1980. Ferns of Nainital (Western Himalayas) An updated list *Fern Gaz* 12. 83-92.

Wieffring, J. H., 1964. A preliminary revision of the Indo-pacific species of *Ophioglossum* (Ophioglasceae). *Blumea* 12: 321-337.

## Ecology and Phytogeography

Awasthi D.K. & P.C. Pande, 1984. A note on phytogeographical distribution of ferns and fern-allies of Almora (W.H.). *Jour Bomb nat Hsit Soc. 81* (1) : 739-741.

Bir S.S., C.K.Satija & Perminder Bir, 1982. Studies on the Indian Ferns: Ecology, distribution and phytogeography of Polypodiaceae. In Aspects of Plant Science. Today & Tomorrow's Printers and Publishers, New Delhi : 1-24.

Dhir K.K. & Y. Saiki, 1984. Phytogeograshic observations on Himalayan ferns. *Nova Hedwigia* 39 : 169-174.

Dixit, R. D. & Bal Krishna, 1990. A Phytogeographical analysis of the species of the family : Thelypteridaceaece- V In : Frontiers in Botanical Research a national symposium Chandigarh Abs. Vol. 69.

Dixit, R. D. & Bal Krishna, 1990. Phytogeographic analysis of the endemic Pteridophytes of India. Conservation priorities. *Indian Fern J.* 7 : 49-53.

Kaur S. & S. Chandra, 1985. Indian Thelypterioid ferns, their nomenclature, distribution and phytogeography. *New Botanist 12* : 83-116.

Khullar S. P., 1982. Distribution of epipytic ferns in the Himalayas in Contemporary Trends in Plant Science (ed.) S. C. Verma Kalyani Publishers, Ludhiana and New Delhi : 202-207.

Niranjan A. R. S. & S. K. Roy, 1978-79 Ecological distribution of some ferns of central Himalayas *Jour Sci Res B H U 29* : 227-32.

Pande, P. C., J. K. Pathak & S. D. Bhatt 1991 Ecological of the wetland Pteridophytes of Kumaun Himalaya. In Ecology of the Mountain waters (eds.) S. D. Bhatt & R.K. Pande. Ashish Publishing House, New Delhi. 174-197.

Pangtey Y.P.S. & G.S.Rawat, 1987. Some ecological observations on the fern flora of Nainital *J. Econ Tax Bot 9* : 147-150.

Punetha, N., 1988. Dry matter production and leaf growth in certain Western Himalayan ferns. *Net conf Pterid* NBRI Abst Vol : 17.

Punetha, N., 1989. Ecological and Photogeographical observations on the fern flora of Phithoragarh district (north west Himalaya). Proc. Ind. Acad. Sci. (Pl. Sci) 98 (4) : 327-333.

Raizada M. B., 1935.The genus Psilotum in India. *Indian For 61* : 654-657.

Singh. Y. & Bir, S., S. 1989. Distribution, ecology and phytogeography of Asplenoid ferns of India Int. Symp. Present& Future Perspectives, Pterid uni Rajasthan, Jaipur. Abst. Vol. 146-147.

Singh, Y & Bir. S. S. 1989. Distribution, ecology and phytogeography of Asplenoid ferns of India. *Indian Fern. J.*, 6 : 268-284.

Singh, S. & G. Panigrahi, 1987. On the taxonomy and distribution of *Ceratopteris thalictroides* (L.) Ad Brongn. (Parkeriaceae) in India. *J. Econ. Taxo. Bot.* 10 : 425-431.

## Cytology and Genetics

Gibby, M., 1985. Cytological obsevations on Indian subcontinent and Chinese *Dryopteris and Polystichum* (Pteridophyte : Dryopteridaceae). *Bull Brit Mus nat Hist.* 14 : 1-42.

Khullar, S., 1989. Cytotaxonomy of *Polystichum* in West Himalaya. In *Flora Malesiana* Symposium Leiden (The Netherlands) Abs : 37-38.

Khullar, S. P. & S. C. Gupta, 1978. Cytotaxonomy of the genus *Polystichum Plant Systematics & Evolution 129* : 269-276.

Khullar, S. P. & P.N. Mehra, 1972. a. Cytotaxonomy of Western Himalayan Ferns 3 Polypodiaceae (S.S.) Nucleus 15 : 156-162.

Khullar S. P. P. N. Mehra 1972 a. Cytotaxonomy of W Himalayan Ferns 1 Schizaeaceous series. *Res Bull Panjab Uni* (n s) 23 : 189-204.

Mehra, P. N., 1961. Chromosome numbers in Himalayan Ferns. *Res. Bull.* (ns). *Panjab Uni* 12 : 139-169.

Punetha, N, 1989. Cytological observations on Ferns of Kumaon (N. W. Himalayas) in Plant Sci Res in India. Trevedi *et. al.* (eds). Today & Tomorrow's Printers & Publishers, New Delhi : 459-465.

Punetha, N. B. S.Kholia & Abha Sen, 1988. A note on the cytology of *Polystichum luctuosum* (Kunze) T Morre: Indian Fern J. 5. 125-126.

Punetha N. &. A. Sen 1988. Cytology of *Christella multiaurita* Punetha *Curr Sci. 57 (8)* : 439-440.

Punetha, N & A. Sen, 1989. Cytological observations on some species of *Pteris* L. from Kumaon. North-West Himalaya. *Proc. Indian. Acad Sci.* (Plant Sci). 99 (2): 131-134.

Verma S.C. & D. S. Loyal. 1960. Chromosome counts in some ferns from Nainital Curr Sci. 29 : 67-70.

## Spores

Punetha, N., 1989. Studies on atmospheric fern spores at Pithoragarh (N. W. Himalayas) with particular reference to distribution of ferns in the Himalyas. Internal. Symp Present and Future prespectives pterid uni. Rajasthan Jaipur Abst. vol. 96-98.

Punetha, N. and B. S. Kholia, 1988. Seasonal variation of fern spores in the air of Pithoragarh town. Him J. Env. Zool. 2 : 110-113.

## Morphogenesis and Experimental Studies

Punetha, N., 1986. Apical inhibition in the fronds of gleichinioid ferns. 9 Bot Conf *J. Indian Bot Soc 65* (Supplm) : 39.

Punetha, N., 1987. Effects of ultraviolet rays on *Lygodiium flexuosum (L)* . Sw.*New. Botanist* (New Delhi) *14* : 67-71.

Punetha N. 1987. Apical dominance an apical inhibition in the fronds of *Lygodium Japonicum* (Thunb) Sw. *Phytomorphology* 37 (4) : 299-301.

Punetha N., 1987. Apical dominance and Apical inhibition in the fronds of *Lygodium japonicum* (Thunb) Sw. *Proc 74th Indian Sci Cong* Part III Abstract : 96.

Punetha N., 1988. Apical inhibition in the fronds of Gleichenioid ferns*Phytomorphology* 38 (2,3) 187-192.

Punetha N., 1988. Leaf development *Hypolepis punctata* (Thbg) Mett ex Kuhn. Proc 75th Indian Sci Congr Part III Abstract. 111.

Punetha N & S. Kaur, 1987. Growth inhibition in apices of fronds of *Dicranopteris linearis* (Burm) underw. *Indian Fern J. 4* : 86-94.

## Ethnobotany

Khullar, S. P., 1989. Bracken in the Western Himalaya . In Bracken 89 Fourth International conf on Bracken : Biology, Economic significance and Control. Abstracts University of Sydney & Australian Institute of Agricultural Science : Sydney.

Pande P.C.& Pangtey Y. P. S., 1987. Studies on Ethnobotany- I on some less known edible and economic ferns of Kumaun region of Western Himalaya. *J. Econ Tax Bot.* 11 (1) : 81-85.

## Horticulture

Pande P. C., 1988. Himalaya Ke Fern-Kitane Akarshak ? In Ornamental Horticulture in Himalayas (eds) S. D. Bhatt & A.K. Singh, New Delhi : 20-25.

Sah, B.h., P.C. Pande and G.C. Joshi, 1986. Ornamental ferns in relation to the economy of the Kumaun Hill region. All India Seminar on Development of Entrepreneur ship in Hill Areas, Almora (Abst.)

## Parasites

Pande, P. C., 1988. *Cuscuta reflexa* Roxb. A parasite on *Onychium contiguum* Wall ex Hope. *Indian J For* 11 (1) : 91-92.

Pande, P. C., P. S. Basera & H.C. Pande 1989. Occurrence of *Cuscuta reflexa Roxb.* on ferns. *Geobios New Reports 8* : 150-152.

Pande, P.C., R. S. Dashila & M.M. Kandpal, 1986. New cryptogamic hosts of *Cuscuta reflexa* Roxb. *J. Econ Tax Bot 8 :* 221-223.

Sen, D. N. 1959. *Cuscuta reflexa* - A parasite on *Equisetum ramosissimum* Desf var *atisissimum* A Br. *Sci. Cult* 24 : 435-36.

Singh, T.C.N., 1932. On the occurrence of *Cuscuta* on ferns. *Proc Indian Sci Congr. 19* : 332.

## General

Pande, H.C. & P.C. Pande, 1989. An interesting phenomenon in *Elaploglossum stelligerum* (Wall ex. Baker) Moore ex Alston & Bonner. *Indian Fern J.* 6 : 284.

Punetha, N., 1979. Freaks in certain ferns., abstract in *2nd all Inida bot Conf J. Indian Bot Soc 58* (suppl) 34.

Punetha, N. 1988. The flianean crozier. A comparative study *Phytomorphology* 38 (4): 347-356.

Singh, T.C.N. 1931. On the forking of the fronds in certain ferns Nainital *Proc Indian Sci Congr 18* : 269.

After the middle of the present century the Pteridologists from different universities and Institutions, viz. Allahabad University, Allahabad; B.H.U., Varanasi; B.S.I. Allahabad; B.S.I., Dehra Dun; N.B.R.I., Lucknow; Panjab University, Chandigarh; Punjabi University Patiala; University of Jodhpur; University of Rajasthan, Jaipur frequently visited the Kumaun Himalaya which provided them sufficient material for their studies on Morphology, Taxonomy, Systematics, Anatomy, Cytology, Genetics, Palynology, Reproductive biology, Morphogenesis and Experimental studies. Their research in the above mentioned fields cosnstitute a part of Peteridophytes of Kumaun Himalaya.

Thousands of species of plants / animals, all over the world, face threats to their existence, due to one or other reasons and require immediate conservation. A thorough study of "Rare and Endangered Plants" is needed to attain the proper knowledge or to arrive at a definite conclusion. The following publications give useful hints on "Rare and Endangered Pteridophytes" and may render help in locating more references on threatened plants especially on pteridophytes.

Bir S. S., 1987. Pteridophytic flora of Indian Rare and Endangered elements and their conservation. *Indian Fern J.* 4 : 95-101.

Chandra, S. 1982. Check list of ferns endemic to India. *Nova Hedwigia* 36 : 241-247.

Chandra, S. & S. Kaur. 1984. Additions to the ferns endemic to India. Indian Fern J 1 : 83-87.

Chatterjee, D., 1940. Studies on the endemic flora of India and Burma. *Jour Asiat Soc Beng* 5 : 1-69.

Datt, Anukul, 1983. Occurrence and distribution of some rare ferns in India In an Assessment of Threatened Plants of India. BSI, Howrah : 323-327.

Dhir, K. K. &. Y., Saiki, 1984. Phytogeographic observtions on Himalayan ferns. *Nova Hedwigia* 39 : 169-174.

Dixit, R. D., 1983. Rare and interesting Pteridophytes of India - II. *In An assessment of Threatened plants of India*. BSI, Howrah : 328-334.

Dixit, R.D. 1986. Tree fern- An urgent need of conservation. *Indian Fern.* J 3 : 42-45.

Dixit, R. D. B. Krishna, 1989. Additions to the endemic pteridophytes of India and their survey-cum- conservtiona strategies In. Plant Sci Res in India Trivedi et. al. Today & Tomorrows Printers & Publishers, New Delhi : 7-13.

Dixit, R.D. & Balkrishna, 1989. Studies in the family Thelypteridaceae-II Endemic species and their conservation. Eleventh All India Confr. Sect. VII, Triuchurapalli 43.

Jain, S. K. & A. R. K. Sastry, 1980. *Threatened Plants of India A state of Art Report*. New Delhi.

Kaur, S., 1989. Conservations of Fern resources of India-Present Status and Future perspectives In *Plant Sci. Res. in India* Trivedi *et. al.* (Eds). Today & Tomorrows, Printers & Publishers, New Delhi: 1-5.

Pande P. C. & A. K. Singh 1988. Depleting Vascular Cryptogames of Kumaun Himalaya and their management. *Mountain Research & Development 9* (4) (accepted).

Pangtey, Y. P. S. & S. S. Samant, 1988. Rediscovery of three rare plants from Kumaun Himalaya. *J. Bomb. Nat. Hist. Soc.* 85 : 456-457.

Pangtey, Y.P.S. & S. S. Samant, 1988. Observations on the threatened rare and endangered flowering plants and ferns in the flora of Kumaun Himalaya. *Advances of Forest Res. in India* 3 : 65-74.

Pangtey, Y. P. S. & S. S, Samant, 1990. A note on the rare and endangered plants of Kumaun Himalaya (Western Himlaya) In : Frontiers in Botanical Res., National Symp. Chandigarh, Abst : 72.

Pangtey, Y.P.S., S. S. Samant & R.S. Rawal, 1988. Rediscovery of a rare fern *Macrothelypteris ornata* (Wall. ex. Bedd). Ching. (Thelypteridaceae) from Kumaun Himalaya after a century. *J. Bomb. Nat. His. Soc. 85* : 660-661.

Pangtey, Y. P. S., S. S. Samant and R. S. Rawal, 1989. A note on *cythea spinulosa* wall. ex. Hook (Cytheaceae) in Western Himalaya. J. Econ. *Tax. Bof. 13* (1) : 106-109.

Samant, S.S. & Y.P. S. Pangtey 1989. Observations on the rare ferns of Kumaon Himalaya (Western Himalaya) India. *Internat Sump. Present and future perspectives Pterid.*, Univ. Rajasthan , Jaipur. Abst. Vol : 138.

Fraser-Jenkins (1984) observed that the taxonomy of Indian ferns and fern-allies suffers from studeis on poor herbarium specimens. A herbarium worker or plant collcetor therefore, must be acquainted with the 'art of botanical collections' and preservation of specimens. The following publications may be helpful for the said purpose.

Bir., S. S. 1962. Taxonomy of the Indian members of family 'Aspleniaceae' *Bull Bot Surv India* 4 : 1-16.

Fraser-Jenkins C.R., 1984. An introduction to fern genera of the Indian subcontinent *Bull Br Mus nat Hist (Bot) 12* (2) : 37-76.

Holttum, R.E., 1968. A Revised Flora of Malaya Vol II. Malaya Govt. Printing Press, Singapore.

Lawrence, G.H.M., 1977. *Taxonomy of Vascular Plant,* London.

Pandeya, S.C., G.S. Puri & J.S. Singh, 1968. Research Methods in Plant Ecology Asia Publishing House, London : 55-75.

Reichestein T., 1981. Das Sammeln Von. Farnen. *Farnblatter* 6 : 10-17

Researches on Pteridophytes have frequently been reviewed by various authors from time to time and may be of utility for all those who wish to take up studies in Pteridophytes. A few very important references are given as follows:

Bir, S. S., 1973. Cytology of Indian Pteridophytes. In *Glimpses in* Plant Research. Nair P. K. K. (ed). Vikas Publishing House Pvt. Ltd. New Delhi 1 : 28-119.

Bir, S.S., 1976 a. Taxonomy of Indian Pteridophytes. *In Recent Advances in Botany* (ed) Kachroo P. Bishen Singh Mahendra Pal Singh, Dehra Dun India : 70-115.

Bir, S.S. 1977 a. Anatomy of Indian Pteridophytes. In *Frontiers of Plant Sciences Prof P. Parija Felicitation Volume* (ed) Padhi B. Utkal University, Bhubaneshwar : 365-400.

Bir, S. S. 1977 b. Pteridophytic flora of India : A review of achievments and future challanges in the systematics and taxonomy. *Bull Bot Surv India 19* : 323-329.

Bir, S. S., 1982. Classification and taxonomy of Pteridophytes with particular reference to the higher ferns. In *Aspect of Plant Science (ed)* Bir S. S. Today and Tomorrow's Printers and Publishers, New Delhi 6 : 183-245.

Bir, S.S. 1983. Cytogenetics of Pteridophytes. In *Genetical Researches in Pteridophytes* (eds) Jaiswal et. al. ICAR, New Delhi : 46-60.

Bir, S.S., 1987. Pteridology in India. *Indian Fern J* 4 : 104-150.

Bir, S. S., 1988. Evolutionary trends in pteridophytic flora of India *Presidential Address 75th Indian Sci. Congr. Session Pune Section* VI Botany : 1-56.

Ching, R. C., 1978. the Chinese Fern Families and genera- Systematic arrangement and Historical Origin. *Acta. Phstotax. Sinica* 16 (3) : 1-119; 16(4) : 16-37.

Devi, S. 1981. References Mannual of Fern Spores, EBIS, NBRI, Lucknow.

Devi, S., 1988. Spores of Pteridophytes. *Indian Fern J.* 5 : 28-57.

Dixit, R. D., 1977. Studies in Pteridophytes in the Botanical Survey of India. *Bull Bot Surv India 19* : 309-314.

Dixit, R. D., 1984. A Census of Indian Pteridophytes Flora of India. Series IV, BSI, Howrah.

Erdtman G., 1957. Pollen and SporeMorphology / Plant Taxonomy Vol II Gymnospermae Pteridophyta Bryophyta Almqvist Wiksell. Stockholm.

Fraser-Jenkins, C. R., 1984. An introduction to fern genera of the Indian subcontinent, *Bull Br. Mus nat Hist.* (Bot) 12 (2) : 37-76.

Kaur, S. 1989. Economic Exploitation and conservation : emerging areas in the study of ferns and fern allies. *Indian Fern J. 6* : 23 : 29.

Kaur, S. & F. Raza., 1983. Bibliography of Indian Pteridology. NBRI, Lucknow.

Kramer, K. U., 1982. Systematics of the pteridophytes. In *Progress in Botany* (eds). H. Ellenberg et. al. Springer-Verlag, Berlin., Heidleberg. 4 : 394-405.

Kramer, K. U. 1984. Systematics of Pteridophytes. In Progress in Botany. (Eds) Ellenberg E. et. al. springer-Verlag, Berlin, Heidleberg 46 : 329-343.

Kramer, K. U., 1986. Systematics of the Pteridophytes. In *Progress in Botany* (eds) Ellerberg E. et. al. Spriner-Verlag, Berlin, Heidleberg 48 : 334-346.

Kramer, K. U. 1989. Systematics of Pteridopjhytes, in : *Progress in Botany*, Vol. 50 pp. 295-311.

Love, A & D. Love R. E. G. Pichi-Sermolli, 1977. Cytotaxonomical Atlas of Pteridophyta J Cramer, Vaduz (Germany).

Maheshwari, P & R. N. Kapil, 1963. Fifty years of science in India, Progress in Botany 5 Pteridophyta *Ind. Sci. Congr. Assoc.*, Calcutta : 37-46.

May, L. W. (1978) The economic uses and associated folklore of ferns of fern-allies., Bot. Rev. 44 : 491-538.

Mehra, P. N. & N. P. Chowdhury., 1957. 'Pteridology' Progress of Science in India section IV Botany 4. *Nat Inst Sci India Publ*. New Delhi : 32-41

Miller, J. H., 1968. Fern Gametophytes as experimental material. *Bot Rev* 34 : 361-440.

Page, C. P., 1979 a. The diversity of ferns : An ecological perspective. In *The Experimental Biology of Ferns* (ed) A. F. Dyer. Academic Press London Exp. Biot 14 : 9-56.

Pichi Sermolli, R.E.G., 1965. Index Filicum supplement quartum pro annis 1934-1960. *Regnum Vegetabile* 37 1-XIV. 3-370.

Ollgaard Benjamin, 1987. A revised classification of the Lycopodiaceae *S. Lat. Opera Botanica*. 92 : 153-178.

Rice E.L., 1979. Allelopathy - an update. *Bot. Rev.* 45 (1) : 15-109.

Sharma, B.D., 1988. Tracheary elements in Pteridophytes. *Bionature* 8 (2) : 113-134.

Sharma, B.D., Suther O.P. & Bohra D. R. 1986. Indian fossil pteridophytes Advances and challengers. *Indian Fern J.* 3 : 82-88.

Sheffiefld, E. &. Bell P. R., 1987. Current studies of the Pyteridophyte life cycle. *Bot. Rev* 53 (4) : 442-490.

Stewart, R. R., 1982. Flora of West Pakistan : History and exploration of Plants in Pakistan and Adjoining Areas. Islamabad IV-186 p.

Tryon, Rolla, 1986. The Biogeography of species with special reference to ferns. *Bot. Rev* 52 (2) : 117-156.

Verma, S. C., 1970. Pteridophytes. In the Indian Botanical Society and Botany in India 1920-1970 (ed) B.M, Johri. University of Delhi : 28-31.

Verma, S. C., 1978. Cytotaxonomy of Indian ferns. In Symposium Modern Trends in Plants taxonomy National Botanical Gardens, Lucknow; 27-28.

Verma, S. C., S. P. Khullar, P. Singh, & P. Sharma 1987. *Pteridology in India- A Bibliography*. Bishen Singh Mahendra Pal Singh Dehra Dun, India.

Verma, S.C. & S. P. VIJ., 1975. Indian Cytologists. All India Congr. Cyto & Genet, Udaipur : 1-15.

Wagner, W. H. Jr. 1974 b. 25 years of Botany-Pteridology 1947-1972. *Ann Miss Bot. Gard 61* : 86-111.

White, R. A., 1971 a. Experimental and developmental studies on the fern sporophytes. *Bot. Rev.* 37 : 509-540.

White, R. A., 1971 b. Experimental studies of the sporphytes of fern. *Bio. Sci. 21* : 271-275.

A wide range of information on fern and fern-allies is now available to all Pteridologists, through the publication, "Indian Fern Journal" published by the Indian Fern Society (for details contact Prof. S. S. Bir, Secretary Treasurer, Pteridophytic Taxonomic Laboratory, Department of Botany, Punjabi University, Patiala-147002).

A set of guidelines / recommendation for teaching and researches in Pteridology were outlined by the renowned Pteriodologists of India, during concluding session of the symposium, "Biology of Indian Pteridophytes" held at the Department of Botany, Punjabi University., Patiala form January 10th through 11th, 1986 (Cf. Bir, 1986). These guidelines may be hopeful in giving a direction to the Pteriodological researches in India and abroad.

BIR, S. S., 1986. A write-up about Naitonal Symposium on "Biology of Indian Pteridophytes" *Indian FernsJ* 3 i-x.

## Acknowledgements

We are deeply indebted to Prof. S. S. Bir (Patiala) and Prof. S.P. Khullar (Chandigarh) for the suggestions, guidance and constant encouragements. Our sincers thanks are due to Prof. R. E. Holttum (England). Prof. K.U. Kramer (Switzerland), Prof. D. S. Loyal (Chandigarh), Prof. T. Reichstein (Basel), Prof. B. D. Sharma (Jodhpur), Dr. S. Chandra (Lucknow) for sending us relevant literature. Thanks are also due to Sri MukuPant (Almora) for his help rendered during the preparation of the manuscript. Financial assistance from CSIR, New Delhi is gratefully acknowledged.

# Chapter 5

# Organic Productivity and Nutrient Content of Poplar Plantation in Tarai Belt of Kumaun Himalaya

*L.S. Lodhiyal*
*R.P. Singh*

Poplars in their natural range occur interspersed throughout the forests of temperate regions of northern hemisphere between the southern limit of around latitude 30°N and northern limit of latitude 45°N. There are 35 species of poplars currently recognized in the World. *Populus* is widely planted above 28°N latitude in India in Jammu and Kashmir, Punjab, Haryana, North Bengal, Arunachal Pradesh and Uttar Pradesh along roads, canals, in agriculture fields, towns, parks, orchards and home gardens. There are six *Populus* spp. indigenous to India growing along water courses in the higher hills, in valleys and also on hill sides exposed due to landslides etc.

In India most of the indigenous *Populus* spp. grow in areas having mean minimum and maximum temperature at 6°C and 10°C, respectively. They grow well on low lying and moist loamy soils but may be planted on river beds with sandy soils and in areas with clayey loams in forest soils. The soil pH best suited to poplars ranges between 5.0 to 6.5. Besides its adaptation to certain geographical limits, the poplars prefer good aerated soils. Soils saturated with $Na^+$ assumes solid structure due to dispersed clay particles at pH higher than 8.2 and are considered unsuitable for its cultivation.

Poplar (*Populus deltoides*), a commercial exotic tree species has been planted over a large scale from 1976 onward in the Tarai belt of Kumaun Himalaya. Poplar has many uses, and it can be harvested in a short period. Poplars are multipurpose tree and its wood is used in many way, such as, paper pulp, ply wood, match wood, packing cases, fibre board, veneer and light constructional timbers. It is a fast growing deciduous tree which produces a large amount of litter that decomposes within one year and improves the soil fertility.

This species can be planted as pure crop as well as in association with agricultural crop in agroforestry system. Whenever, this species is planted in farmer's agriculture field, it gives them extra economic income besides the agronomic yields. However, the dry matter production has to increase significantly to obtain an economically feasible production level even when energy prices will rise (Frissel et al. 1978).

A large scale plantations of different species have already been tried to fulfill growing demands of fuelwood and industries but the use of these species are rather limited. Therefore, in order to increase the productivity of existing forests to meet the demands of the industries, the need to replace the slow growing broad leaved species by fast growing species of poplar plantation is considered an important solution in Kumaun Himalaya.

**Study Site**

The present study site is located between 29°03′–29°12′ N lat. and 79°20′-79°23′ E long. at an altitude of 290 m. The climate of Tarai region is sub-tropical. Of the annual average rainfall 1593 mm (average of 1985-1989), about 86 per cent occurs from mid-June to September. The mean monthly temperature ranges from 14.4°C (January) to 31.3°C (June) (Lodhiyal 1990).

**Geology**

The Tarai area is characterised by sedimentary deposits of tertiary age consisting of lower sandstone of old Tarai deposits. The soil is alluvial loamy, deep and fertile; free from quartz and boulders. The sediments in the region are largely derived from gabbro, chlorite, schist, quartzite, sandstone, basalt, shales, slates, gneisses, diorites and granite washed away from the Himalayan (Valdiya 1980).

## Materials and Methods

### Natural Vegetation

The natural vegetation, which was replaced by Poplar (*Populus deltoides* Marsh) plantations, is referred as moist Tarai sal (*Shorea robusta*) forests and alluvial savannah wood land forest type (Champion and Seth 1968).

The present plantation had 666 tree $ha^{-1}$, plant to plant and row to row distance being 3 and 5 m, respectively. The dry matter production of 3-year-old plantation was compared with other fast growing *eucalyptus* species. The total basal area was 12.74 $m^2$ $ha^{-1}$. Trees in 0.25 hectare area of sample plot were measured for dbh (1.37 m) and distributed in respective dbh classes. Shrub vegetation were absent. The analysis of herb layer was accomplished by using ten randomly placed 1 × 1 m quadrats.

Based on the field data the following diameter classes of poplar trees were recognised, i.e., 7.5-10.0 cm, 10.0-12.5 cm, 12.5-15.0 cm, 15.0-17.5 cm and 17.5-20.0 cm. Twelve trees (3 individuals of each diameter class) were harvested. The roots were dug out to 1 m depth within an area extending 1 m in radius around the base of each tree. The tree was then felled into the space adjoining individuals. After morphometric measurements; the stump root was detached. The above ground part was separated into bole wood, bole bark, branch, twig and foliage parts. The root system was categorized into stump roots, lateral roots and fine roots. The fine roots were estimated by digging out three 25×25×25 cm monoliths before exposing the stump roots and lateral roots. Fresh weight of all components were determined in the field. Pre-weighed composite samples from upper, middle and lower strata (about 500 g fresh weight) of each component were brought to the laboratory and oven dried at 60°C constant weight.

The fresh weight multiplied by an appropriate fresh : dry weight factor gave the dry weight for different components. The harvest data were subjected to the regression analysis to relate the dry weight of each component with diameter at breast height.

The regression equation used was Y = a + bx where, Y dry weight of the component (Kg tree$^{-1}$), X dbh (cm), a is the Y intercept and b is the slope or regression coefficient. A straight line relationship was found between dbh (X) and weight (Y) for all components. A good positive relationship was found between these two variables, so a linear regression equation was used.

Shrubs were absent in the stand. The biomass of herb layer was determined when it was peak during September-October (1987). The harvested material was separated into aboveground and belowground components.

Mean dbh value for a diameter class, was used in regression equation to get an estimate of mean biomass (by components) for that diameter class. Stand biomass was calculated by summing up the biomass values across diameter classes. The sum of biomass of trees and herbs gives the total standing biomass of the vegetation of the site.

The litter input was measured by placing 15 litter traps randomly on the forest floor of the plantation. Samples from each litter trap were collected in polythene bags and brought to the laboratory at one month intervals during the study period. Samples were separated into leaf and wood litter components, cleaned, weighed and oven-dried at 60°C to constant weight. Twenty five individuals of different diameter classes were marked with white paint at breast height (1.37 m) in September 1987 in order to assess diameter increment at annual interval. The marked trees were measured for the increase in diameter in September, 1988. Mean diameter increment for each diameter class was then calculated.

Using the allometric equations, the biomass for different component (bole wood, bole bark, branch, twig, foliage, stump roots, lateral roots and fine roots) of trees in mean diameter classes was calculated separately, from the dbh measurement of 1987 ($B_1$) and 1988 ($B_2$). The net increase in biomass ($\Delta B = B_2 - B_1$) gave the year's biomass accumulation. The sum of $\Delta B$ values for different components yielded net biomass accretion in the trees. To the foliage, biomass accumulation was added to the annual leaf fall to represent the leaf production. Wood litter fall values were added to the biomass accumulation in twigs. The death of main roots is rare in actively growing trees but considerable mortality occurs in fine roots. Following Kalela (1954), Orlov (1955, 1968) and Ogino (1977), it was assumed that mortality in fine roots is equivalent to one-fifth of leaf litter fall, but recent studies indicate that the fine root mortality is markedly higher than one-fifth of litter production. Therefore, the present estimate of root production in this study is an under estimation. However, the same approach (as used here) was followed in estimating root production of plantations with which the *Populus deltoides* plantation was compared. In the present study, the assumption that fine roots mortality is equivalent to one-fifth of leaf litter fall as taken into consideration for calculating the fine root mortality, however, be a gross under estimation (Harris, Santantonio and Mc Ginty, 1980). This value, added to biomass accumulation of fine root component, yielded fine root net production. The net production for each of the components was summed across diameter classes which give the total net production of trees for a plantation. The biomass (aboveground and belowground) of herbs on the site was determined during their peak growth in September, 1987. This was assumed equal to net herb production. The sum of net production values of trees and herbs yielded the total net primary production of the site.

Samples of different tree components (bole wood, bole bark, branch, twig and foliage) from the lower, middle and upper stratum of trees were collected separately from 15 trees of all available dbh classes of the plantation. The samples for roots (Stamp root, lateral roots and fine roots) were collected by digging out roots upto 1 m depth. Similarly, samples for herbs (aboveground and belowground) were collected. Three composite samples of each component of the trees and herbs were oven dried at 60°C and the samples analysed for nutrients. For each replicate a composite sample of 0.5 g of plant material was analysed for total nitrogen after digestion in 10 ml concentrated sulphuric acid using 5 g catalyst mixture (potassium sulphate and cupric sulphate in ratio 9 : 1) with a quick digestion unit. The total nitrogen was determined by micro-kjeldahl technique (Peach and Tracy 1956; Misra 1968). Phosphorus and potassium were extracted by wet ashing of 0.5 g plant material, in an acid mixture consisting of 10 ml $H_2SO_4$ + 3 ml $HNO_3$ + 1 ml $HC10_4$ following the method of Jackson (1958). Phosphorus was determined by spectrophotometry and potassium by flame photometry (Jackson 1958).

The total amount of nutrient in the vegetation was obtained by summing the amount in the different components. The amount of nutrients estimated in both strata were summed to obtain total nutrient content.

## Results

Allometric equations, variables and parameters relating biomass of different components to the diameter at breast height for 3-year-old poplar plantation is given in Table 1.

**Table 1 :** Allometric relationship between biomass of tree components (Y = kg $tree^{-1}$) and diameter at breast height (D, cm) for Poplar clone D–121 plantation Y = a + bD.

| *Component* | *Intercept* | *Slope* | $r^2$ | *t* |
|---|---|---|---|---|
| Bole wood | –0.0939 | 2.7312 | 0.994 | 47.6151 |
| Bole bark | –0.0129 | 0.3617 | 0.994 | 47.6109 |
| Branch | –0.2786 | 0.7803 | 0.944 | 14.8043 |
| Twig | –0.0792 | 0.2822 | 0.881 | 9.8202 |
| Foliage | –0.1569 | 0.6219 | 0.954 | 16.5470 |
| Stamp root | 0.0043 | 0.6343 | 0.952 | 16.2047 |
| Lateral root | –0.1579 | 0.3283 | 0.912 | 11.6604 |
| Fine root | 0.0149 | 0.0706 | 0.824 | 7.8090 |

All equations are significant at P > 0.01.

## Biomass

Biomass of tree estimates (kg $ha^{-1}$) across different dbh classes in present poplar plantation is given in Table 2.

**Table 2 :** Componentiwise biomass (kg $ha^{-1}$) in different diameter classes of Poplar clone $D_{121}$ plantation.

| *S. No.* | *Diameter class (cm)* | *Bole wood* | *Bole bark* | *Branches* | *Twigs* | *Foliage* | *Stump root* | *Lateral root* | *Fine root* |
|---|---|---|---|---|---|---|---|---|---|
| 1 | 7.5–10.0 | 400.1 | 53.0 | 110.5 | 40.3 | 89.1 | 93.3 | 45.9 | 10.6 |
| 2. | 10.0-12.5 | 899.8 | 119.1 | 250.0 | 91.0 | 201.1 | 209.7 | 104.1 | 23.8 |
| 3. | 12.5-15.0 | 2805.3 | 371.5 | 783.5 | 284.9 | 629.0 | 653.4 | 326.4 | 73.8 |
| 4. | 15.0-17.5 | 21628.3 | 2864.1 | 6054.9 | 2200.5 | 4857.4 | 5036.0 | 2527.0 | 568.2 |
| 5. | 17.5-20.0 | 2668.7 | 353.4 | 748.7 | 272.2 | 600.2 | 621.2 | 312.7 | 70.0 |

The total tree biomass was 60.0 t $ha^{-1}$, of this, aboveground and belowground accounted for 82.3 and 17.7 per cent respectively. The share of different component of tree was in the following order : bole wood > branch > foliage > bole bark > twig in aboveground and stamp roots > lateral roots > fine roots in belowground roots (Table 3).

The herb layer biomass was 2.3 t $ha^{-1}$, of which aboveground accounted for 82.6 per cent and belowground 17.4 per cent (Table 3).

Total biomass of vegetation was 62.3 t $ha^{-1}$, of which tree layer shared about 96.0 per cent (Table 3).

**Table 3 :** Componentwise net primary productivity ($ha^{-1}$ $yr^{-1}$) in different diameter class of poplar clone $D_{121}$ plantation.

| *S. No.* | *Diameter class (cm)* | *Bole wood* | *Bole bark* | *Branch* | *Twigs* | *Foliage* | *Stump root* | *Lateral root* | *Fine root* |
|---|---|---|---|---|---|---|---|---|---|
| 1 | 7.5–10.0 | 225.3 | 29.8 | 63.5 | 23.3 | 165.7 | 52.3 | 27.1 | 28.5 |
| 2. | 10.0-12.5 | 412.9 | 54.7 | 118.0 | 42.7 | 303.6 | 95.9 | 49.6 | 52.7 |
| 3. | 12.5-15.0 | 1022.6 | 135.4 | 292.2 | 105.7 | 751.9 | 237.5 | 122.9 | 130.5 |
| 4. | 15.0-17.5 | 6637.9 | 879.0 | 1896.6 | 686.0 | 4881.8 | 1541.7 | 797.9 | 845.9 |
| 5. | 17.5-20.0 | 615.9 | 81.6 | 175.9 | 63.6 | 452.8 | 143.1 | 74.1 | 78.3 |

## Productivity

The total net primary productivity of poplar tree was 24.4 t $ha^{-1}$ $yr^{-1}$, of this different component accounted in the following order: bole wood > foliage > branch > bole bark > twig in aboveground and stump roots > lateral roots > fine roots in belowground part (Table 4). The herb layer productivity was 2.3 t $ha^{-1}$ $yr^{-1}$, of which above and belowground part accounted for 82.6 and 17.4 per cent respectively (Table 4). Total productivity of vegetation was 26.7 t $ha^{-1}$, of which, tree layer shared about 91 per cent (Table 4).

**Table 4 :** Total biomass ( t $ha^{-1}$) and net primary productivity (t $ha^{-1}$ $yr^{-1}$) in different state of poplar clone $D_{-121}$ plantation

| *Component* | *Biomass (t $ha^{-1}$)* | *NPP (t $ha^{-1}$ $yr^{-1}$)* |
|---|---|---|
| **Tree layer** | | |
| Bole wood | 28.4 (47.3) | 8.9 (36.5) |
| Boly bark | 3.8 (6.3) | 1.2 (4.9) |
| Branch | 7.9 (13.2) | 2.5 (10.2) |
| Twig | 2.9 (4.8) | 0.9 (3.7) |
| Foliage | 6.4 (10.7) | 6.6 (27.0) |
| Stump root | 6.6 (11.0) | 2.1 (8.6) |
| Lateral root | 3.3 (5.5) | 1.1 (4.5) |
| Fine root | 0.7 (1.2) | 1.1 (4.5) |
| Total | 60.0 (100) | 24.4 (100) |
| **Herb layer** | | |
| Above ground | 1.9 (82.6) | 1.9 (82.6) |
| Below ground | 0.4 (17.4) | 0.4 (17.4) |
| Total | 2.3 (100) | 2.3 (100) |
| Total Vegetation | 62.3 | 26.7 |

Values in parentheses indicate per cent contribution.

**Litter fall**

Total litter fall was 4.53 t $ha^{-1}$ $yr^{-1}$, of this leaf litter was the sole component. The total leaf litter fall, about 50 per cent occurred in the month of October 1987.

**Nutrient**

**Nutrient Concentration**

The per cent nutrient (NPK) concentration varies with tree components. The nutrient concentration was found in the following order : N > K > P (Table 5).

**Nitrogen**

The nitrogen concentration in different tree component varies in the following order : foliage > twig > bole bark > branch > fineroots > lateral roots > stump root > bole wood. The maximum N-concentration is recorded in foliage (2.450 ± 0.037) and minimum is noticed in bole wood (0.228 ±0.043) (Table 5).

**Phosphorus**

The phosphorus concentration in different tree component varies in the following order : foliage > twig > bolebark > branch > fineroots > lateral roots > stump roots > bole wood. The maxium P-concentration was found in foliage (0.196±0.006) and minimum in bole wood (0.037±0.005) (Table 5).

**Potassium**

The potassium-concentration was in the following order : foliage > twig > bole bark > branch > fine roots > lateral roots > stump roots > bole wood. The maximum K concentration was registered in foliage (1.105 ± 0.017) and minimum is found in bole wood (0.143 ± 0.016) (Table 5).

**Nutrient content**

**Nitrogen**

**Tree layer :** The standing state of nitrogen was 416.36 kg $ha.^{-1}$ The different components shared it in the following order : foliage

Table 5 : Nutrient concentration (%) in different component of Poplar clone $D_{121}$ plantation.

| Component | Nutrient concentration (%) | | |
|---|---|---|---|
| | N | P | K |
| Bole wood | 0.228 ± 0.043 | 0.037 ± 0.005 | 0.143 ± 0.016 |
| Bole bark | 1.108 ± 0.022 | 0.072 ± 0.004 | 0.505 ± 0.025 |
| Branch | 0.728 ± 0.035 | 0.074 ± 0.007 | 0.266 ± 0.026 |
| Twig | 1.412 ± 0.024 | 0.142 ± 0.006 | 0.625 ± 0.025 |
| Foliage | 2.450 ± 0.037 | 0.196 ± 0.006 | 1.105 ± 0.017 |
| Stump root | 0.489 ± 0.029 | 0.048 ± 0.007 | 0.255 ± 0.012 |
| Lateral root | 0.535 ± 0.038 | 0.050 ± 0.006 | 0.271 ± 0.014 |
| Fine root | 0.653 ± 0.036 | 0.054 ± 0.007 | 0.279 ± 0.018 |
| **Herb layer** | | | |
| Aboveground | 1.570 ± 0.019 | 0.092 ± 0.008 | 1.350 ± 0.014 |
| Belowground | 1.660 ± 0.020 | 0.070 ± 0.021 | 1.100 ± 0.009 |

(37.5 per cent) > bole wood (15.5 per cent) > branch (13.9 per cent) > bole bark (10.0 per cent) > twig (9.8 per cent) > stump root (7.8 per cent) > lateral root (4.3 per cent) > fine root (1.2 per cent) (Table 6).

**Herb layer :** The nitrogen stored in herbs was 35.64 kg ha$^{-1}$, of this, aboveground and belowground accounted for 85.9 per cent and 14.1 per cent respectively (Table 6). Total nitrogen in vegetation was 452.0 kg ha$^{-1}$, of this, tree layer shared 92.1 per cent (Table 6).

## Phosphorus

**Tree layer :** The standing state of phosphorus was 41.07 kg ha$^{-1}$. Different components of tree shared it in the following order : foliage (30.8 per cent) > bole wood (25.6 per cent) > branch (14.2 per cent) > twig (10.2 per cent) > stump root (77 per cent) > bole bark (6.6 per cent) > lateral root (4.0 per cent) > fine root (1.0 per cent) (Table 6).

**Herb layer :** The phosphorus content in herbs was 2.09 kg ha$^{-1}$. Of this, aboveground and belowground accounted for 85.6 and 14.4 per cent. Total phosphorus in vegetation was 43.16 kg ha$^{-1}$, of this, tree layer shared about 95 per cent (Table 6).

## Potassium

**Tree layer :** The potassium content in trees was 197.40 kg ha$^{-1}$.

The different components of trees shared it in the following order: foliage (35.7 per cent) > bole wood (20.6 per cent) > branch (10.8 per cent) > bole bark (9.6 per cent) > twig (9.1 per cent) > stump root (8.5 per cent) > lateral root (4.6 per cent) > fine root (1.1 per cent) (Table 6).

**Herb layer :** The potassium content was 31.09 kg ha$^{-1}$, of this, aboveground shared 84.7 per cent and belowground recorded 15.3 per cent (Table 6). Total potassium in vegetation was 228.49 kg ha$^{-1}$. Of this, tree layer shared about 86.4 per cent (Table 6).

**Table 6 :** Standing state of nutrients (kg ha$^{-1}$) in different components of poplar clone D$_{-121}$ plantation.

| *Component* | *Nutrient (kg ha$^{-1}$) (%)* | | |
|---|---|---|---|
| | *N* | *P* | *K* |
| **Tree layer** | | | |
| Bole wood | 64.76 ( 15.5) | 10.51 (25.6) | 40.62 (20.6) |
| Bole bark | 41.68 (10.0) | 2.71 (6.6) | 19.00 (9.6) |
| Branch | 57.85 (13.9) | 5.85 (14.2) | 21.31 (10.8) |
| Twig | 40.79 (9.8) | 4.12 (10.2) | 18.05 (9.1) |
| Foliage | 156.24 (37.5) | 12.64 (30.8) | 70.46 (35.7) |
| Stump root | 32.41 (7.8) | 3.18 (7.7) | 16.87 (8.5) |
| Lateral root | 17.75 (4.3) | 1.65 (4.0) | 8.99 (4.6) |
| Fine root | 4.88 (1.2) | 0.41 (1.0) | 2.10 (1.1) |
| Total | 416.36 (100) | 41.07 (100) | 197.40 (100) |
| **Herb layer** | | | |
| Aboveground | 30.62 (85.7) | 1.79 (85.6) | 26.33 (84.7) |
| Belowground | 5.02 (14.1) | 0.30 (14.4) | 4.76 (15.3) |
| Total | 35.64 (100) | 2.09 (100) | 31.09 (100) |
| Total Vegetation | 452.0 | 43.16 | 228.49 |

## Discussion

The reliability of estimates of production for a site depends mainly on the accuracy in determinations of the annual biomass increment of trees. The regression equations indicates that the biomass of each component and of the total tree is significantly related to the bdh. By using the same Vernier Calliper at exactly the same location on the tree, systematic errors in successive measurements of dbh of marked trees are reduced. The dry matter production of poplar trees includes the biomass and net primary productivity of the stand. The total tree biomass is 60.0 t $ha^{-1}$, of this, above-ground accounts for 49.4 t $ha^{-1}$. The bole wood and foliage are the important component for economic and soil fertility (nutrient cycling) point of view, respectively. There is significant difference in organic matter content with dbh (P > 0.01). Shelton et al. (1982) reported 35-40 t $ha^{-1}$ stem wood of standing crop for the thinned plantations. This is found lower than the value reported 84.0–189 t $ha^{-1}$ (Lodhiyal et al., 1992), 85.9–129.0 t $ha^{-1}$ (Kaul et al., 1983) and 175 t $ha^{-1}$ (Kaul and Sharma, 1983), and higher than 8.7-33.8 t $ha^{-1}$ (Lodhiyal and Singh 1993), 30.0 t $ha^{-1}$ (Frison, 1974) and 48.0 t $ha^{-1}$ (Zsuffa and Balatinecz 1975). The tree layer biomass of present study poplar plantation is higher than values ranged from 25.9-47.1 t $ha^{-1}$ of age 3 to 4 year-old and lower than 67.5 t $ha^{-1}$ of age 5-year-old *Eucalyptus* hybrid plantations (Bargali et al. 1992).

The net primary productivity (NPP) of tree layer is 24.4 t $ha^{-1}$ $yr^{-1}$, higher than 15.4-22.4 t $ha^{-1}$ $yr^{-1}$ (Lodhiyal 1990) and 12.2-20.4 t $ha^{-1}$ $yr^{-1}$ for 3 to 8 years-old *Eucalyptus* hybrid plantations in the adjacent region of Central Himalaya (Bargali et al., 1992). The average annual NPP reported for poplar plantations are 12.0 t $ha^{-1}$ $yr^{-1}$ (Shelton et al., 1982) and 13.0 t $ha^{-1}$ $yr^{-1}$ (Olson 1975).

The ratio of biomass to NPP [the biomass accumulation ratio (BAR)] has been used to characterize the production conditions in communities by Whittaker (1966) and Whittaker and Woodwell (1969). It expresses the quality of biomass retained per unit of net production. The biomass accumulation ratio (BAR) of *Populus deltoides* plantation is (2.45) higher than similar age of *Eucalyptus* hybrid (2.21) plantation. The NPP/FSC (net primary productivity/ foliar standing crop) indicates that the production per unit leaf mass is lower in poplar (3.81) than *Eucalyptus* hybrid (4.55) plantation.

The litter fall is affected by tree density and age of stands. It is important factor for maintaining the nutrient cycling of the system. The litter fall is 4.53 t $ha^{-1}$ $yr^{-1}$ (6.66 trees $ha^{-1}$) in the present study plantation compared with 5.35 t $ha^{-1}$ $yr^{-1}$ (leaf + wood litter) in 5-year-old (400 trees $ha^{-1}$) poplar plantation. The litter fall in present study plantation is compared with 4.91 t $ha^{-1}$ (6-year-old) and is found higher than the range 1.67-3.66 t $ha^{-1}$ $yr^{-1}$ of 3 to 5 years-old stands of *Eucalyptus* hybrid plantations (Baragali et al., 1992).

Absolute nutrient removals could be reduced considerably by not removing foliage or small branch components from the site, an approach suggested by other authors (Hansen and Baker 1979). In the present study plantation, foliage accounted for maximum nutrient concentration (2.45 per cent N, 0.196 per cent P and 1.105 per cent K) and minimum nutrient concentration is registered in bole wood (0.228 per cent N, 0.037 per cent K and 0.143 per cent K). The nutrient concentration is found higher in comparison to other fast growing tree viz., *Eucalyptus hybrid* (maximum 1.41 per cent N, 0.08 per cent P and 0.950 per cent K in foliage and minimum 0.08 per cent N, 0.01 per cent P and 0.06 per cent K in bole wood) (Bargali et al. 1992). The total nutrients (NPK) content in tree layer was 416.4 kg ha$^{-1}$ (N), 41.1 kg ha$^{-1}$ (P) and 197.0 kg ha$^{-1}$ (K), of this, foliage shared about 31-37 per cent. The total vegetation nutrients was (452.0 kg ha$^{-1}$ N, 43.16 kg ha$^{-1}$ P and 228.49 kg ha$^{-1}$ K) found higher than *Eucalyptus hybrid* (3-year-old) stand (Bargali et al. 1992).

**Acknowledgements**

Financial support from the Council of Scientific and Industrial Research, Government of India, New Delhi is gratefuly acknowledged.

## References

Bargali, S.S., S.P. Singh and R.P. Singh, 1992a. Structure and function of an Age Series of Eucalypt Plantation in Central Himalaya. I Dry Matter Dynamics. Annals of Botany, 69 : 405-411.

Bargali, S.S., S.P. Singh and R.P. Singh, 1992b. Structure and function of an Age Series of Eucalypt Plantations in Central Himalaya. II Nutrient Dynamics. Annals of Botany 69 : 413-421.

Champion H.G. and S.K. Seth, 1968. A revised survey of Forest Types of India. Government of India Publication, New Delhi.

Frissel, M.J., C.P. Van Goor, D.W. de Hoop and J.J. Oli, 1978. Feasibility study of short rotation forestry for energy purposes. Report to the 2nd Coordination Meeting of Contractors of the EC Programme on Solar Energy, Brussels.

Frison, G., 1974. Poplar plantation with a two year rotation. Cellulosae Carta 25 : 10-21.

Hansen, E. and J.B. Baker, 1979. Biomass and nutrient removal in short rotation intensively cultured plantations. In Proceeding Symposium on Impact of Intensive Harvesting on Forest Nutrient Cycling : 130-151.

Harris, W.F., D. Santantonio and McGinty, 1980. The dynamic belowground ecosystem. In Fresh perspectives from ecosystem analysis. (ed.) by R.H. Waring. Oregon State University Press, Cornvallis : 119-130.

Jackson, M.L., 1958. Soil Chemical analysis. Prentice Hall Inc. USA : 498.

Kalela, E.K., 1954 Mantysiemen poiden Ja paustojen juurisuhiteista. Acta For Ann. 61 : 1-16.

Kaul, O.K., D.C. Sharma and V.N. Tandon, 1983. Biomass distribution and productivity in Poplar plantation. Indian Forester Vol. 109 (11) : 822-828.

Kaul, O.M. and K.K. Sharma, 1983. Biomass production systems of Poplars and Willows in India. Indian Forester 109 (9) : 645-654.

Lodhiyal, L.S., 1990. Structure and functioning of poplar plantations in Tarai belt of Kumaun Himalaya. Unpublished Ph.D. Thesis, Kumaun University, Nainital, India.

Lodhiyal, L.S., B.S. Rana and S.P. Singh, 1992. Biomass and productivity in an age series of short rotation *Populus deltoides* plantations. Tropical Ecology, 33 (22) : 214-222.

Lodhiyal, L.S. and R.P. Singh, 1993. Biomass, productivity and litter fall in 9-year-old cottonwood plantation in Tarai plains area of Central Himalaya. Advances in Forestry Research in India, Vol. IX.

Misra, R., 1968. Ecology Work Book. Oxford and IBP Publishing Company, Calcutta: 244.

Orlov, A.J., 1955. The role of feeding roots of forest vegetation in enriching with Organic matter. Pchvovedenie, 6 : 14.

Orlov, A.J., 1968. Development and life duration of pine feeding roots. In Methods of *Productivity Studies in Root Systems and Rhizosphere Organisms.* M.S. Gholarov, V.A. Kovda, L.N. Novich Kuval Vanova, L.E. Rodin and V.M. Sveenikova. Nauka, Leningrad : 139-165.

Ogino, K., 1977. A beech forest at Ashia : biomass, its increment and net production. In (eds) *Primary Productivity of Japanese Forests, Productivity of Terrestrial Communities.* T. Shidei and T. Kira (Editors), Univ. Tokyo Press : 172-186.

Olson, J.S. 1975. Productivity of Forest ecosystems. Productivity of World ecosystems. Nat. Acad. Sci. Washington D.C. : 166.

Peach, K. and M.V. Tracy, 1956. Modern methods of plant analysis. Vol. I Springer-Verlag, Berlin.

Shelton, M.G., L.S. Switzer, L.E. Nelson, J.B. Baker, and C.W. Muellar, 1982. The development of cottonwood plantations on alluvial soils. Technical Bulletin 113, Mississippi State University.

Vaidiya, K.S., 1990. Stratigraphic scheme of the sedimentary units of the Kumaun Lesser Himalaya. In (eds) *Stratigraphy and Correlation of the Lesser Himalayan Formations.* K.S. Valdiya and Bhatiya. Hindustan Publishing Corporation, Delhi : 7-48.

Whittaker, R.H., 1966. Forest dimensions and production in the Great Smoky mountains. Ecology 47 : 103-121.

Whittaker, R.H. and G.M. Woodwell, 1969. Structure, production and diversity of the Oak-pine forest at Brooke-haven, New York, Journal of Ecology 57 : 155-174.

Zsuffa, L. and J.J. Balatinecz, 1975. Poplar pulpwood production with a one-year rotation. Populier 12 : 6-8.

# Chapter 6

# Glimpses of History of Agriculture in Uttarakhand in the Central Himalaya

*B.P. Joshi*

The history of agriculture is inextricably intermixed with the progress of the civilization which enabled the mankind to form a society. However, the history of Indian agriculture, particularly the Himalayan agriculture has not attracted much attention of the historians. The subject of writing the history of agriculture is difficult and I am not unaware of the limitations but it is hoped that learned scholars in future, will improve upon the text.

The history of agriculture goes back to the Neolithic period (7500 BC) in the Middle East and later archaeological evidence of agriculture (2300 BC) is found in the Indus valley. These facts apply to the agrological history of the Western Himalaya as well. Apart from the Buddhist scriptures, records on the Indian agriculture are available in the travelogues of Megasthenes (300 BC), Fa-Hian (399-415 AD), Hiun Tsang (629-645 AD), Al-beruni (1031 AD) and a chronological description appeared in 12th century (*Krishi Parashar* 950-1100 AD, *Krishi Sukti* 1000 AD). *Arthashastra of Kautilya* (300 AD) also finds mention of agricultural economics in ancient India.

*Himavat*, todays Uttarakhand (*Kedarkhand* and *Manaskhand*) in a wider sense comprises of 9 districts of the Central Himalaya of Uttar Pradesh such as, Almora, Pithoragarh, Nainital, Chamoli, Garhwal, Tehri, Uttarkashi Dehradun and Udhampur. It is a range of lofty mountains and deep valleys, except for the narrow submontane strip of Bhabhar and Tarai which existed from the time of Pauranic history (*Skandh-puran*).

Although, the region has undergone repeated organization and reorganization under political turmoil, the entire tract in the ancient times seems to have been described as Kumaun. However, in 1839 the Kumaun province was divided into 2 districts viz., Kumaun and Garhwal. Later, the division was composed of 3 districts of Kumaun, Garhwal and Tarai, and subsequently the 2 districts—Kumaun and Tarai were redistributed as Almora and Nainital (total area 51000 $km.^2$, cultivated area 9000 $km^2$ and population 4.78 million, in 1981). It is territorially larger and numerically stronger than some member countries of the UN and Seven Sisters of NE India.

The history of agriculture of a country or region cannot be considered without the study of contemporary socio-political history and thus the history of agriculture in the Himalaya cannot be separated from the mainstream of the historical events in the subcontinent, although Uttarakhand has its own ancient and medieval history. A critical review of the old agronomic practices and the people may help in the identification of the reasons for the ecological holocast in the fragile Himalaya and clues for the environmental regeneration.

People of Uttarakhand are an intermixture of various ethnographical groups with rich and variegated cultural heritage, nevertheless, the history remains illdefined and disconnected. Before the Aryan invasion (2000 BC), Kirat, Saka, Naga and Khas dynasties appear to have ruled the Western Himalaya. The people have a great heritage of centuries–inheritors of great builders of temples, known for their gallantry, honesty, hard work and intelligence (raji aborigines, immigrants and bhotia).

Hill agriculture involves great labour and considerable skill in tilling the terraced fields with ages-old implements. Though fortunately free from serious floods, draughts and blight, land-slides and leeching of the soil may present continuous hazards. Women are the backbone of the entire rural economy including agriculture. During the period of great famine in India (1868), the hills remained unaffected and grain was exported to the plains. It seems that the agro-ecosystem in the Himalaya was not much disturbed till 1910. Ramsay comments 'the hill-farmer is comparable to the best farmers of India who worked in perfect harmony with his environment for centuries'.

During the early period, the cultivators were unstable and migratory. They used to desert the land because the right on the land was sovereign which did not permit prosperity in agriculture and growth of the industry. Every class of cultivation from intensive cropping (upto 6000') to shift agriculture (forest-farming) was in vogue. Cultivation in the higher precipitous slopes was poor. General pattern of crop rotation followed since ages was well suited in the Himalaya and thus the land fertility was maintained. The situation changed with the introduction of HYV, exotic crops and fertilizers etc. The valleys are highly fertile where every patch of land is intensively tilled. Atkinson (1882) described Someshwer and Bhimtal valleys as most prosperous and beautiful in the whole of Asia. Khas invaded Western Himalaya from the Central Asia and spread in the low-lands for livestock grazing and it seems they were good husbandry-men.

Hills were prosperous till 1762 AD and in fact, there is no evidence of import of grains from plains till the advent of 1900 AD. Agricultural commodities exported were turmeric, ginger, red pepper, buckwheat, ghee, honey, walnuts, wax, soap-nuts, tea, sheep, goat, Tibetan pony (Gunt), agro-forestry products (timber, species and condiments, drugs and narcotics, tans and dyes, gums and resins, musk and herbs etc., 1830). About 24966 quintals of wheat and barley was exported to Tibet (1883) and 2490 quintals of wool was imported (1872).

Different dynasties which ruled Uttarakhand were Katyuri for 700 years (upto 700 AD), Chand for 1000 years (757-1790 AD), Rohilla for 9 months (1743 AD) Gurkhas for 25 years (1790-1815 AD) and British for 135 years (1815-1947 AD). In certain parts of the region, Khas (869-1065 AD) and other princepalities ruled from time to time. Frequent fueds between the different hill tribes rendered the land waste and fallow thus adversely affecting the agriculture. Hwen Thsang (7th century) on his visit to the Western Himalaya observed that the region is very prosperous, soil is fertile and the main corps are barley, chua (*Hordeum coetesta*) and phaphar (*Fagopyrum tataricum*). People reared sheep and horses and were chiefly engaged in war and husbandry. Early ancestors of humped cattle, *Bols indicus (B. primigenius, B. acutiforons* and *B. nomadicus)* and wild buffaloes (*Bubalus arnoe*) were probably reared in the Himalayan tracts.

People of Uttarakhand with a curious blending of Buddhist and Brahmanical cult worship *Bhumia (Khetrapal)* as a tutelary god of fields in each village for prosperity-germination of seed and protection of crops; *Saim* a deity for the preservation of the forests and *Chammu (Badan, Kalbisht)* for increased livestock production. It reflects on the antiquity of agriculture in the Himalaya.

Although, the land under plough in the hills due to geographical peculiarities is very limited (19 per cent in 1972), various phrases for the description and use of land and its measurement for time immemorial indicate the systemic pattern of agriculture practiced, e.g., irrigated lowland (*Talaon*), unirrigated upland (*Upraon*), swampy land (*Semar*), a sheet of plain cultivated land (*Tano, Tok*), tilled valley (*Gair*), fallow wasteland (*Banjo*), cultivated steep slopes (*Relo*) etc; similarly, sheep pen (*Khor*) and cattle sheds (*Goth, Kharak*) were in use. The unit; for the measurement of the land such as *Nali* (area requiring 2 kg of seed for sowing (240 sq yds) and *Bissi* (40 yds less than an acre) and *Bhara* (6 *Bissi*), *Nali* equal to 16 *Muthhi* and *Bissi* equal to 20 *Nali* etc., are indicative of an established agrarian system in the Himalaya.

Main crops in the Himalaya are necessarily the same as in the North Indian plains except that the high altitude varieties are early, temperate and hardy. Rice is the main crop sown, of which a large number of varieties have been described (*Basmati, Duda, Hansraj, Timila, Jamal, Anjan, Sal, Rotuwa, Katyuri, Dhania, Kirmuli, Motiya, Jogyana, Chunkuli, Salam, Paltya* etc.) and about four wheat varieties are found mention. Among the coarse grains, Kodo (*Paspalum scrobiculatum*), Mandira (*Optismenus frumentaceus*), Koni (*Setaria tlalica*), Mandua (*Fleusine)*, soyabean (*Glcyine soja*) and various pulses oilseeds are common produce. *Chura* (*Bassia butyraea*) is used for making vegetable butter of pomade, is of pharmaceutical importance.

## Alpine Agriculture (Bhotia agriculture)

Fiscal divisions in the northern alpine valleys, bordering Tibet is the area called Bhot (Byans, Chaundas, Johar, Darma, Niti etc.) comparises mainly of the snowy ranges. The narrow valleys are cultivable but the tract largely consists of barren rocks, forests and snow peaks. Bhotia are industrious, hardy and Mongoloid people who monopolised Tibetan trade (1962). They are more interested in commerce than in agriculture. The later being of secondary importance. A single crop is obtained during the period of habitation and the yield is scanty. Major crops are buckwheat (*F. esculentum*), phaphar (*F. tataricum*), chua (*A. frumentaceus*), Bhotia wheat or *Nephal* (*Tritfoum vulgare*), celestrial barley (*Hordoum Himalayanse, H. harastihylem*), awnless monostrous barley (*H. aegieras*), oat (*Avena sativa*) and wild oat (*A. fatua*).

Bhotia villages are built in the hollows and ravines to avoid avalanches and furious gales. They take their animals (sheep, goat, cattle, yak, ponies) to the alpine pastures of meadows (*Bugyals*) for flushing and steaming up of their livestock. However, recent geo-political developments have changed the agro-ecosystem in this area. About 4800 quintal of wool was exported to the plains from Kumaun (1909).

**Submontane Agriculture (Tarai and Bhabhar)**

These are submontane tracts in the south which have been integral part of the kingdom of Uttarkhand. Bhabhar is water-less forest land, 5-15 miles in breadth which as uncultivated till very late and Tarai is a marshy savanah about 10 miles in breadth and 90 miles in length. These tracts provided excellent pastures for live-stock where forest farming was practised. The land remained un-inhibited owing to thick forests, hostile malarious climate, bandits and wild beasts (454 cheetah were killed between 1892-1901). Early settlements probably started in 1385 AD by Tharu and Boksha tribals. Due to rich forest wealth and rich soil, it was under constant assault by the plains landlords and was annexed many times from the hills. British after taking this area finally determined the rights of hill people in this region. Trail commented 'if the plain farmer can compete with the hill farmer in this difficult region, the later should make room for the former'. Some forests of Tarai were as luxuriant as the best ones in America (Atkinson, 1882). Extensive canal system for irrigation was laid down after the mutiny (1861). Inspite of the rich fertile soil, the mode of agriculture was traditional (3 rice crops were sometimes sown viz., *Gaja* (April-May), *Bijhura* (June) and *Rasauta* (July).

## Agrarian Management

History of Uttarakhand can also be divided into ancient, mediaval and modern periods and can be discussed under following heads in relation to agriculture as well. However, complete details are not available.

### 1. Chand Dynasty

It may be stated that various Himalayan kingdoms were hardly interested in rural development. Records indicate that the earliest land settlement in Kumaun was done by Bhim Raja of Sor (Pithoragarh) in the year 1518 AD. Agricultural community was divided into different classes such as proprietors (*Thatwans*), cultivators (*Khayakars*), those who paid the land revenue in cash (*Sirtans*) and 126 slaves (*Kainis*) etc., having different rights on the land they tilled. The land tax was payable in 36 different forms (cash, grain, bhet and coolie etc.) and were realised through a number of middlemen. Apart from it, commerce, forest-produce and livestock (*Gheekar* 25 paise per buffalo). Cash tax was called *Sirthi* and that paid in kind was *Kut*. There were 68 types of taxes levied on the farmers, viz., 32 items of ministerial fees (*Battis Kalam*) and 36 of revenue (*Chattis Kalam*), leaving very little beyond subsistence with the cultivators. The tax was 1/3rd of the production for ordinary and 1/2 for the fertile land. Villagers in order to avoid tax started 'roof-top- cultivation'

### 2. Gurkhali Period

Gurkha rule in Uttarakhand was ruthless and cruel (1790-1815 AD) that the villagers deserted the dwellings and the terraced land become barren and fallow. Old revenue rules were completely changed (1812 AD) and the taxes were heavily increased. The state revenue declined owing to bad management, poor administration and military rule. It resulted in decreased agricultural production. Gurkha were selling slaves (Rs. 10 to Rs. 150) when the price of a pony was Rs. 250 The prices of agricultural commodities fell down. It appears that land settlement was done during this period (1805).

### 3. Mohammadan Period

In fact, Uttarakhand was never ruled by Muslim kings except for a brief period (1743 AD) by Rohillas marked by loot and destruction. They redistributed the administration of Himalayan lowlands. Farishta (1623 AD) writes about province of Kumaun extending from Tibet in the north to Sambhal in the south which was prosperous and rich in copper and gold.

**4. British Period**

British conquered Kumaun in 1815 AD, although they captured the lowlands in the year 1805 and freed hill people from the tyrany of Gurkhas. This was the period of agricultural development under priority, in the administration of Trail, Bettan and Ramsay. They rationalized the land revenue after meticulous survey, assessment and mapping of the land. The land was classified (areable, waste and forest etc.). British provided stability and security to the farmers, laid irrigation channels and cleared the areas infested with tigers (624 tigers were killed in Kumaun between 1860-1880 AD). Tarai and Bhabhar was permanently merged in the hills to provide pasturage and winter harvest to the hill people.

These fiscal reforms increased the price of land and the prices of the agricultural commodities increased by 400 per cent and more waste-land was tilled (1821). Agricultural land was further classified as irrigated, first class dry and second class dry. For assessment, produce of the irrigated land was equal to the double of the second class land was equal to 1/3rd more than the second class. They made rules suited to the terrain and the people. However, it resulted into an increase of 81.43 per cent in the land revenue (from Rs. 0.75 per *Bissi*). During this period, the prices of grains were low and the produce was about 5.5 quintals of rice, 4.5 quintals of wheat, 3 quintals of mandua and 2 quintals of urd per acre. Record show 5673 villages, 3 million *Bissi* of total and 0.2 million *Bissi* of cultivated area (1863 AD).

British province of Kumaun was divided into small units-subdivisions, having a cluster of villages (*Patti* and *Paragnah*). Changes were made in the tenural rights on the land which were with the sovereign rules. Exotic crops and livestock cross-breeding programmes were started. Tea cultivation was started in 1846, and in the year 1880, 9000 quintals of tea was produced in 3342 acres of land. East India Company in the year 1800 indicated the prospects of hemp (*Cannabis sativa*) cultivation for the manufacture of superior quality of fibre.

Sericulture in Uttarakhand was originally introduced by Indra Chand, ruler of Chand dynasty in the year 797 AD by importing silk-moth from China through Tibet and Nepal. Whatsoever, Captain Hutton (1856) recommended silk industry schemes in the hills. Later, all these projects died out.

Exploitation of the Himalaya by British people resulted into Forests Revolution (1911 to 1917), *Kuli Begar* revolution (1921) which later merged into the main national revolution for freedom.

Post independence era of agricultural development is marked with the *Zamindari abolition* and mechanization of submontane regions.

## Inferences

A review of the agricultural history of the Uttarakhand Himalaya revealed that the 'Green' and 'White' revolutions have no relevance for these regions probably because of the constraints such as socio-political changes (Indo-Tibet trade restrictions), difficulties in the mechanization of agriculture, continued fragmentation of the already small land holdings, denudation of the vegetal cover, lack of inputs, discontinued winter migration to the foot-hills for pasturage and harvesting and summer migration to the alpine meadows for pasturage has increased the biotic pressure, population increase, and lack of strategic planning (tea, brewery fibre, herbs, honey, silk, unconventional livestock-feral farming etc.).

## References

Anon 1975. *Encyclopedia Britanica.*, Vol. 11 & 13., William Bantons Pub., London., pp. 516 & 562.

Anon 1975. *Smarika : Almora Janpad.*, Indra Printing Press, Almora, p. 116.

Anon 1981. *Bharat : Varshik Sandarbh Granth.*, Publication Div., Information & Broadcasting Ministry, Govt. of India, New Delhi, p. 689.

Atkinson, E.T. 1882. *The Himalayan Gazetteer.*, Vol. 1, 2 & 3, 2nd edn., Cosmo Pub., New Delhi. p. 1981.

Awasthi, D. 1987. A Hill Demand, *India Today*, New Delhi, July 15, p. 87.

Corbett, J. 1954. *The Temple Tiger*, Oxford Univ. Press, New Delhi, p. 183.

Jameson, W. 1845. Report on Culture and Manufacture of Tea in Kumaun and Garhwal., *J. Agri. Hort., Bengal*, Vol. II & IV.

Joshi, B.P. 1964. Himalayan Pastures : A Blessing to Sheep Breeders., *Indian Livestock, ICAR, New Delhi.* p. 18.

Joshi, B.P. 1986, High Altitude Animal Agriculture in the Central Himalaya : Strategies for 2000 AD., *5th National Convention of Indian Society of Veterinary Medicine, Trichur, Kerala*. pp. 65.

Joshi, B.P. 1987, New Strain of Goat for the Semi-temperate Climatic Regions of the Central Himalaya., *Pashudhan* 2(6) : 1.

Joshi, 1990. Animal Agriculture in Central Himalaya. *Himalaya : Environment, Resources & Development* (U.G.C.) Almora Book Depot. pp. 480.

Joshi, S.D. 1929. Report on the Preliminary Agricultural Survey of the Almora District., *The Supdt. Government. Press, United Provinces,Allahabad.* p. 12.

Khanka, S.S. 1990 Lopsided Industrialization in Kumaun Div.–a Study. *Yogana* 8 : 24.

Nautiyal, J.C. and P.S. Babor, (1985). Forestry in the Himalayas : How to Avert the Environmental Disaster, *International Science Review* . 10, 27.

Pande, D. 1984. *Kurmanchal Kesari Badri Datt Pande Janm Shatabdi Smarika* (edited.), Shakti Prakashan, Almora. p. 516.

Pant, C.P. andN.G. Gratz, 1979 Outlook on Agriculture., *WHO Chronicle* 10, 111.

Pathak, C.S. 1985. *Pahar-2* (edited); Peoples Association for Himalayan Area Research, Nainital p. 224.

Paul, S. 1985. *Fuelwood in India; Prospects and Propositions.*, India House Development, New Delhi, p. 98.

Randhawa, M.S. 1980. *A History of Agriculture in India.*, Vol. 1, 2 & 3, ICAR, New Delhi, p. 410, 512 & 422.

Roychaudhari, S.P. 1964. *Agriculture in Ancient India.*, New Delhi. p. 250.

Sankhala, K. 1987. A Great Himalayan Plunder, *Frontline*, New Delhi, May 30. p. 50.

Sanskritayanas, R. 1958. *Kumaun*, Gyanmandal Ltd., Kabir Chaura, Varanasi, p. 359.

Singh, J.S. 1985. *Environment Regeneration in Himalayan: Concepts and Strategies*, (edited). OHEA, p. 468.

Venkataraman, G. 1987. Overburdened Ecosystem., *Frontline*, New Delhi, May 30. p. 54.

Vaishnav, Y.D. 1983. *Land and People of Himalayan District*, Almora Book Depot, Almora, p. 262.

# Chapter 7

# Irrigation Development and Agriculture in Garhwal Himalayas

*K.N. Joshi*

In India, the importance of irrigation emnates from pervasive character of its impact on agricultural economy, 'Satpati' (1984). As a traditional protective input it ensures a secure harvest, acts as an insurance against inadequate and inconsistent monsoon, and as such constitutes a 'sine-qua-non' for agricultural stability. However, the advent of new crop technology has considerably enhanced the curuciality of irrigation as a basic productive input. Briefly, it possesses significant area effect, crop pattern effect and yield effect. In this light, present paper aims to :

(a) evaluate State Governments efforts in developing irrigation during last 10 years, and

(b) to assess the area, crop pattern and yield effects of irrigation development in hill agriculture.

Although, a number of important rivers and their tributaries have their origin in the Himalayas, but the agriculture in the hills is mainly rainfed. Nearly 80 per cent of the regions gross cropped area depends exclusively on rain, which is unevenly distributed both in respect of time and space. Water is not available in most parts of the region and more than 70 per cent of the surface water remains unexploited and under developed. The scarcity of irrigation facilities is evident by the fact that in 1983-84 only 16.84 per cent of the net sown area (NSA) of the region was irrigated. What is more pinching is that, most of its share is limited to rainy season.

Therefore, irrigation would have been given the highest priority in any programme for increasing agricultural production and self reliance.

## I. Financial Investment on Irrigation

Till Third Five Year Plan, public investment on irrigation in the Hills was very insignificant, but afterwards the State Government has paid more attention towards the development of agriculture in the Hill region. As a result, irrigation development had been given top priority and more funds were allocated. During Fifth and Sixth Five Year Plans, the State Government has invested Rs. 7.73 and Rs 16.54 crores respectively in the region. On an average Rs 15.46 lakhs and Rs 33.09 lakhs were invested in each district of Garhwal region, showing an increase of more than double during Sixth Plan over Fifth Plan. Perhead investment, as shown in Table 1, also show an increase from Rs 39.3 in Fifth Plan to Rs 68.1 in the Sixth Plan (70 per cent).

**Table 1 :** Per-head Public Investment on Irrigation (in Rs.)

| Districts | *1974-75* *Actual* | *1983-84* *Actual* | *Vth Plan* *Planned* | *VIIth Plan* *Actual* |
|---|---|---|---|---|
| 1. Pauri | 1.09 | 9.5 | 21.51 | 45.31 |
| 2. Tehri | 2.26 | 10.68 | 32.96 | 53.97 |
| 3. Chamoli | 0.69 | 18.66 | 38.31 | 105.47 |
| 4. Uttarkashi | 1.75 | 32.24 | 95.40 | 191.59 |
| 5. Dehradun | 2.63 | 12.94 | 58.11 | 47.60 |
| 6. Garhwal Region | 1.96 | 14.02 | 39.28 | 68.10 |
| 7. U.P. | 12.9 | 20.3 | 58.9 | 94.7 |

Source : 1. Fifth & Sixth Five Year Plans of Pauri, Tehri, Chamoli, Uttarkashi and Dehradun Districts.

2. Fifth and Sixth Plans of Uttar Pradesh.

Above table reveals the fact that during 1974-75 to 1983-84 perhead government investment on irrigation increased remarkably, by more than 7 times, in Garhwal region. Though, the total amount is still very meagre. In relation to it government investment in the State increased by 1.5 times over its high base figure. Similarly, during Fifth Plan period perhead investment or irrigation was Rs 39.28 which increased substantially to Rs 68.10 in Sixth Plan period, whereas corresponding figures for the state were Rs 58.9 and Rs 94.7 respectively. This considerable increase in investment on irrigation is seen particularly because some major projects, viz.

Maneribhali, Tehri Dam, Vishnu Prayag, Utyasu and Sweet projects were surveyed, investigated and construction started. Though, owing to regions geographical conditions these large projects will not benefit the area so extensively as they will benefit the western plains of the State.

Apart from it, the Central and State Governments provide loans and subsidies to individual farmers and village communities from time to time for constructing and improving the privately owned irrigation works. These works also add to the irrigated area of the region (Joshi, 1987).

## II. Development of Means of Irrigation

Generally canals, gravity channels, water lifting and storing devices are considered suitable in a hill region. Increasing government investment during planning period developed irrigation network and irrigating capacity in the region. Table 2 depicts the growth of minor irrigation works in Garhwal region.

**Table 2 :** Development of Irrigation works in Public Sector

| *Districts* | *1973-74* | | *1982-83* | |
|---|---|---|---|---|
| | *Length of canals in kms.* | *No. of Tube-wells & Pump-sets* | *Length of canals in kms* | *No. of Tube-wells & Pump-sets* |
| 1. Pauri | 220 | — | 435 | 8 |
| 2. Tehri | 120 | — | 215 | 10 |
| 3. Chamoli | 98 | — | 180 | 4 |
| 4. Uttarkashi | 106 | — | 375 | 13 |
| 5. Dehradun | N.A. | 26 | 368 | 75 |
| 6. Garhwal Region | 544 | 26 | 1573 | 110 |
| 7. U.P. | 51238 | 12564 | N.A. | 19257 |

*Source :* 1. Shankhiya Patrika–Districts Pauri, Tehri, Chamoli, Uttarkashi and Dehradun 1975 and 1984.
2. Statistical Diary of U.P. 1975 and 1984.

Table 2 shows that in 1973-74, 544 km canals and 26 tubewells were operating in Garhwal region (excluding Dehradun), which, inclusive of Dehradun, increased to 1573 km canals and 110 tubewells and pumping sets in 1982-83. This shows an attractive progress. Regarding tubewells the plains of Dehradun and Pauri districts are only physiographically suitable, but in the hill valleys water lifting pump sets are now keeping pace and their number is increasing. Besides government efforts, a number of guls (gravity channels) tubewells and pumpsets and hauzes are also functioning in the region under private ownership. These means are also adding significantly to minor irrigation works. In 1982-83 4036 km guls, 794 tubewells and pumpsets and 12953 hauzes were working in Garhwal region.

## III. Growth in Irrigated and Cropped Areas

Relative growth in irrigated area and cropped area is an important indicator of area effected and generated by irrigation development. Regarding irrigation facilities the position of hill region in general, and Garhwal region in particular, is not satisfactory. In 1980-83 only 16.21 per cent of the net sown area in Garhwal region was irrigated, about 2.5 per cent higher than the base figures of 1971-74. In this regard Dehradun stood at the top while Chamoli at the bottom. Uttarkashi and Tehri also possess higher rank but others two agged behind.

**Table 3 :** Proportion of NIA and Compound Growth Rates of Irrigated and Cropped Area (From 1971-74 to 1980-83)

| Districts | *1971-74 NIA as % of NSA* | *1980-83 NIA as % of NSA* | *CGR of NIA* | *CGR of GIA* | *CGR of NSA* | *CGR of GSA* |
|---|---|---|---|---|---|---|
| 1. Pauri | 8.13 | 8.06 | 0.04 | 0.9 | 0.12 | −0.12 |
| 2. Tehri | 13.15 | 15.28 | 1.1 | 1.1 | −0.24 | −0.20 |
| 3. Chamoli | 5.0 | 5.44 | 0.02 | 0.9 | −0.53 | −0.33 |
| 4. Uttarkashi | 14.65 | 16.08 | 3.2 | 4.7 | 2.02 | 1.44 |
| 5. Dehradun | 32.35 | 40.93 | 4.5 | 2.7 | 0.41 | 0.95 |
| 6. Garhwal Region | 13.72 | 16.21 | 2.0 | 2.2 | 0.08 | 0.14 |
| 7. U.P. | 41.63 | 55.83 | 2.9 | 4.1 | 0.06 | 0.62 |

Regarding relative growth in irrigated area and sown area it is learnt that net and gross irrigated area grew at a faster rate than the net and gross sown areas in the region as well as in the state. Net irrigated area (NIA) and Gross irrigated area (GIA) recorded a compound growth rate of 2.0 and 2.2 per cent respectively, while NSA and GSA remained almost stagnant, recording only 0.08 and

0.14 per cent growth respectively in the region. The point to be highlighted is that due to considerable growth in NIA and GIA in Dehradun and Uttarkashi districts, NSA and GSA too have grown fast in these districts. But in other districts of the region, except Pauri in the case of NSA, cropped area changed reversely for three main reasons, (i) growth of irrigation was almost negligible, (ii) population growth in the past has not left any arable land without plough, and (iii) farmers are sometimes forced to put the marginal agricultural lands as fallow due to regular soil erosion in the region.

## IV. Changes in the Cropping Pattern

Because of the extremely small agricultural holdings, which area not also extensively irrigated, the cropping pattern in the Himalayan region has not undergone any substantial change during past decade. Briefly, it did not show any significant diversification yet. Major portion of the cropped area remains under foodgrain crops and very little under non-food commercial crops, though they are more profitable.

**Table 4 :** Distribution of cropped area among different crops in Garhwal Region (1982-83)

| *Crops* | *Pauri* | *Tehri* | *Chamoli* | *Uttar-kashi* | *Dehradun* | *Garhwal Region* | *U.P.* |
|---|---|---|---|---|---|---|---|
| 1. Paddy | 18.5 | 14.4 | 26.4 | 23.2 | 18.3 | 19.3 | 20.5 |
| 2. Mandua | 25.8 | 19.0 | 24.0 | 17.5 | 4.7 | 19.3 | 0.6 |
| 3. Sanwa | 15.4 | 20.3 | 5.4 | 8.0 | 5.0 | 12.3 | 0.5 |
| 4. Maize | 1.6 | 1.4 | 0.4 | 0.9 | 14.4 | 3.6 | 4.4 |
| 5. Wheat | 29.7 | 33.4 | 32.3 | 30.2 | 32.3 | 31.5 | 33.6 |
| 6. Barley | 6.0 | 4.3 | 4.9 | 1.7 | 2.6 | 4.3 | 2.5 |
| 7. Other cereals | — | — | — | — | — | — | 6.1 |
| 8. Pulses | 0.8 | 1.6 | 0.3 | 1.4 | 4.6 | 1.7 | 12.0 |
| 9. Non food crops | 2.2 | 5.6 | 6.2 | 16.8 | 18.1 | 8.0 | 19.8 |
| Total | 100.00 | 100.00 | 100.00 | 100.00 | 100.00 | 100.00 | 100.00 |

Source : Uttar Pradesh Ke Krishi Ankare 1982-83.

Table 6 reveals that in 1982-83, 92 per cent of the cropped area was under foodgrain crops and only 8 per cent of it was under non-food crops in Garhwal region. It is important to note that from 1971-72 to 1982-83 the area under non-food crops recorded an increase of

3 per cent. Since the proportions of NIA to net sown area (NSA) in Dehradun and Uttarkashi is comparatively high, therefore, proportion of cropped area under non-food crops in these districts is also found remarkably high, but other districts having small proportions of NIA register less land under non-food crops. Thus slowly developing irrigation could not substitute food crops by non food crops rapidly, though they are considered more remunerative.

As a result of irrigation development irrigated area under main crops such as paddy, wheat and maize grew in the region as well as in its constituent districts, as depicted in Table 6.

**Table 6** : Compound Growth Rates of Irrigated and Sown Areas of Main Crops (From 1971-74 to 1980-83)

| *Districts* | *CGR of IA under Paddy* | *CGR of IA under Wheat* | *CGR of IA under Maize* | *CGR of SA under Paddy* | *CGR of SA under Wheat* | *CGR of SA under Maize* |
|---|---|---|---|---|---|---|
| 1. Pauri | 0.56 | 0.61 | 8.31 | −1.48 | −3.14 | 0.14 |
| 2. Tehri | 0.97 | 2.41 | 0.57 | −3.53 | −1.56 | 0.55 |
| 3. Chamoli | 1.75 | 0.05 | 1.55 | 0.65 | −2.53 | −10.10 |
| 4. Uttarkashi | 5.25 | 4.31 | −3.0 | 2.28 | −0.51 | −1.64 |
| 5. Dehradun | 3.02 | 3.60 | 0.43 | 2.22 | 1.97 | 1.54 |
| 6. Garhwal Region | 2.0 | 2.45 | 1.09 | −0.56 | −1.45 | 0.81 |
| 7. U.P. | 5.84 | 4.83 | 2.70 | 1.50 | 2.88 | −2.39 |

The relative growth trends of irrigated area and sown area under main crops in Garhwal region are revealing. Irrigated area under paddy, wheat and maize is increasing in the region, but growth in the case of paddy and wheat is faster than maize. Maize is not an important crop in the hilly districts in reference to area coverage but generally sown in irrigated fields. It is also worth noticing, that growth rates of irrigated area under concerning crops are significant in Dehradun and Uttarkashi districts than others. In this regard a clear cut disparity may be observed between Garhwal region and the state. As far, the growth of sown area under paddy, wheat and maize is concerned, irrigation development in Garhwal region did not show much impact on sown area, under these crops.

## V. Growth in the Yield of Main Crops

Irrigation enhances the yield per hectare as it enables the farmers to switch over to improved technology. This change in the cultivation under irrigation leads to greater productivity. Studies (Kalhon *et al.* 1965) on the relative contribution of various component elements in the growth of output established that increase in per hectare yield is attributable to the adoption of new agricultural technology emerging from rapid expansion of irrigation facilities. Agriculture in Garhwal region also reveal this truth, as the productivity, H.Y.V. coverage and fertilizer consumption are increasing in response to irrigation development.

**Table 6 :** Compound Growth Rates of Productivity, Area under H.Y.V. and Consumption of Fertilizers.

(From 1971-72 to 1982-83)

| *Districts* | *Productivity of Paddy* | *Productivity of Wheat* | *Productivity of Maize* | *HYV Area under Paddy* | *HYV Area under Wheat* | *HYV Area under Maize* | *Per hectare consumption of fertilizers 1971-72* | *1982-83* |
|---|---|---|---|---|---|---|---|---|
| 1. Pauri | 1.52 | 0.51 | – 1.33 | 4.32 | 10.19 | 7.75 | 1.18 | 1.98 |
| 2. Tehri | 2.85 | 1.21 | – 0.78 | 15.23 | 9.23 | 16.80 | 1.10 | 1.45 |
| 3. Chamoli | 2.31 | – 0.16 | – 0.88 | 10.71 | 13.74 | 3.91 | 1.40 | 3.09 |
| 4. Uttarkashi | 3.35 | 0.74 | 0.75 | 14.97 | 8.36 | 11.95 | 2.70 | 6.45 |
| 5. Dehradun | 1.29 | 3.05 | 4.65 | 13.54 | 17.17 | 11.45 | 17.80 | 21.62 |
| 6. Garhwal Region | 2.23 | 1.16 | 3.11 | 11.92 | 10.38 | 11.03 | 4.05 | 5.99 |
| 7. U.P. | 3.07 | 4.08 | 0.48 | NA | NA | NA | 20.00 | 57.62 |

Table 6 reveals that productivity of main crops viz. paddy, wheat and maize is increasing, though at a slower rate, in the region. During 1971-72 to 1982-83, productivity of paddy, wheat and maize grew at a compound rate of 2.23, 1.16 and 3.11 per cent respectively in the region as against 3.07, 4.08 and 0.48 per cent respectively in the state. Per hectare yield of paddy has only recorded a smooth growth in each segments otherwise per hectare yield of wheat in Chamoli and maize in Pauri, Tehri and Chamoli have decreased. The other important fact is that growth in per hectare yield in Dehradun and Uttarkashi is found higher than other districts, due to better growth in irrigation in these districts. Coverage of High Yielding Varieties (HYV) seeds in each segment is increasing substantially, though compound growth rates of HYV paddy, wheat and maize are observed higher in Dehradun, Uttarkashi and Tehri districts than others. It is because they have comparatively higher proportion of net sown area 'Joshi' (1987).

Per hectare consumption of fertilizers is also responding to irrigation development in the region and its constituent districts. Though per hectare consumption of fertilizers is increasing in the region, but it is still very low when compared with that of the plains of U.P. In 1982-83, per hectare consumption of fertilizers in Garhwal region was only 6 kg as against 58 kg in the state. Dehradun and Uttarkashi with better irrigation facilities utilized more fertilizers than others. Thus the utilization of fertilizers in the hills, to a large extent, is determined by irrigation facilities.

## Conclusions

The above discussion leads us to the conclusion that inspite of the herculean efforts made by the Government of U.P. and the people of the region during last decade, irrigation facilities could not be provided to the desired level. Neither sufficient irrigation capacity was created nor created potential was utilized optimally with due care. Defective construction of many schemes, seepage improper distribution of channels, frequent landslides, lack of sufficient funds for quick repairs and unlevelled fields are some other problems retarding growth of irrigation in the hills in general and Garhwal region in particular. Irrigation development in the region could not change the sown area effectively. Scarcity of irrigation compelled the cultivators to stick to unproductive and less remunerative traditional agricultural practices. It could not change the cropping pattern extensively in the region. New technology which consists of the use of HYV seeds, chemical fertilizers, insecticides and huge capital investment have to concentrate up to only irrigated patches. Hence a small proportion of the cropped area is only under the progressive agriculture.

## References

Joshi, K.N., 1987. *Agricultural Development in Garhwal Region since 1971-72*. Garhwal University.

Kalhon, A.S. and S.S. Johl., 1965. Agricultural Growth Rates in Punjab, Panjab University, Chandigarh.

Vaidyanathan A 1965. Growth of crop output in India, 1951-54 to 1958-61.; An analysis by Component Elements. *Journal of the Indian Society of Agricultural Statistics*, Vol. XVII.

Satpathi, T., 1984. *Irrigation and Economic Development*, Ashish Publishing House, New Delhi, p.1.

# Chapter 8

# Dynamic Dimensions of Fruit Crops' Ecofarming in an Hill Environment

*D.C. Pandey*
*J.C. Kuniyal*

The inclination towards the adoption of high input technologies with the measures of mechanization, use of fertilizer and chemical pesticides in crop field operation have been the long term practice of farmers even in the third world countries. But one of the most important drawbacks of this technology is that it remained limiting only to large scale farmers. Successes achieved in soil irrigation, land cultivation and the use of chemicals are, however, accompanied by a loss of some very important protective natural properties (Gerasimer, 1983). Besides being various covert and overt naturo-cultural imbalances, ecological risks on the farm/local/regional levels remained problematic. Owing to the various increasing disorders, 'ecofarming concept' came into existence as it leads to a balanced agro-environmental development. The principal view of ecofarming is mainly based upon the use of extraneous inputs, achieving higher productivity and the protection of ecosystem.

The continuous criticism of high inputs technology in the crop cultivation is due to increasing problems of the means of transport and inhomogeneous distribution of farming inputs. However, India in particular and the hills in general are suffering from other (except high input technology) infrastructural services in the adoption of agro-crops' farming. The inputs, like improved seeds, adequate fertilizers, effective irrigation have not been properly distributed in the wake of ardent physio-climatic and socio-economic determinants. Moreover, though intensive and extensive cultivation measures (i.e., land improvement and reclamation) have been in usual practices. The adoption of these ad hoc and compromising solutions in sight have become more or less discourageous. The changes in techno-organisational levels in accordance with the changes in traditional crop systems is yet to overcome generally in India and particularly in the study region.

## Ecofarming Approach

Development, i.e., 'ecological' or 'specially adopted to locational requirements' is defined as the objective of achieving high, sustained productivity with limited use of extraneous inputs, while at the same time preserving or recreating a balanced ecosystem (Krings, 1990). The eco-rural development levels (local or regional) has been expressed with its methodological considerations as under :

— protective measures for erosion syndrome (by means of horticulture process).

— the enhancement in the growth of 'fruit crops mixed with vegetables' particularly in the rainfed agricultural zone.

— considerable use of compost through sheded and decaying leaves of deciduous type horticultural plants.

— optimum use of rainwater through constructing inward slopy terraces and building up of resistant *medhs* (boundaries) surrounding the fields.

— supply of fuel and fodder requirement of villagers through the self-designed and generated forest ecosystem.

— to economize the earning of farmers.

— to develop balanced dietary habits in corroboration with non-cereals.

— to maximize the subsistence dependency of villagers from food crops to fruit crops.

The main aim of ecofarming lies in the fact in achieving closed cycles of foods, energy and water in the local/regional level. The integration of plant biomass, weeds, compost and rainwater must constantly be made possible in the process of farming. Within this new approach of ecofarming, preservation of soil fertility and stabilization of ecosystem can be achieved.

The production of fruit crops has been evaluated through the computation of efficiency levels. With the revelation of the existing level of the development of fruit crops by means of efficiency in a unit of land, it becomes, relatively easier to pinpoint the potentiality or the restorative growth power of these fruits within the ecological envelops. Efficiency therefore can be regarded as a concrete measuring scale to examine the dynamic dimensions of fruit crops's ecofarming. Before discussing the dynamic dimensions of fruit crops' efficiency, it would be better to throw light on the concept of efficiency to avoid any confusion.

The term 'efficiency' refers to the existing performance of growing crops in a unit of land in relation to the physio-climatic and socio-cultural factors. Efficiency more precisely, is essentially a scientific device to philosophize the inherent fertility, productivity and potentiality of land. There are physio-climatic, techno-organisational determinants which singly or collectively affect the produce of crops. Despite the fact that the performance of crops' growth is primarily subject to the natural calamities, but sometimes even more cultural determinants, such as, choice of crops have far reaching effects on the agricultural efficiency in the forms of deterioration of land in terms of unsystematic practices of agriculture in the hilly terrain.

The ever increasing pressure of population and the daily needs of makind have widened the gap which may hardly be bridged up in coming future. The urgency of the day for people is to yield more and more produce of crops from the limited land resources. But the foodgrain's culture irrespective to the agro-ecologically sensitive areas of the hills can not be stepped up confidentially in the prevalent features of rugged and steep topography, irratic agro-climatic conditions, overgrazing and desertification of forested land, etc. As a consequence, several environment disorders, such as, landslips, landslides and soil erosion etc. have begun to reach upto their full maturity in making the crop fields filthy particularly in the U.P. Himalaya and generally in the whole mountains.

Moreover, in the study area- the U.P. Himalaya, the areal coverage of cereals is on higher side than those of non-cereals. The over practice of the cereal farming will prove nothing but will result in a great loss of inherent soil fertility and its capability on the one hand, and inadequate dietary elements in the intake of daily diet of villagers on the other. To reach upto an equilibrium stage in every respect of ecofarming, replacement of subsistence foodgrains farming into peasant farming and then commercial fruit crops' farming is the urgent need of the existing cropping pattern. This systematic but gradual changes in the traditional cropping pattern will inevitably require the crops which are economically viable and ecologically sound. Cultivation of agricultural food crops above 30° slope (Kuniyal and Om Prakash, 1990); are ecologically very sensitive and prone to the agents of denudation (weathering + erosion). Alongwith this, the productivity of agro-food-crops in the hills is declining day by day. The hilly tract needs serious attention as agronomic yield per hectare (64.1 x $10^5$ k cal/ha/yr) is very low (Kuniyal, 1989; Kuniyal and Om Prakash, 1990). Therefore, cereals are uneconomic as well as ecologically unfit. In such a diversified physio-climatic region which differs from levelled to rugged with steep topography; and tropical hot to Alpine climatic conditions, arable farming of subsistence crops will retrogade the wheels of ecosystems. The other factors, like small size of holdings ( i.e., 1.89 ha), the higher number of fragmentation, thin soil surface, rapid erosion, tilling small holdings (i.e 1.89 ha) though the indigeneous ploughing and usual practice of terrace farming through indigeneous tilling are some of the major and challenging echoes which can be heard everywhere in the hills. Usually a crop field is tilled by twice or thrice times during a year. This continual activity of ploughing makes the soil very loose alongwith various adverse changes in bio-physico-chemical properties of soil. Cultivation before the arrival of monsoons drastically enhances the rate of soil erosion in the U.P. Himalaya region.

To achieve good returns and to maintain ecological balance, hilly farmers should be encouraged for the development of horticulture, vegetables, and edible and medicinal plants in place of crop farming (Kuniyal, 1987); which requires either no or less ploughing. For making the whole region economically viable and agroecologically stable, fruit, culture and vegetable growth amongst horticultural activities may prove as boon endowed by nature offering the ecologically most suitable choice of horticultural development in the U.P. Himalaya.

## Present Methodology

Agricultural efficiency for 9 fruit crops, viz., apple, pear, peach, plum, apricot, chestnut, mango, lemon (includes also other citrus varieties), litchi, which are indispensably be grown in the U.P. Himalaya, have been taken into account. The basic data were colected from the Deptt. of Horticulture, U.P., Lucknow for the year 1991-92. Thereafter, per ha production of fruits and their summation values were reckoned which after addition was divided by the total yield of nine fruit crops of eight districts.

The derived values through the mentioned procedure is termed as Efficiency Location Co-efficient (ELC). The present format to determine ELC was adopted from Bhaskar (Bhaskar, et. al., 1987) with certain modifications. Based on the range of data, the study region has been categorized into four productivity grades, viz., medium (M) and quartile ranges ($Q_3$ to $Q_1$).

## Discussion

The efficiency index-co-efficient is computed as 15.24 for Naini Tal as the highest and 9.07 for Pauri district as the lowest (Table 1). Similarly, the total sown area coefficient (SAC) is derived as 18.09 for Naini Tal and 8.04 for Uttarkashi district as the highest and lowest respectively (Fig. 1). The ELC index values are obtained in the ranges of 15.24 to 14. 18, 14.17 to 12.95, 12.94 to 10.71 and 10.70 to 10.7. The regional average of ELC is recorded as 12.95. Moreover,

**Table 1 :** Fruit Crops efficiency in relation to total area sown in U.P. Himalaya (1991-92)

| *Sl. No.* | *Name of the District/ Region* | *Total per ha produce (in quintals) for nine fruits* | *Location co-efficient as agricultural efficiency (ELC)* | *Total area sown (in ha)* | *Location coeffiicient for total sown area (SAC)* |
|---|---|---|---|---|---|
| 1. | Almora | 25.38 | 11.17 | 27329 | 15.79 |
| 2. | Naini Tal | 34.62 | 15.24 | 31312 | 18.09 |
| 3. | Pithoragarh | 33.03 | 14.55 | 20522 | 11.86 |
| 4. | Dehradun | 23.29 | 10.25 | 19640 | 11.35 |
| 5. | Pauri | 20.60 | 9.07 | 22073 | 12.75 |
| 6. | Tehri | 30.36 | 13.37 | 21864 | 12.64 |
| 7. | Chamoli | 31.38 | 13.81 | 16406 | 9.48 |
| 8. | Uttarkashi | 28.48 | 12.51 | 13909 | 8.04 |
| | U.P.Himalaya | 227.14 | 100.00 | 173055 | 100.00 |

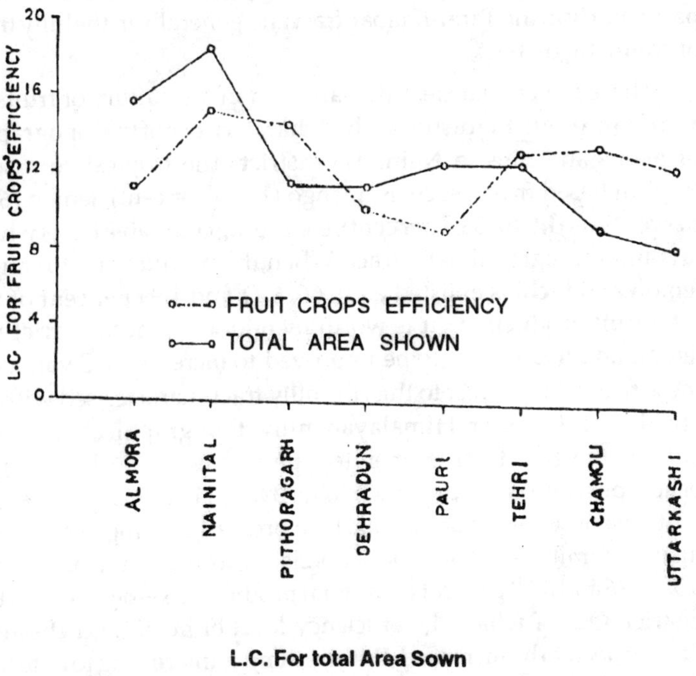


**Fig. 1.:** U. P. Himalaya : Distribution of Total Area Shown and fruit crops' efficiency, 1991-92

the results for fruit crops' efficiency are categorized into four different efficiency grades as under. (Table 1)

### High Fruit Crop Efficiency (15.24 - 14.18)

Naini Tal (15.24) and Pithoragarh (14.55) are the two districts of Kumaun which fall under high fruit crop efficiency. Naini Tal district is relatively situated in the close proximity of marketing facilities. Amongst other facilities, means of transport, high yielding variety of fruit seeds, almost favourable agro-climatic conditions, assured water supply, skilled labour and use of modern techniques, well off conditions of farmers, and establishment of new agricultural research centres such as Pantnagar Agricultural University have mostly favoured the growth of orchard farming particularly in the Tarai-Bhabar tract and generally in the hilly tract of Naini Tal district.

There is a crystal clear demarcation in the culture of fruits in accordance with the distinguished charateristics of the topography of plains and hills in Naini Tal district. The tropical cum sub-tropical based fruits, such as, mango (16.03 per cent), lemon (6.64 per cent) and litchi (3.67 per cent), etc. are the cultivable fruits which favours this agro-climatic tract. When the production of mango, lemon and litchi is enlisted as 16.66, 6.11 and 1.99 per cent of the total fruit production, it is worth mentioning that the variety of lemon and litchi is yet to be improved to increase ELC values of these fruits. In contrast to this, the hilly tract characterised with its Siwalik and Lesser Himalayan hilly topographical features, favours the growth of temperate type fruit crops, such as, apple, peach, pear and chestnut which cover 39.63, 11.08, 5.22 and 4.95 per cent area under them of the total fruit orchard farming in Naini Tal district. Similarly, their production is recorded as much as 53.83, 6.35, 5.26 and 0.79 per cent of the total production respectively in the district. Out of which the efficiency level of peach and chestnut further need to be increased. But one thing is interesting to note that the total fruits growing area coverage and their produce in the hills have relatively higher than that of plains. The main reason for having so is that the productivity level of temperate fruits is higher than that of the tropical cum subtropical fruits in the Tarai and Bhaban tracts.

Accordingly, Pithoragarh district produces apple, lemon and chestnut which occupy 24. 31, 15.40 and 10.18 per cent area along with the production of 42.11, 17.80 and 2.12 per cent respectively of the total production of fruits. The chestnut variety is of inferior type which is locally termed as *Kaithi akhrot,* whereas the superior variety, i.e., *Daithi akhrot, chestnut* appears to be prosperous in conserving the issues of crop ecology and ecofarming in many ways.

— Oils may be derived from chestnut (which is generally mixed with mustard in the *okhal*) that is helpful in meeting the oil and fat requirement to some extent.

— Chestnut may assure economically high profit to the villagers.

— To a larger extent, the fuel demands of villages may be fullfilled by the older and dead chestnut plants.

— Chestnut is a deciduous vegetative type amongst other temperate fruits which sheds their leaves in the winter season. In the passage of time, these old leaves become the inherent source of humus in the soils. Therefore, soil fertility is maintained.

— Like other fruits, it does not perish quickly due to its hard cover, hence does not require facilities like cold storage or close proximity to the market.

**Medium Fruit Crop Efficiency (14.17-12.95)**

Chamoli and Tehri are the two districts having ELC figures of 13.81 and 13.37 respectively. Apple, lemon and chestnut are the major fruit crops in Chamoli district which contributes to 24.26, 21.80 and 14.51 per cent area of the net sown area in the district. The produce for apple, lemon and chestnut is recorded as 48.46, 21.99 and 3.61 per cent respectively. The other subsidiary fruit crops which may yield high produce are peach and pear. Lemon variety citrus juicy fruits (i.e., mosambi, *narangi, malta,* etc.) are complimentary in restoring and recouping ecofarming. Moreover, lemon or other sweet citrus are the abundant supplier of vitamin 'C' in human diet. Besides, citrus are further valuable because they are basically grown for commercial purposes. Therefore, farmers may earn appropriate income. However, citrus fruits require certain measures to keep them fresh. Facilities like 'godown', cold storage, accessibility to roads, availability of skillful workers, high investment and immediate marketing services are of utmost importance.

Joshimath (consisting of surrounding villages of Sunil and Rabigram etc.), Holy, Adibadri in the upper Alaknanda catchments; Lohajung (in the way of Roopkund marg) and Chibaia (near Devthali) in the Dauli catchment and Gwaldam record significant apple production in Chamoli district.

Tehri is another district occupying medium fruit crop efficiency. Apple (31.92 per cent), mango 10.03 per cent) and lemon (9.96 per cent) are the important fruits amongst which apple is extensively grown in the upper reaches whereas mango and lemon varieties are intensively grown in the Bhagirathi and Mandakini valleys. These fruits enlist as many as 42.53, 17.21 and 13.55 per cent production of the total fruit output in the district.

**Low Fruit Crop Efficiency (12.94-10.7)**

Uttarkashi in the Garhwal region and Almora in the Kumaun register low fruit crop efficiency accounting for 12.54 and 11.17 ELC indexes. The SAC legends for these districts encompass as much as 8.04 and 15.79 respectively. This is indicative to the fact that agricultural efficiency is not proportionable to SAC values in Almora district. There are tremendous potentials to develop fruit culture in Almora district.

To have a micro level glimpses of predominating fruit crops, it was noticed that apple, chestnut and lemon based varieties extend upto about 45.29, 11.14 and 9.67 per cent area under the total horticultural crops in Uttarkashi district. Amongst these crops apple is not only the prominent crop but it also registers the highest produce (i.e., 64.82 per cent) in Uttarkashi.

In the Almora district, the principal crops are apple, lemon and mango comprising about 29.24, 13.46 and 10.05 per cent share of the total orchard farming respectively. Their produce is accounted as much as 39.21, 11.76 and 16.89 per cent respectively. Except these, the other fruits worthmentioning in Almora district are pear, peach, plum and apricot, etc.

**Very Low Fruit Crop Efficiency (10.70-9.07)**

Dehradun and Pauri are the two districts of the Garhwal region encompassing as 10.25 and 9.07 ELC results in the U.P. Himalaya. Their SAC figures are 11.35 for Dehradun and 12.75 for Pauri.

The major fruits crops of Dehradun district are mango, apple and litchi accounting for 22.48, 18.99 and 14.56 per cent of the total area under fruits. Their production is 26.20, 40.61 and 7.04 per cent for mango, apple and litchi of the total fruit output. Litchi seems to be less productive as its per hectare production is very low.

Accordingly, in Pauri district, apple, lemon and mango are the principal crops covering approximately 29.99, 13.59 and 10.03 per cent of the total harvested area under fruits. The produce of these fruits is recorded as 42.25, 14.07 and 17.26 per cent respectively.

**Conclusion**

To sum up, the ELC computed for the U.P. Hills are 13.66 and 11.54; whereas the former value is detected above the region's average (i.e., 12.95) and the latter one falls below the regional standard in the U.P. Himalaya. In other words, Garhwal is such a region where 10.89 per cent ELC values is yet to be increased to bring it upto the very same level of SAC.

When the distributional pattern of major fruits (amounting more than 10 per cent area of the total gross cultivated area under fruits) such as, apple (30.63 per cent), lemon (12.02 per cent), mango (11.18 per cent), etc. are associated to their yields (i.e., apple - 46.90, lemon - 12.39 and mango- 14.49), it has been concluded that in accordance with the percentage share of net sown area, their productivity figures constitute a positive relation. But there are certain fruits such as chestnut, peach, pear, litchi, plum and apricot which are not so productive in spite of having relatively larger areas under them. Such a heedless view of either planner or villagers towards these subsidiary fruit crops will definitely degenerate the ecological status of ecofarming rather than regenerating and restoring it in the Himalayan region. Amongst, land use systems, foodgrain cultivation has to be readapted to local requirements by including carefully selected elements of ecological farming (i.e., protective measures through horticultural process, enhancement of fruits mixed with vegatables, optimization of rain water and maximization of fruit crops in rainfed zone, etc.)

## References

Bhaskar, K. S. *et. al.*, 1987. Agricultural efficiency of Vidharbha Region (Maharashtra). *NGJI*. Vol. 33, pt. 2 (June).

Gerasimov, I. P., 1983. *Geography and Ecology*. Progress Publishers, Moscow.

Katyal, S.L., 1972. Twenty-five years of Research on fruit crops, *Indian Farming*, vol. 22, No. 6 (Sept.).

Krings, Thomas, 1990. Ecofarming in Africa : Experiences in East Africa- Possibilities and Models in the Sudan zone, *Applied Geography and Development*, vol. 36.

Kuniyal, J. C. 1987. Crop concentration and diversification in Naini Tal District, U.P. Himalaya, *NGJI*, vol. 33, pt. 2 (June).

Kuniyal, J. C. and Om Prakash, 1989. Irrigation Intensity and Agricultural Development : A case study of the Eastern Kumaun Himalaya. In D. C. Pandey and P. C. Tiwari (eds.) *Dimensions of Development Planning*, Criterion Publication, New Delhi.

Kuniyal, J. C. and Om Prakash, 1990. Rationality of Resource Utilization for sustained Growth in Central Himalaya. In P. C. Pandey (Ed.) *Current Trends in Development Planning*, Criterion Publication, New Delhi.

# Chapter 9

# New Farm Technology and Hill Peasantry in India

*D.C. Pande*

New Farm technology is the adoption of High Yielding Variety Programs (HYVS) in the sphere of agriculture with an objective to raise the quality and scale of food grains production. The technology is usually manifested in combining optimally modern farm inputs such as improved seeds, fertilisers, pesticides, irrigation and mechanical devices with land farm coupled with scientific cropping pattern and crop rotation. Thus the new technology in its composite form is understood alongwith its biochemical and mechanical aspects.

India did not achieve the new agricultural technology on its own all of a sudden. Instead only a persistent concern and effort of not less than sixty years has led her in the development of an indigenous expertise placing her ultimately right on the top with a very comfortable food grains output situation of almost 172 million tonnes per annum. However, needless to say here is that our task is not over and the agricultural technology will have to grow steadily with each invention and discovery if a record foodgrains output of 250 million tonnes has to be achieved to suffice the need of 10,000 million Indians by the turn of the century.

As referred above a number of programs and study circles have been responsible in shapingout new farm technology in modern India. A few of these which find a mention over here are : Grow More Food Campaign (GMF, 1942) Balwant Raj Mehta Committee Report (1952) and Ford Foundation Team of Experts (1959).

Although modern farm practices have proven merits for the farming community of today, however, their universal application sometime becomes doubtful. This controversy becomes still louder as we descend down from a big to a small farm holding. When a batch of noted experts considers modern agriculture of not much avail for the small farmers (Bhalla, 1972; Shah & Singh, 1970; Dwivedi, 1972; Mishra and Puri, 1984), however the other group ascribes the positive results of new farm technology also for the small peasantry (Saini & Bardhan, 1961; Griffin, 1974; Raj Krishna, 1976; Sharma, Roy & George, 1979). This controversy should be resolved with a facts finding approach through organising empirical studies about small farming sector all over the country. This very attitude prompted the author to explore the behaviour of hill peasantry in response to new agricultural devices in Kumaun Hills. For the sake of logistic convenience Almora district was selected as our study sample.

## Sample Profile

District Almora extends from 79.20° to 81.31° east longitude and from 28.59° to 30.49° north latitude. With a geographical area of 5385 sq. km. The district divides into two physiographic units, i.e., the greater and lesser Himalayas. The altitudinal zones show a variation from 600 to 800 meters. The low altitudinal zones in south west and south east, viz. Syaldeh, Bhikiasain Bageshwar, Garur and Takula are demarcated by 600 to 1200 m of contours while the maximum elevation zone Kapkot Block in the north extends upto 4500 m. There are 14 development blocks and the total number of populated villages is 2967. According to 1991 census the total population of the district was 7,53,373 out of which the farm population was 169814 (23 per cent). Considering that 93.72 per cent of the total population is rural, the pressure upon the rural peasantry is scrupulously high. This problem becomes more heart rendering as one recognizes exceedingly small cultivable area and the high degree of subdivision and fragmentation of an average land holding in the district. Figures of the Statistical Diary of Kumaun for 1982-83 and Finance Board, U.P. 1982 reveal that in the district out of a total number of 1,04,838–1,01,843 holdings are less than 5 hect. in area, while one third holdings are less than one hect. in area. The sample study conducted for 4 out of 14 development blocks of the district reveal that a farming family owns operational holdings ranging between 10 to 20 and, further, approximately 82-62 per cent farm holdings are less than 50 nalies (2.5 acres) in area.

The reason behind the acute fragmentation is the hill terraced cultivation besides the practice of traditional social institution of the law of inheritance. In the hill topography terracing becomes an essential evil and as such one should not expect the number of holdings to confine to a certain small limit what soever programme of land reform is enforced. It is also our experience from the study of the district that atleast in 22 per cent of such terraces which lie beyond a contour of 2250 m agriculture losing all rationales of economic cultivation reduces to a mere convention. New agricultural technology faces a real challenge to deliver goods to the doomed cultivators of these settlements. The real Oasis of hope in the cultivation map of the district is the flat fertile land tangled in the Saryu-Ramganga micro catchment. It comprises three development blocks of Garur, Bageshwar and Chaukhutia.

**Table 1** : The sample design

| *Study Area* | *Study Sample* | *Reason for the Selection of a Sample Unit* |
|---|---|---|
| District Almora with Development Blocks | Kosi-Saryu-Gomti microcatchment with Development Blocks | Sensitive to modern agricultural practices |
| — | — | — |
| 1. Syaldeh<br>2. Chaukhutia<br>3. Garur-Baijnath<br>4. Kapkot<br>5. Bageshwar | 1. Garur–Baijnath | Wheat-paddy-maize intensive modern agriculture technology implemented/demonstrated by IGADA |
| 6. Salt<br>7. Bhikaisain<br>8. Tarikhet | 2. Bageshwar | Rich in cereals and local varieties of tropical fruits, demon strations led by Vivekanand Lab. |
| 9. Dwarahat<br>10. Lamgara<br>11. Hawalbagh | 3. Hawal Bagh | Specialises in vegetable production of high class variety. |
| 12. Takula<br>13. Bhasiachhana<br>14. Dhauladevi | 4. Lamgara | Orchard plantation |

These blocks outline the fate of the district in matter of food grains production and may safely be put under the 'fine grain zone' of the district. Besides this 'fine grain zone' other blocks of the district on the basis of their agro-climatic character may be grouped under 'vegetable-orchard zone' and 'millet zone'. 'Vegetable orchard zone' comprises 5 blocks, viz., Lamgara, Tarikhet, Dhauladevi, Dwarahat and Hawalbagh while the 'millet zone' comprises the development blocks of Bhikiasain, Salt and Syaldeh. This zone is partially or fully arid sporting some local coarse varieties of grains and deserves a special attention when any policy of modernizing agriculture is formulated for the district.

**New Technology : Issues**

A study of the land use pattern for the district reveals cultivable land in the total reporting area dwindling from 69170 ha in 1983-84 to 62892 ha in 1984-85 and to 62892 ha in 1985-86; the corresponding figures for gross sown area also show a declining trend over the same period from 219663 ha to 179173 ha and finally to 175256 ha. Net sown area also represents a marginal fall during this time span. This speaks about decline in 'land intensity' in the district during the observed period. This fact arises serious concern in view of the already scanty arable area in the district. The decline in land intensity may be explained on the basis of the sharp rise in the cost of cultivation in recent years compelling the cultivators to withdraw from the remote and far flung holdings. This increasing indifference to professional cropping has resulted in enlargenning 'current fallows' in the district which may even endanger the creation of 'barren lands' in years to come. Modern agricultural strategy may provide a soothing solution to this burning problem by motivating the farmers in such sub-dwellings to plant improved varieties of orchards. Oil seed cultivation mainly that of sunflower and castor seed may also prove of high commercial value for the cultivators in such arid belts.

The district grows paddy, maize, pulses and the millets such as Mandua, Jhangura, Bhatt in the Khariff-Zaid season while Wheat, Gram, Mustard peas and Potato are grown as the major Rabi crops. In the irrigated tracts the cropping pattern is paddy based (sown over 80 per cent stretch) in the Khariff and wheat based (sown over 80 percent stretch) in the Rabi season. Rain fed areas are millet intensive where the cropping pattern varies from Mandua based to Bhatt based. In the grain belt one comes across fairly stable cropping pattern. However, intra-village differentials cannot be ruled out and in some exceptional cases cropping pattern may even emerge as lentil based or mustard based. We do not recommend any substantial change in the fairly stable cropping pattern in the hills for suggestion sake, excepting that, enough room should gradually be provided for the oil seed cultivation in the existing cropping pattern and that ways and means be devised to enable mashroom cultivation over moist cold tracts of the district.

## Seed Factor

After the land factor seed occupies next important place so far as farm turnover is concerned. It was a pleasant surprise for the author, to note that in a constrained agriculture like Almora, cultivators in majority were found using hybrid and improved varieties of seeds. The percentage of the users of such types of seeds varied from 60 to 80 per cent in the valley region while far uplands it stood slightly lower at 50. For the uplands, which are rainfed, hybrid seeds are not recommended. 'Vivekanand Parvatiya Krishi Anusandhanshala' (a research and extension unit of ICAR based at Almora since 1936) in such an eventuality has devised improved varieties of the traditional seeds which can be applied both in irrigated and unirrigated conditions of the soil. Vivekanand Lab has also continuously demonstrated the new agricultural technology in real practice for motivating the rural people. For this purpose the lab had adopted Chaunsali village. Another land mark in the agricultural advancement in the district is Indo-German Association for the Development of Agriculture (IGADA) which functioned from 1966 to 1978. During its tenure of 13 years the agency concentrated its activities on improvements in the Garur–Baijnath block. To its credit it owes some very productive varieties of Paddy, lentils and oilseeds. The Directorate of Fruit Utilisation (1953) runs a model government orchard at Chaubatia in Tarikhet block. Later in 1963 the Directorate set up a U.P. Hill Horticulture Board. Under the guidance of the Directorate two big orchards : one at Chaubatia and other at Doonagiri have been set up and three small nurseries respectively at Bageshwar, Karmi (Kapkot Block) and Matela (Lamgara block) installed. Besides, two progeny orchards are functioning at Garurabanj (Dhauladevi block) and Dol (Lamgara block). All these units together devote in producing quality fruit plants, seedlings and vegetable seeds to meet the demand of the region.

**Table 2 :** Some improved varieties released by Vivekanand Lab

| Variety | | Year | Variety | | Year |
|---|---|---|---|---|---|
| Gehun | 401 | 1973 | Rajmash | 63 | 1978 |
| Gehun | 404 | 1973 | Masoor | 1 | 1981 |
| Gehun | 421 | 1978 | Matar G C | | 1981 |
| Madua | 204 | 1978 | Soya | 1 | 1982 |
| Madira | 8 | 1978 | Gahat | 1 | 1981 |
| Matar | 1 | 1978 | Bauni Bean | | 1973 |
| Makka | 43 | 1979 | Lata Bean | 12 | 1973 |
| Dhan | 398 | 1980 | Lata Bean | 17 | 1973 |
| Dhan | 206 | 1983 | Shakarkand | 19 | 1974 |
| Makka | 54 | 1962 | Groundnut | | 1983 |
| Makka | 42 | 1972 | Dhan– | 16 | 1984 |
| Makka | 1 | 1973 | Barley– | 1 | 1984 |
| Makka | 16 | 1989 | Gehun | 616 | 1986 |
| Amber Pop corn | | 1981 | Makka | 41 | 1986 |
| Madua (Regi) | 101 | 1974 | Makka | 88 | 1988 |
| Madua | 204 | 1978 | Dhan | 163 | 1987 |
| Madira | 8 | 1978 | Madira | 29 | 1988 |
| Jawar | 1 | 1973 | Madira | 21 | 1989 |
| | | | Madira | 124 | 1989 |

Note : Each variety carries VL as prefix to its name

## Fertiliser Application

While the use of the better seeds is quite convincing, the situation at the front of fertiliser use is not as assuring. Very low consumption figures for NPK combine during past few years proves this fact.

**Table 3 :** Fertiliser distribution : comparative analysis ('000mt)

| Variety | Nitrogen | | Phosphorous | | Potash | |
|---|---|---|---|---|---|---|
| | Kumaun | Almora | Kumaun | Almora | Kumaun | Almora |
| 1980-81 | 22.4 | 2.41 | 0.77 | N.A. | 3.6 | N.A. |
| 1981-82 | 26.7 | 0.00 | 9.10 | N.A. | 3.6 | N.A. |
| 1982-83 | 31.1 | 3.42 | 13.15 | 0.19 | 5.7 | 0.054 |
| 1983-84 | 34.4 | 0.41 | 10.88 | 0.30 | 3.6 | 0.078 |
| 1984-85 | 24.2 | 0.34 | 11.9 | 0.33 | 5.0 | 0.091 |
| 1985-86 | 21.8 | 0.13 | 8.7 | 0.19 | 4.8 | 0.020 |
| 1986-87 | 22.5 | 0.30 | 7.8 | 0.15 | 2.12 | 0.060 |

*Source :* Krishi Nideshalaya U.P. 1985, Statistical Diary Almora 1986, Economic Analysis Kumaun Mandal 1988.

Low levels of fertiliser consumption in the district may be attributed to hopelessly low levels of soil fertility, high costs of the fertilisers, general ignorance among the cultivators and absence of package programs relating to input dozes. Another factor which draws attention in this regard is that most of the cultivators are not applying the recommended dozes of fertilizers, or, the crop and soil differentials are not taken care of. In most of the cases actual doze is scrupulously smaller than the recommended doze.

The following table gives an account of the dozes of fertiliser as recommended by the Vivekanand Lab for some important crops.

**Table 4 :** Levels of fertilisers recommended for new varieties (kg/ha)

| *S.No.* | *Crops* | | *N* | $P_2O_5$ | $K_2O$ |
|---|---|---|---|---|---|
| 1. | Rice | Irrigated | 80-100 | 60 | 40 |
| | | Rainfed | 40 | 30 | 20 |
| 2. | Maize | Rainfed | 100-120 | 60 | 40 |
| 3. | Mandua | Rainfed | 40 | 20 | — |
| 4. | Soyabean | Rainfed | 20 | 80 | 40 |
| 5. | Rajmash | Rainfed | 20 | 60 | 30 |
| 6. | Wheat | Irrigated | 80 | 60 | 40 |
| | | Rainfed | 40 | 20 | — |
| 7. | Barley | Rainfed | 30 | 15 | — |
| 8. | Lentil | Rainfed | 10 | 40 | — |
| 9. | Peas | Rainfed | 20 | 60 | — |
| 10. | Gram | Rainfed | 20 | 60 | — |
| 11. | Toria | Rainfed | 40 | 30 | — |
| 12. | Rai | Rainfed | 50 | 40 | 30 |

*Source :* Annual Report, VPKA, Almora : 1988

## Irrigation dogma

Means of irrigation are as important as other aspects of modern agriculture. 20 per cent of the sown area commands irrigation–facilities in the district. Irrigation is a point of discontentment among the cultivators with oddly situated land holdings : either elevated or tukked up amidst spherically clustered land holdings of other persons. In such a predicament irrigation becomes a problem instead of a means of development. Similarly a marginal cultivator, most distantly situated from the water source, may also find himself in great difficulty when the water flow of a canal or a 'gool' becomes so feeble before reaching him that he is not in a state to irrigate his field. Hence due care and caution are required so as to ensure proper water flow so that a cultivator does not go dry even in an irrigated zone. The irrigation potential built up in the district during past few years finds a description in the Table 5.

**Table 5 :** Irrigation sources : a comparative analysis

| *Year* | *Length of canals* | | *Pumpsets (Nos)* | | *Tanks (No)* | | *Gool (km)* | |
|---|---|---|---|---|---|---|---|---|
| | *Kumaun* | *Almora* | *Kumaun* | *Almora* | *Kamaun* | *Almora* | *Kum.* | *Alm.* |
| 1981-82 | 2238 | 470 | 2651 | 40 | 2224 | NA | 3652 | 2066 |
| 1982-83 | 2450 | 473 | 2825 | 40 | 8868 | 5320 | 3652 | 2066 |
| 1983-84 | 2709 | 656 | 2938 | 44 | 9505 | 5538 | 4129 | 2136 |
| 1984-85 | NA | 656 | NA | 44 | NA | 5748 | NA | 2176 |
| 1985-86 | NA | 669 | NA | 44 | NA | 5677 | NA | 2202 |

*Source :* Statistical Diary 1985, 1986, Kumaun Region

## Some Observations

The comparative analysis of yield rates of local and High Yield Variety (HYU) seeds reveals that the HYV's of paddy and wheat yield 2 to 3 times more output compared to local varieties. This also resulted into more income accrual to the farmers. It was found that majority of the farmers in the study area were using either HYV or the improved seeds. However, farmers were not very particular in the year to year renewal of the HYV seed which is a prerequisite for the success of modern agriculture. Only 50 per cent to 60 per cent of the HYV users were adopting this practice. If the farmers can be persuaded for a cent per cent renewal of HYV seeds, then adoption of modern agriculture ensures a bright future for the study area in years to come. Farmer has also come across similar experiences for likewise regions (Farmer, 1980).

Regression and correlation analysis revealed that HYV technology increases gross output and yield. This leads to a greater demand for family and hired labour. Sadhu and Mahajan (1985) and Farmer (1980) have also reached to similar conclusions. The conclusions of our study area are interesting from this point of view that use of hired labour varies substantially from valleys to upland in a 4 : 1 ratio. Hired labour is mostly employed for cereal cultivation in valleys, while in mid and up-lands it is usually used for picking and marketing fruits.

Vegetable cropping pattern does not show obvious favour for a particular crop. However, potato, onion, gourd, cabbage, marrow, brinjal and leaf vegetables are some prime vegetables grown in the district. Lately, off-season vegetables are reaching into the market beans, marow, raddish and cauliflower are some important vegetables of this category. This speaks about cultivators' inclua-tion towards double and multiple cropping, an essential dimension of modern agriculture. This is indeed landable and governmental agencies must devise ways and means to sustain the interests of the vegetable growers by keeping their hopes.

Land utilization pattern shows a moderate cropping intensity. For the entire district it is 188 per cent. This moderate cropping intensity may be further enhanced through introducing short maturing species in the valley areas.

An encouraging and important fact discovered for the study area is that most of the cultivators have great faith in the demonstration programmes conducted by the extension agencies and research organisations such as Indo German Association for the Development of Agriculture (IGADA) and Vivekanand Hill Agricultural Research Lab, Almora. Programmes organised under IRDP have also proved useful for the rural folks. Roy et. al. (1979) express the same view in their studies.

The application of new agricultural technology has improved the socio-economic status of the big and medium farmers of the area who have reaped the benefits of this technology. They have better access to resources and timely and accurate information about new technology.

In the irrigated areas in the valley the new technology has been generally adopted and two crops in a year, viz., HYV paddy-wheat or vegetables and HYV maize and Soyabean are reaped. There are good prospects to increase cropping intensity and income of farmers. Meanwhile massive research efforts are being undertaken by Vivekanand Laboratory, Almora to develop new varieties and cropping patterns for rainfed agriculture.

**Constraints to New Agricultural Technology in Hills**

It was observed that new agricultural technology when actually implemented did not produce the expected results in many of the reported cases. The reasons for such failure are both man made or natural. In certain cases there was a feeling among the non-beneficiaries that only the enlightened and the richer sectors of the farmers were generally favoured by the block level and other extension officials. In some other cases lack of timely irrigation and uncertainties due to the occurrence of pests and diseases, low seed multiplication and absence of technical progress were some of the constraints to the agricultural productivity.

For the irrigated tract it was found that the tillers did not share water equitably. While some farmers had excess water supply to their fields, other situated at the tail end of the source complaint about scarce and irregular water supply to their fields although they were bearing the irrigation costs and charges in full.

Extension services have not been fully effective in the study region because of inadequacy of personnel connected with extension work, poor working conditions and disincentives built in the system of posting and lack of awareness among the field staff about the social and cultural background of the people particularly in upland areas (Kakkar, 1989).

There is also lack of farmer oriented extension programme. A proper identification of the farmers would show that most of them are women, whereas, there is a woeful dearth of women extension workers and lack of involvement of women in the decision making process in agricultural development.

**Suggestions for the Future Promise**

In the study area nobody holds special training in Agricultural Science, while in the city area several boys and girls hold agricultural diploma. If this work force is appointed in the entire area for the extension of new agricultural technology, the rural women, can acquire enough knowledge in this direction.

Farmers should grow more than two varieties of each crop, so that if any one variety is infected by any pest or disease the other one is saved. Thus the loss will be reduced.

The horticulture department should maintain a continuous rapport with the orchardists and extend them technical guidance in the matter of laying out and planning orchards. This is important from this point of view because many respondents in the field study did not seem convinced about the quality and long life of the plants obtained from the nurseries. In many cases plants died within 4 or 5 years of plantation.

Proper training programmes should be organised to impart training to women in (a) improved agricultural practices adaptable to that area and building up of local and HYV seed stores, (b) application of fertilisers having optimum combination with farm manure, and (c) the use of insecticides and pesticides at the proper time and manner. This type of training is yet to start in hills. "Mahila Mandals" in the region may play a decisive role in building up an opinion among the women folk in favour of modern agriculture.

The technological gaps in the crop production for the different sub-regions should be filled and large scale field testing of the technology should be carried out.

The research and development programme should be worked out for both irrigated as well as unirrigated areas. In the initial stage one research station should be established in each zone of the hill region of Kumaon Hills at different altitudes. Comparative advantages of new technology must be worked out and demonstrated among the village farmers.

## References

Bhalla, G.S., 1972. Changing Structure of Agriculture in Haryana–A Case Study of the Impact of Green Revolution, 1969-70. Deptt of Economics, Panjab Univ. Chandigarh.

Dwivedi, R.C., 1972. *New Strategy of Agricultural Development in India*. Lyall Book Depot, New Delhi.

Shah, S.L. and L.R. Singh, 1970. Increasing Income Disparities Due to the New Technology of Agriculture in North West U.P. *IJAE*, Vol. 25.

Griffin, K, 1974. *The Green Revolution : An Economic Analysis*. United Nations Research Institute for Social Development, Geneva.

Sharma, J.S., S. Roy, and P.S. George, 1979. *Analysis of Indian Food grains Production and Consumption Data*. Research Report, New Delhi.

Farmer, B.H., 1980. *Green Revolution, Technology and Change in Rice Growing Areas of Tamil Nadu and Sri Lanka*. The Eng. Language Book Society & MacMillan.

Mahajan, R.K. & A.N. Sadhu, 1985. *Technological change and Agricultural Development in India*. Himalaya Pub. House Bombay.

Raj Krishna, 1979. Small Farmers Development. In Desai B.M. (ed.) *Intervention for Rural Development : Experience of Small Farmers Development Agency*, IIM, Ahmadabad.

Ray, et.al. 1979. *Policy Planning for Agricultural Development*. Tata McGraw Hill, New Delhi.

Kakkar, Anita T., 1989 The Contribution of New Agricultural Technology in the Development of Agriculture in District Almora. Unpublished Ph.D. Thesis, Deptt. of Economics, Kumaun University, Nainital.

## Chapter 10

# Status, Utilization and Potentials of Water Resources in Kumaun Lesser Himalaya

*P.C. Tewari*
*Bhagwati Joshi*

Water is a most valuable natural resource for mankind on the earth surface. It plays a very crucial role in domestic, agricultural and industrial sectors of activities. In a large measure, the spatial pattern and productivity of various resource development processes and activities, and the evolution and growth of human settlements are determined and regulated by the availability and potential of surface and ground water. The significance of water resources in the regional and economic development of an area, therefore, hardly needs to be emphasized. Hence, the geohydrological studies with respect to the availability and potential of surface and sub-surface water and planning and management of water resources assume paramount importance, keeping in view the sustainable development of natural resources in the Himalayan watersheds.

The ever increasing demand of water for agricultural and industrial development and for domestic requirements could be met out by synthesizing the intensive and extensive systems of water conservation and management in Himalaya. Intensive system in this context is referred to as micro level planning and management of water resources to meet village level requirements, whereas, the extensive system of water utilization involves the macro level conservation and management of water resources.

The Himalaya is geotectonically unstable and ecologically sensitive, and the fragile geoecological components of the region are now passing through a process of rapid anthropogenic transformation under the impact of unscientific and irrational resource-development processes being operated in the region. The reckless deforestation and the accelerated pace of soil erosion has reduced the water generating capacity of land to a considerable extent, and brought about major geohydrological changes in the entire region (Tewari, 1995).

The Upper Kosi Catchment encompassing a geographical land surface of 480.15 sq km. in Kumaun Lesser Himalaya was selected for the case illustration, in this investigation. The watershed has its spatial extension between 29°31′ and 29°55′N latitudes and 79°30′ and 79°45′ East longitudes in district Almora (Fig. 1). Kosi is one of the major river of Ramganga (West) system of Kumaun, which drains an area of about 1475 sq. km. between its source in the north-west of Kausani in district Almora and Ramnagar in district Nainital. The Upper Kosi Catchment is constituted by the northern most part of Kosi drainage basin which extends between Pinath, i.e., the source of the river, and Kwarab where the main tributary of the river, Suyal joins it on its left bank. The watershed has an average altitude ranging between 1080 m and 2720 m above the mean sea level. The watershed is one of the densely populated and agriculturally prosperous tracts of entire Kumaun. Consequently, the activities of cultivation and grazing are extended over areas as large as nearly 31 and 10 per cent of the total geographical area of the catchment respectively, whereas, the vegetal cover has been shrinking progressively, and is now estimated to confine only to 39 per cent of the total land surface of the basin.

The Upper Kosi Catchment, like other watersheds of the Lesser Himalaya in Kumaun, presents a paradoxical situation with respect to the actual potential of water resources and their utilization. The large surface flow system available in the region has great potential for various uses, but owing to very poor management, a very little proportion of this water potential has been utilized so far. The results of this mismanagement and underutilization of water resources can clearly be seen in terms of acute shortage of drinking water in majority of areas and a very insignificant proportion of irrigated land in the catchment.

The entire watershed was divided into ten micro basins following the stream ordering techniques, for the study of water volume and status of water resource utilization (Fig. 2). The broad drainage basin characteristics of all the ten micro-watersheds have been presented in Table 1. While, the detailed study of stream and spring discharge, water generating capacity of land and anthropogenic impact on water resources was carried out only in Shail Gad micro-watershed, i.e., the experimental watershed of the Upper Kosi Catchment (Fig. 1 and 2 and Table 1).

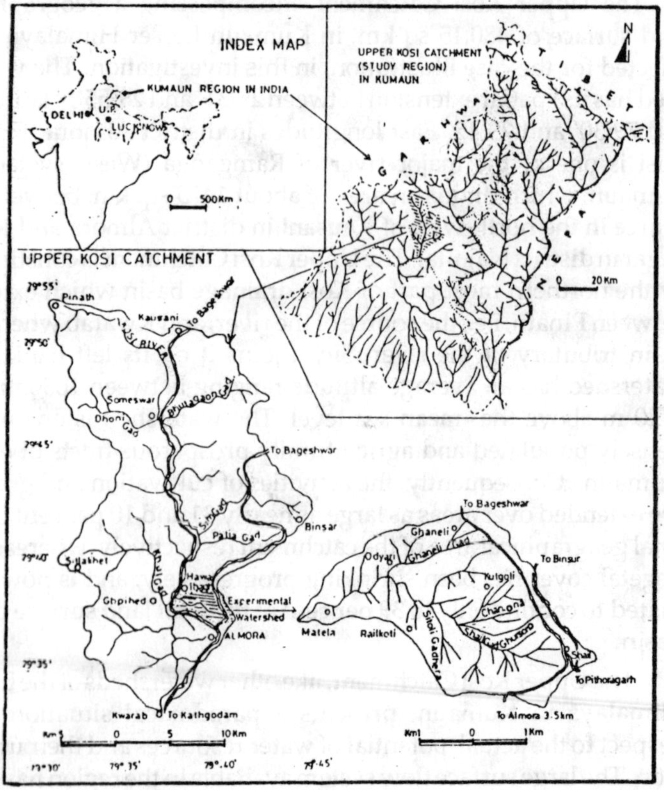


**Fig. 1. : Location of the study area**

Fig. 2 : Upper Kosi catchment : Surface drainage and micro watersheds

## Hydrological Parameters of Watershed

The water in any region is contained either in surface bodies or in surrounding atmosphere or in the form of underground water. The hydrological studies are mainly concerned with the quantification of water in these three forms and analysing the process between them. But the recent, developments in the field of applied hydrology are concerned more specifically with the assessment of water resources in terms of water availability and potential, water generating capacity of land to springs and channels, water flow and discharge, geo-hydrological balance, anthropogenic impact on water resources, and also the prediction of forthcoming hydrological and related events.

The quantity of water received by a drainage basin through rainfall and snow can be assessed by the Thiessen method or isohyetal technique or by the representative basin approach (More, 1967). The Thiessen technique estimates the overall amount of rainfall in a watershed by adding up the areas geometrically nearest to the meteorological data recording station whose rainfall figures are taken as representative, whereas in the isohyetal method, the geographical area and the quantity of rainfall between the regional isohyets are added up. The technique of representative sample extrapolates values from small, carefully measured sub-catchment chosen as typical of the main one (Mitchell, 1973).

Different types of evaporation pans can be used for measuring evaporation from a free water surface. Since, a number of practical difficulties are involved in using the evaporation pan methods which ultimately influence the result of the analysis, various empirical formulae have also been derived by hydrologists which are mainly based on the concept of the energy balance. Doing measurements of air-temperature, wind velocity, saturation deficit and length of day, Penman (1963) derived his formula for assessing evaporation from the surface equating energy input and output; Blaney and Griddle (1950), Thornthwaite (1948 and 1954) and Olivier (1961) have also developed somewhat similar equations, applicable in regions of different geoenvironmental conditions. But the factor of terrain, which not only directly control and regulate the process of infiltration and overland flow in a drainage basin but also play a crucial role in determining the entire hydrological system has not been incorporated in any of these techniques.

The stream discharge which mainly depends on channel dimensions and conditions can be assessed approximately from the measurements of cross sectional area current speed. 'On the whole, channels become wider and shallower relative to their depth and adjusted to larger flows as they progress down-stream. Bankfull discharges increase in the same direction in proportion to the square of the width of the channel or of the length of individual meanders and in proportion to the 0.75 power of the total drainage area focused at the point in question (Mitchell, 1973).

The stream flow can be predicted by employing statistical and hydrometeorological techniques incorporating the estimation of infiltration, evaporation and transpiration. In statistical analysis it is assumed that recurrence intervals of extreme events bear a consistent relationship to their magnitudes. In accordance with given hydrological requirements a recurrence interval standard is fixed. The hydrometeorological investigation for predicting the stream flow are based on the estimate of the highest and lowest values which can reasonably be expected from all the meteorological factors involved in causing the extreme conditions and the assessment of their likely effect when acting in concert. More (1967) has suggested that in order to ensure maximum possible reliability the second method of hydrometeorological analysis should be used preferably in combination with the statistical method. Owing to several practical difficulties involved in the application of these methods, particularly the complexity of measuring the indices of stream and channel flow, the hydrological characteristics of a catchment can alternatively be assessed up to high satisfaction level through approximations techniques. The concept of 'unit hydrograph' evolved by Sherman (1932) is an example of approximation techniques. The method of unit hydrograph is based on the postulation that the most important hydrological characteristics of any watershed can be seen and examined from the direct runoff hydrograph resulting from 25 mm of rainfall evenly distributed over twenty four hours. This is obtained by drawing a graph of the total stream flow at a selected point as it changes with time after such a storm, from which the normal base flow caused by groundwater is substracted (Mitchell, 1973).

The proportion of water entering the soil to feed the ground water, running off into the drainage channels, or lost by evapotranspiration is primarily determined by the interrelated factors of gradient, soil infiltration capacity and surface roughness. The land surface determines the quantity of water which can be stored in basin by controlling its overall area, depth and the permeability of its surface. The larger and shallower the basin, the greater the area over which evaporation and downward percolation can take place. Topographic features also control and regulate the quantities and rates of ground water recharge to, transmission, storage within and loss from, the ground water reservoir. The recharge may be defined as that proportion of rainfall which is neither diverted by overland flow nor lost through evapotranspiration. The relative amount of recharge is directly associated with the scale of surface roughness and the magnitude of soil permeability and inversely related to the degree of gradient. To put another way, the higher the permeability and the surface roughness the greater the ground water recharge while, steeper the slope the lower recharge in drainage basin. The permeability is generally a function of soil texture and particle size, whereas, the surface roughness varies with type, height and density of vegetation and flow, size, cross-section alignment and frequency of channels. As stated above, infiltration capacity increases with surface roughness and permeability, and decreases with gradient, it therefore tends to be greatest in the relatively rough, coarse textured, permeable and more gently sloping terrain around streams than on the intervening surface. Permeability is directly determined by granulometry and soil structure. The geological character of a drainage basin such as rock types, dip and fault line also has a considerable impact, rather say strong control on the ground water table and recharge, geohydrological balance, water transmission, occurrence of springs, channels and lakes and on water generating capacity of land. Anthropogenic factors especially the landuse pattern of a watershed has great impact on the hydrological processes of runoff, infiltration and evapotranspiration which ultimately control the water generating capacity of land to springs and channels. All these above mentioned hydrological parameters have great relevance in the water resource management in a drainage basin, as they provide a scientific base for the generation and development of hydrological information system. Though the measurement of these parameters involves several complexities yet a number of field and laboratory methods and techniques lead to meaningful conclusions. For details of various relevant and applied techniques both statistical and hydrometeorological for measuring and analysing the hydrological characteristics of a watershed, resource managers and planner may refer Panman (1963), Haggett and Chorley (1969), Smith (1950), Strahler (1957), Horton (1945), Cowan (1956), Schwab et. al. (1966).

Thus, in order to obtain a precise understanding of the water holding and transmitting characteristics and for the appraisal of the water resource potentials of a catchment the investigation and analysis of slope, surface roughness, infiltration capacity and underground storage capacity in combination with an assessment of the local balance between rainfall and evapotranspiration are essential. Owing to some practical difficulties only few of these parameters, such as, water volume, water flow and discharge in springs and channels, water generating capacity of land and anthropogenic impact on water holding potentials have been studied in the watershed and one experimental micro-catchment. Besides the study of status and utilization pattern, a framework for the management and conservation of the water resources has also been outlined, both in ecological and developmental perspectives of the region.

## The Volume of Water

The main source of surface and sub-surface water is the precipitation. As discussed earlier, the functioning of geo-hydrological cycle of a region depends on the complexities of various factors operating in combination in a watershed, such as, precipitation, evapotranspiration, overland flow, percolation infiltration, deep transfer and sub-surface flow, etc. Since, no comprehensive and scientific investigations have been carried out in Himalaya, so far, for the complete geohydrological appraisal of the watersheds of the region, by key government agencies mainly entrusted with the task of planning, management and monitoring of water resources, the information on baseline hydrological parameters and also on spatio-temporal aspects of water resources, in the region are not available. Even the very general information available on hydrological aspects, of the region, is not based on proper scientific observations and devoid of statistical treatment. Pathak (1983) in his study of uses and potentials of water resources of U.P. Himalaya, has assumed that on an average the surface run-off in the region accounts for 49 per cent of the total rainfall. The amount of surface-flow and infiltration which play a very significant role in maintaining the geohydrological balance of a watershed, vary spatially with varying complexities of physio-cultural factors, such as, surface configuration, vegetation, climate and the pattern of landuse operated by the regional population. An experimental exercise carried out in a micro watershed of the region revealed that the rate of overland flow is the highest (18.00 per cent of the gross rainfall) in cultivated land and was measured lowest (3.2 per cent of the gross rainfall) in forested sites. The results obtained through this experiment are presented in Table 2. Pathak, (1983) in his study of vegetation Hydrology of Kumaun Himalaya while studying the apportionment of gross rainfall into different pathways in the forests of Kumaun Himalaya, derived the result that in the Himalayan forests the amount of surface run-off is as low as 0.70 per cent of the gross rainfall. The precise information regarding the apportionment of gross rainfall on the basis of Pathak (1983) in summarized in Table 3. These experimental exercises carried out in different parts of the Kumaun Himalaya reveal the fact that vegetal cover plays a dominant role in increasing the groundwater recharge in a watershed. In view of the non-availability of detailed information on various hydrological parameters, as noted above, the estimates of total volume of water in all ten micro-catchments of the region was made from the average annual precipitation by Thiessen Polygon method (James and Robert, 1971). For calculating the volume of water in different micro watersheds, drainage basin characteristics were also taken into consideration.

**Table 1 :** Drainage Basin Characteristics of the Micro-watersheds of Upper Kosi Catchment

| | *Micro catchments* | *Width (km)* | *Length (km)* | *Area (km²)* |
|---|---|---|---|---|
| 1. | North Kosi | 7 | 17 | 74.90 |
| 2. | Bhutagaon Gad | 6 | 10 | 34.90 |
| 3. | Dhoni Gad | 7 | 8 | 40.50 |
| 4. | Sim Gad | 9 | 22 | 53.90 |
| 5. | Middle Kosi | 10 | 11 | 68.90 |
| 6. | Nana Kosi | 8 | 11 | 62.05 |
| 7. | Patia-Ron Gad | 6 | 11 | 44.60 |
| 8. | Khul Gad | 8 | 9 | 45.90 |
| 9. | Shail-Jamthara Gad | 5 | 6 | 24.50 |
| 10. | South Kosi | 7 | 7 | 30.10 |
| Upper Kosi Catchment | | 19 | 33 | 480.15 |

**Table 2 :** Surface runoff in different landuse classes

| | | |
|---|---|---|
| 1. | Forested Sites | 3.2 |
| 2. | Degraded Forests | 6.2 |
| 3. | Grazing Land | 11.70 |
| 4. | Horticulture area | 4.00 |
| 5. | Cultivated Land | 18.00 |
| 6. | Barren Land | 9.25 |

**Table 3 :** Apportionment of gross rainfall into different pathways in Kumaun Himalaya forests

| | Pathways | Per cent of Gross Rainfall |
|---|---|---|
| 1. | Gross Rainfall | 100.00 |
| 2. | Canopy Interception Loss | 16.70 |
| 3. | Net Rainfall | 83.30 |
| 4. | Ground Flora and litter Interception | 12.30 |
| 5. | Water Available to Soil | 71.00 |
| 6. | Over land and Surface Run-off | 00.70 |
| 7. | Sub Surface Flow | 56.30 |
| 8. | Infiltration | 70.03 |

Table 4 showing the distribution of water volume in different micro-basins, reveals that the entire catchment has 210.77 million cu. m volume of water, out of which North Kosi accounts for the highest proportion (15.61 per cent) and Shail-Jamthara Gad shares the lowest (5.10 per cent). The northern and central part of the region constituting the micro basins of North Kosi, Bhutagaon Gad, Dhoni Gad, Sim Gad, Middle Kosi and Nana Kosi accounts for more than 65 per cent of the total volume of water of the catchment, whereas, the water volume in the southern part of the region is comparatively lower (Table 4).

**Table 4 :** Micro Catchmentwise Average Annual Volume of Water in the Upper Kosi Catchment

| | Micro catchments | Area ($km^2$) | Per cent of the total area of the catchment | Volume of water in million cu.m. | Per cent of the total water volume of the catchment |
|---|---|---|---|---|---|
| 1. | North Kosi | 74.90 | 15.60 | 32.90 | 15.61 |
| 2. | Bhutagaon Gad | 34.90 | 7.27 | 15.30 | 7.26 |
| 3. | Dhoni Gad | 40.50 | 8.43 | 17.76 | 8.43 |
| 4. | Sim Gad | 53.90 | 11.23 | 23.66 | 11.22 |
| 5. | Middle Kosi | 68.90 | 14.33 | 30.26 | 14.36 |
| 6. | Nana Kosi | 62.05 | 12.92 | 27.24 | 12.92 |
| 7. | Patia Ron Gad | 44.60 | 9.29 | 19.56 | 9.28 |
| 8. | Khul Gad | 55.90 | 9.56 | 20.15 | 9.56 |
| 9. | Shail-Jamthara Gad | 24.50 | 5.10 | 10.74 | 5.18 |
| 10. | South Kosi | 30.10 | 6.27 | 13.20 | 6.26 |
| Upper Kosi Catchment | | 480.15 | 100.00 | 210.77 | 100.00 |

## Existing Status of Water Resources Utilization and Potentials

As stated earlier despite huge water resources, the entire

region is facing great scarcity of water for domestic as well as for agricultural purposes. Only about 7 per cent of the total water resources of the region are being used at present, mainly for household requirements, irrigation and running of water mills. Out of total 322 villages 310 have been provided with water supply. But it was noticed during field surveys that in most of the villages the water supply is irregular and partial. In 63 villages of the region, the water supply schemes have blocked and become inoperative mainly owing to lack of proper maintenance. As many as 197 villages of the catchment are facing accute shortage of water for domestic purposes. But the striking feature is that about 75 per cent of these villages have been provided domestic water supply in the official records of the concerned department. This is because of the fact that most of the water supply schemes in these villages have become inoperative due to several reasons. Out of the total geographical area of the region 28.53 per cent in faced with the problem of water scarcity for household purposes. Table 5 reveals that the micro basins of Middle Kosi, Nana Kosi, Patia-Ron Gad, Khul Gad, Shail-Jamthara Gad and South Kosi constituting the southern part of the catchment have the great scarcity of water, compared to northern part of the region. Detailed information regarding the total requirement of domestic water supply in the catchment, including Almora town, as per recommended standard of 120 and 90/Litre/Person/Day for urban and rural areas, respectively, is given in Table 6. The table reveals the fact clearly that owing to demand and supply gap resulting from the mismanagement and under utilization of water resources, the actual quantity of water supplied to Almora town is about three times less than the minimum requirement.

The total cultivated land of Upper Kosi Catchment is 14880 ha which accounts for 30.99 per cent of the total geographical area of the region. Out of this total arable land only 1364 ha or 9.16 per cent is under irrigation with a gross irrigated area of 2772 ha. Water resources for irrigation are managed in the area, mainly through the construction of small canals, locally known as 'guls', lift, hydrums and tanks. Canals and tanks are the main sources of irrigation as these two individually share 38.05 and 36.80 per cent of total irrigated land of the watershed, respectively (Table 7). The actual

**Table 5 :** Water scarcity area in Upper Kosi Catchment

| Micro-water sheds | Water Scarcity Area in km² | Per cent of total Area | No. of village in the water scarcity area |
|---|---|---|---|
| 1. North Kosi | 10.00 | 13.35 | 8 |
| 2. Bhutagaon Gad | 7.00 | 20.05 | 11 |
| 3. Dhoni Gad | 4.00 | 9.88 | 19 |
| 4. Sim Gad | 16.00 | 29.68 | 14 |
| 5. Middle Kosi | 15.00 | 21.77 | 27 |
| 6. Nana Kosi | 13.00 | 20.95 | 31 |
| 7. Patia-Ron Gad | 13.00 | 29.14 | 21 |
| 8. Khul Gad | 15.00 | 32.68 | 25 |
| 9. Shail-Jamthara Gad | 20.00 | 81.63 | 20 |
| 10. South Kosi | 24.00 | 79.73 | 16 |
| Upper Kosi Catchment | 137.00 | 28.53 | 197 |

**Table 6 :** Status of domestic water supply in Upper Kosi Catchment

| Total No. of Villages | Population | Water requirement L/Day | Villages provided drinking water | No. of Villages having full water supply | No. of Villages having partial water supply | No. of Villages where the scheme have blocked | Water supplied L/Day | No. of Urban Centre | Water Requirement L/Day | Water supplied L/Day |
|---|---|---|---|---|---|---|---|---|---|---|
| 322 | 99,663 | 89,69,670 | 310 | 166 | 101 | 43 | NA | 1 | 27,24,600 | 9,08,200 |

NA Data Not Available

proportion of irrigated land to net sown area varies between 0.86 and 20.37 per cent in different micro-basins of the region (Table 7). Paddy, wheat, barley and potato are the principal irrigated crops of the valley. Paddy and wheat share the highest proportion of the irrigated area which account for 51.01 and 43.25 per cent of the gross cultivated land under irrigation. The gross irrigated land in the region ranges between 16 ha and 882 ha in different parts of the catchment (Table 8).

The irrigated agriculture is confined only to low lying areas where the regular supply of water for irrigation is available, whereas, the cultivated land lying in the upland which accounts for about 90.84 per cent of the total cultivated area of the region, is never irrigated, due to non-availability of water. Therefore, on the whole, the productivity of agriculture is very low and unreliable. The requirement of water for irrigation at its optimum level for every hectare of cropping vary to a considerable extent, mainly owing to consumptive use of crops and effectiveness of irrigation. The

**Table 7 : Sourcewise irrigated area in Upper Kosi Catchment**

| *Sources* | *Canals* | | | | *Tanks* | | | *Hydrams* | | | *Lifts* | | *Total* | |
|---|---|---|---|---|---|---|---|---|---|---|---|---|---|---|
| *Name of Microwatersheds* | *Length in Km* | *Irrigated area in ha* | *Irrigated area in per cent* | *Nos.* | *Irigated area in ha* | *Irrigated area in %* | *Nos.* | *Irrigated area in ha* | *Irrigated area in%* | *Nos.* | *Irrigated area in ha* | *Irrigated area in %* | *Irrigated area in ha* | *Irrigated area per cent of Net Cultivated area* |
| 1. North Kosi | 19.00 | 235 | 81.32 | 76 | 50 | 17.30 | 2 | 4 | 1.38 | – | – | – | 289 | 15.92 |
| 2. Bhutagaon Gad | 5.00 | 30 | 28.57 | 95 | 75 | 71.43 | – | – | – | – | – | – | 105 | 14.79 |
| 3. Dhoni Gad | 3.50 | 36 | 44.45 | 66 | 45 | 55.35 | – | – | – | – | – | – | 81 | 6.75 |
| 4. Sim Gad | 2.80 | 12 | 16.44 | 81 | 61 | 83.56 | – | – | – | – | – | – | 73 | 5.18 |
| 5. Middle Kosi | 12.00 | 153 | 35.25 | 102 | 70 | 16.13 | 2 | 12 | 2.76 | 8 | 199 | 54.86 | 434 | 20.37 |
| 6. Nana Kosi | 3.50 | 22 | 17.74 | 105 | 75 | 60.48 | 2 | 9 | 7.6 | 1 | 18 | 14.52 | 124 | 4.74 |
| 7. Patia-Ron Gad | 1.00 | 14 | 8.23 | 98 | 65 | 38.23 | – | – | – | 4 | 91 | 53.54 | 170 | 11.15 |
| 8. Khul Gad | – | – | – | 61 | 40 | 100.00 | – | – | – | – | – | – | 40 | 2.19 |
| 9. Shail-Jamthara Gad | 2.5 | 17 | 42.5 | 27 | 13 | 32.50 | 2 | 10 | 25.00 | – | – | – | 40 | 5.52 |
| 10. South Kosi | – | – | – | 19 | 8 | 100.00 | – | – | – | – | – | – | 8 | 0.86 |
| Upper Kosi Catchment | 49.30 | 519 | 38.05 | 730 | 502 | 36.80 | 8 | 35 | 2.57 | 13 | 308 | 22.58 | 1364 | 9.16 |

**Table 8 : Cropwise irrigated area in Upper Kosi Catchment**

| Crops | | Paddy | | Wheat | | Barley | | Maize | | Massor | | Oilseeds | | Potato | | Others | | Gross irrigated Area | |
|---|---|---|---|---|---|---|---|---|---|---|---|---|---|---|---|---|---|---|---|
| Micro-Watershed | | Irrigated area in ha | Per cent of Gross Irrigated Area | Irrigated area in ha | Per cent of Gross Irrigated Area | Irrigated area in ha | Perc ent of Gross Irrigated Area | Irrigated area in ha | Per cent of Gross Irrigated Area | Irrigated area in ha | Per cent of Gross Irrigated Area | Irrigated area in ha | Per cent of Gross Irrigated Area | Irrigate4ed area in ha | Per cent of Gross Irrigated Area | Irrigated area in ha | Per cent of Gross Cropped Irrigated area | Gross Irri-gated area in ha | Per cent of Gross Cropped Area |
| 1. | North Kosi | 286 | 20.23 | 268 | 22.35 | 4 | 20.00 | 1 | 16.67 | 9 | 18.75 | – | – | 15 | 26.78 | 4 | 16.00 | 587 | 21.18 |
| 2. | Bhutagaon Gad | 98 | 6.93 | 9376 | 7.76 | 3 | 15.00 | 1 | 16.68 | 7 | 14.58 | 1 | 25.00 | 8 | 14.28 | 2 | 8.00 | 213 | 7.68 |
| 3. | Dhoni Gad | 88 | 6.22 | 56 | 4.67 | 1 | 5.00 | 2 | 33.33 | 7 | 14.58 | 1 | 25.00 | 7 | 12.50 | 3 | 12.00 | 165 | 5.95 |
| 4. | Sim Gad | 90 | 6.36 | 48 | 4.00 | – | – | – | – | 5 | 10.42 | – | – | 3 | 5.36 | 2 | 8.00 | 148 | 5.34 |
| 5. | Middle Kosi | 425 | 30.05 | 405 | 33.78 | 8 | 40.00 | 2 | 33.33 | 14 | 29.17 | 2 | 50.00 | 18 | 32.14 | 8 | 32.00 | 882 | 31.82 |
| 6. | Nana Kosi | 130 | 9.19 | 110 | 9.97 | 2 | 10.00 | – | – | 4 | 8.33 | – | – | 2 | 3.58 | 4 | 16.00 | 252 | 9.09 |
| 7. | Patia-Ron Gad | 180 | 12.73 | 156 | 13.02 | 2 | 10.00 | – | – | 2 | 4.17 | – | – | 3 | 5.36 | 2 | 8.00 | 245 | 12.45 |
| 8. | Khul Gad | 52 | 3.68 | 30 | 2.50 | – | – | – | – | – | – | – | – | – | – | – | – | 82 | 2.96 |
| 9. | Shail-Jamthara Gad | 55 | 3.89 | 27 | 2.25 | – | – | – | – | – | – | – | – | – | – | – | – | 82 | 2.96 |
| 10. | South Kosi | 10 | 0.72 | 6 | 0.50 | – | – | – | – | – | – | – | – | – | – | – | – | 16 | 0.57 |
| Upper Kosi Catchment | | 1414 | 51.01 | 1199 | 43.25 | 20 | 0.73 | 6 | 0.22 | 48 | 1.73 | 4 | 0.14 | 56 | 2.02 | 25 | 0.90 | 2772 | 100.00 |

irrigation requirement is the consumptive use of crops, minus the precipitation available or the consumption by plants plus the losses that occur in the way. The consumptive use may be expressed as the amount of water required by the plants to produce maximum yields. The water requirement for irrigation, also includes the losses due to evaporation, percolation, surface waste and non-productive consumptions which altogether account for 50 per cent of the water released for irrigation of a particular crop (Pathak, 1983).

## Geoecological Aspects of Water Resources

Investigations on ecological aspects of water resources play most significant role in the management and monitoring of environmental components for ecologically sustainable regional development and planning of a region. Besides the study of water potential, the ecological part of geohydrological research includes the study of anthropogenic impacts on the water resources and their role in transforming water flow and water generating capacity of land to channels and springs. In order to have a scientific, detailed and comprehensive study of water resource potential of the region and to measure the degree of impact of human transformation process of natural components on the geohydrology, a detailed and comprehensive study was carried out in the experimental watershed, i.e., the Shail Gad.

## Water Flow in Springs and Streams

In hilly Terrain, springs and channels are the only available water resources for domestic as well as agricultural purposes. Almost all the perennial stream sources and springs in the watershed are on dip-slopes and they are generally located on the fault line or fracture of hard impermeable rocks, such as gneissic structure, etc. In the extreme north and north east, the Kalimat reserved forest belt and village Ghaneli which lie on anti-dip slopes are almost devoid of springs and perennial streams. This shows the control of geological formation of the experimental watershed on the availability and potential of sub-surface water. Structural lines are so strong in regulating the hydrological system of the region that the entire Kalimat reserved forest falling within the watershed and large part of Patal Devi area and village Shail which lie above Almora-Bageshwar motor road, between Dhar Kee Tuni and Ghaneli

**Table 9 :** Spring and stream discharge capacity in Shail Gad Watershed

| *Sub-watersheds* | | *Spring* | | | *Streams* | | *Total* | |
|---|---|---|---|---|---|---|---|---|
| | *Total Geographical Area (ha)* | *Density Km²* | *Water Discharge L/Day* | *Water Generating Capacity L/ha/Day* | *Water Discharge L/Day* | *Water Generating Capacity L/ha/Day* | *Water Discharge L/Day* | *Water Generating Capacity L/ha/Day* |
| 1. Shail Watershed | 185.25 | 5.94 | 25,400 | 137.11 | 35,000 | 188.93 | 60.400 | 326.04 |
| 2. Ghurson-Channa Watersheds | 301.95 | 2.99 | 54,800 | 181.49 | 82,800 | 92.73 | 82,800 | 274.21 |
| 3. Kutgoli Watershed | 102.15 | 1.96 | 11,800 | 115.52 | 26,000 | 254.53 | 37,800 | 370.04 |
| 4. Ghaneli-Odyari Watershed | 220.15 | 1.81 | 600 | 2.72 | 14,000 | 63.59 | 14,600 | 66.32 |
| 5. Naikhota-Railkote Watershed | 204.50 | 2.45 | 2,650 | 12.96 | 15,000 | 73.35 | 17,650 | 86.31 |
| Shail Gad Watershed | 1014.00 | 3.06 | 95,250 | 93.93 | 1,18,000 | 116.37 | 2,13,250 | 210.30 |

Dhar, have no perennial water source, except two springs, one at Shail (Sunarkhola) and other near Kalimat Forest Chauki. There are total 49 springs in the area, out of which 18 have almost dried and the stream density is 3.74 km of stream length per sq. km.

Various water discharge measuring instruments were used to monitor water flow of streams and springs in the region. This exercise was carried out during winters, in 1991, and the water flow of all perennial channels and spring was measured throughout the experimental watershed (Table 9). The density of perennial springs in the region was computed to be 3.06 springs per sq. km. Shail subwatershed lying in the extreme east of the region has on an average 5.94 spring per sq.km. which is the highest in the area, whereas, per square kilometer spring density was found lowest (1.81 springs) in Ghaneli-Odyari sub-drainage area (Table 9).

The flow of water in the springs of the region ranges between 100 L/day to as high as 15,000 L/Day, and the total spring water flow was measured to be 95,250 L/Day in the watershed. Ghurson-Channa Sub-drainage area has recorded the highest spring flow of 54800 L/Day which is closely followed by Shail area (25,400 L/Day). The lowest spring discharge (600 L/Day) was measured in Ghaneli-Odyari sub-watershed. The total water discharge of the master stream at its mouth near Kosi was estimated to be 1,25,000 L/Day. Which is very high as far as the geographical area of the watershed is taken into consideration. The stream flow of different sub-watersheds of the region ranges between 14,000 L/Day in Ghaneli-Odyari and 35,000 L/Day in Shail sub-drainage area (Table 9 and Fig. 3).

## Water Generating Capacity of Land

In order to obtain a quantitative picture of spatial variation in the water generating potential of land to streams and channels the total water flow of different sub-watersheds was divided by their respective catchment area, and thus the water discharge capacity of streams and springs was transformed into hectare/litre water per day. The experimental watershed has on an average, 210.30 L/ha/Day water generating potential with a minimum of 66.32 L/ha/Day in Ghaneli-Odyari and maximum of 370.04 L/ha/Day in Kutgoli sub-drainage area (Table 9). It is obvious from Table 9 that the availability of water is very high in Shail, Ghurson-Channa and

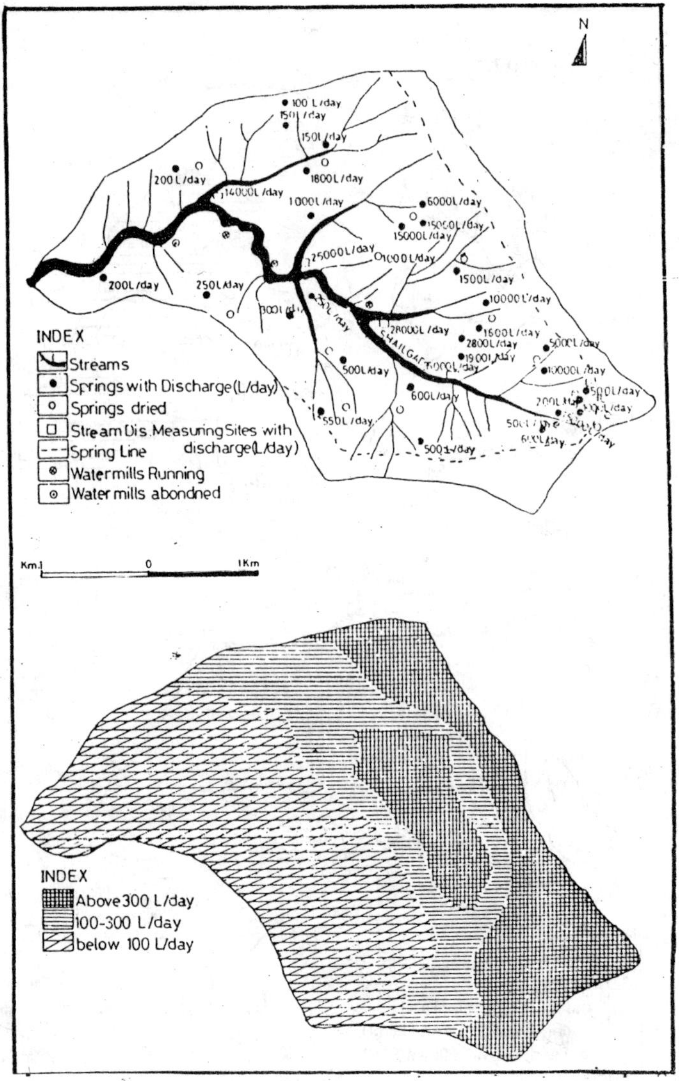


**Fig. 3. :** Shail Gad Microwatershed—(A) Water flow in springs and streams. (B) Water generating capacity of land

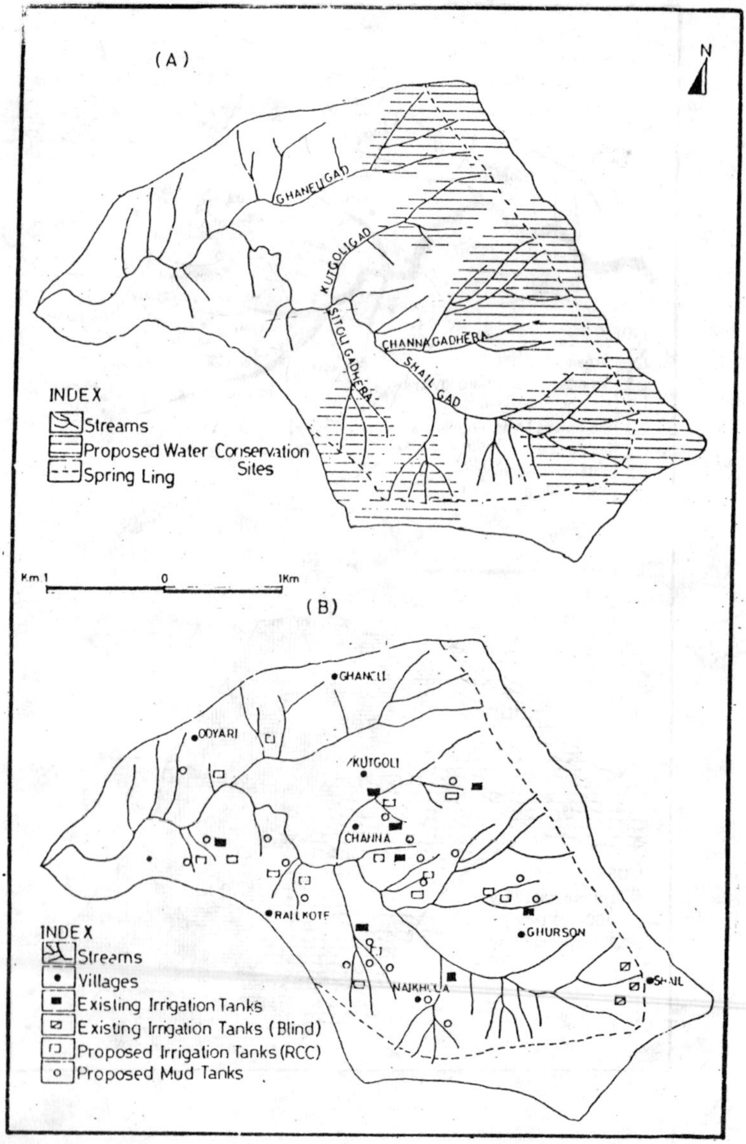


**Fig. 4. : Shail Gad micro-watershed : (A) Proposed water conservation sites, (B) Existing and proposed water harnessing measures**

Kutgoli area while it is considerably low in other areas of the region. Since, the Almora-Bageshwar motor road which roughly makes the upper limit of all perennial springs and streams in the region, the entire settlement area and most of the agricultural terraces of Shail and Patal Devi area which lie above this state highway are almost devoid of water, except one 500 L/Day capacity spring at Shail which is the only source of drinking water for about twenty families (Fig. 3). Consequently, the entire Shail village and Patal Devi area has great scarcity of drinking water, all the year round, except during monsoon. But there is abundance of water resource below the road in these areas, from where the water is supplied to neighbouring villages and also to some localities of Almora town. Similarly, the Ghaneli and Odyari area is also faced with accute water shortage. But the most striking fact regarding the water resources of the region is that, as also stated above, despite critical water problem, Shail and Ghurson-Channa area has been supplying 40,000-60,000 litre per day water to Almora town, and at the same time village Shail faced with the most accute water problem in the region is said to have been receiving less then 4 Litre/Person/Day water from Almora water supply project. This state of affairs brings out the fact clearly that the high water potential of the region is being managed very poorly and inefficiently causing great hardship to the inhabitant population and resulting in underdevelopment of the region.

**Anthropogenic Impact on Water Resources**

Along with the knowledge of spatial pattern of water potential of a region, a resource manager as well as an environmental planner should have a very good understanding of the magnitude of impact of human transformation process of landscape, such as, deforestation , agriculture, construction of roads etc. on the geohydrological balance. He should also be well versed in translating all relevant environmental monitoring techniques and water resource management strategies in increasing the water generating capacity of land and harnessing its available potential for the ecologically sustainable development of an area. Large scale deforestation mainly for fuel and fodder and poorly and unscientifically managed agricultural practices are some of the major human transformation processes of landscape causing serious geohydrological disturbances in the region. These unscientific transformation processes have been resulting in reduced groundwater recharge through increased overland flow and declined rate of infiltration. Consequently, the water generating potential of land to channels and springs has reduced dramatically, in all the disturbed areas, during the last 20-30 years. As stated earlier, out of the total 49 springs surveyed in the experimental watershed, 18 have completely dried due mainly to reduced-ground water generating capacity of land owing to considerable decrease in infiltration. Besides, as has been reported by the villagers the flow of all remaining springs had decreased to a considerable extent, during the last 3-4 decades. As a result of reduced discharge of streams, out of four water mills in the catchment area, three lying in the upper part of the watershed, have been abandoned for the last 10-15 years. One more important reason which seems to be equally responsible for diminishing stream and spring discharge in the region, is the dominance of pine in the existing forests as well as in the regeneration process of forests in the region.

The need of sustainable development in Himalaya makes it imperative to adopt a comprehensive and integrated approach, for the planning, management and monitoring of water resources by maintaining the fragile balance between productivity functions and conservation practices. Necessary scientific measures should be taken for increasing the water generating capacity of land to springs and channels. Rainwater harvesting measures may also be taken to bring more land under irrigation. For this purpose six water conservation sites have been identified in Shail Gad micro-watershed for their ecological treatment and conservation, particularly, through planting water conserving trees. It would be ecologically desirable that these water conservation pockets should be developed as water sanctuaries and be fully protected and reserved. RRC tanks are the most suitable for the upper parts of watershed for harnessing spring water, whereas, in lower areas small guls can be constructed directly from streams and channels to cultivated terraces. Fifteen sites for the construction of RRC tanks have been identified besides some existing ones (Fig. 4B). In addition, temporary small reservoirs can also be constructed with village level technology, at several locations along low discharge streams for winters and dry summers. Twenty locations for the construction of such small reservoirs have also been proposed after detailed survey of the area and consulting their beneficiary population in different villages.

## References

Blaney, H.E. and W.D. Griddle, 1950. *Determining Water Requirements in Irrigated Areas from Climatological and Irrigation Data,* U.S. Department of Agriculture, soil conservation service.

Cowan, W.L., 1956. Estimating Hydraulic Roughness Coefficients. *Agricultural Engineering,* 37 : 473-5.

Haggett, P. and R.J. Chorley, 1969. *Network Analysis in Geography,* Edward Arnold, London.

Horton, R.E., 1945. Erosional Development of streams and Their Drainage Basins : Hydrophysical Approach to Quantitative Morphology, *Bulletin of the Geological Society of America,* 56 : 275-370.

James, L.D. and R. Lee Robert, 1971. *Economics of Water Resources Planning,* TMH Edition.

Mitchell, B., 1973. *Terrain Evaluation,* Longman London.

More, R.J., 1967. Hydrological Models in Geography. In R.J. Chorley and P. Haggett (eds.) *Models in Geography,* Methuen, London.

Olivier, H., 1961. *Irrigation and Climate.* Edward Arnold, London.

Pathak, H.G., 1983. Water Resources of U.P. Himalaya : Their Uses and Potentials. In O.P. Singh (ed.) *The Himalaya : Nature, Man and Culture,* New Delhi : 161-169.

Pathak, P.C., 1983. *Vegetation Hydrology and its Impact on Nutrients Cycling in Selected Forest Ecosystems of Kumaun Himalaya,* Unpublished Ph.D. thesis Kumaun University, Nainital.

Penman, H.L., 1963. *Vegetation and Hydrology,* Common Wealth Bureau of Soils, Harpenden, Technical Communication, No. 53.

Schwab, G.O., R.K. Frevert, T.W. Edminster and K.B. Barnes, 1966. *Soil and Water Conservation Engineering.* J. Wiley, New York.

Sherman, L.K., 1932. Stream-flow from Rainfall by Unit Hydrograph Method. *Engineering News Record,* 108 : 501-5.

Smith, K.G., 1950. Standards for Grading Texture of Erosional Topography. *American Journal of Science,* 648 : 655-68.

Strahler, A.N., 1957. Quantitative Analysis of Watershed Geomorphology. *Transactions of the American Geophysical Union,* 38 : 913-20.

Thornthwaite, C.W., 1948. An Approach Towards a Rational Classification of Climate, *Geographical Review,* 38 : 85-94.

Thornthwaite, C.W., 1954. A Re-examination of the concept and Measurement of Practical Transpiration. In R.J. Mather (ed.) *The Measurement of Potential Evapotranspiration : Problems in Climatology.* Seabrook, New Jersey : 200-9.

Tiwari, P.C., 1995. *Natural Resources and Sustainable Development in Himalaya.* Almora : 129-146.

# Chapter 11

# Status of Forest Resources and Sustainability of Rural Eco-system in Gomti Basin, Kumaun Lesser Himalaya

*B.S. Bisht*
*P.C. Tewari*

Forests are the most valuable natural resources of the Himalayan region of India. Besides providing fodder, fuel, timber, and raw material for different forest based industries, vegetal cover plays the most significant role in recycling and balancing the delicate and sensitive ecological system of the region. Forest protects the hill slopes against soil erosion, land slides, land slips and increase the effectiveness of precipitation by regulating and controlling overland flow, maintaining geohydrological balance and increasing humidity by transpiration.

The diversified physio-climatic conditions obtaining in the Gomti watershed are reflected in a variety of natural vegetation in the region. Diversity of surface relief with varying aspects of slopes and their exposures resulting in sharp variation in temperature and rain fall, mainly control and condition the growth, distribution and density of forests in the entire region.

## The Study Area

The Gomti Basin encompassing a geographical land surface of 362.8 sq km. in district Almora of Kumaun Lesser Himalaya was selected for this investigation. The watershed has its spatial extention between 29° 46$^{-1}$ and 30° 2$^{-1}$ North Latitudes and 79° 28′ and 79° 47′ East Longitudes with mean elevation ranging from 960 m to 2560 m from the mean sea level (Fig. 1). Out of the total geographical area of the basin 54.5 per cent and 7.2 per cent is respectively under forest and pastures, and as much as 17.1 per cent is under cultivation. As per the 1991 census records, the total population of the watershed was 70231 which is distributed in 261 villages.

**Objectives**

The study has been focussed on achieving the following objectives :

(i) To assess the status and availability of forest resources.

(ii) To identify the fodder and fuelwood surplus and deficit situations in the catchment. (iii) To examine the trend and magnitude of grazing pressure of livestock on forest and pastures of the region. (iv) To assess the impact of the degradation of forest resources on the regional agro-ecosystem through socio-economic approach.

**Research Methods**

The research problem was approached by catering the following research components :

(i) The entire catchment was divided into six micro watersheds following the stream ordering technique. The calculation of forests, pastures, grassland area was done by using Survey of India Topographical Sheets, land record maps, forest maps, field surveys and mapping also through secondary information collected from various government offices.

(ii) Fuelwood and fodder consumption at household level was quantified in each micro watershed by household survey and measurement in representative villages. The livestock were converted into cattle units on the basis of their fodder requirement of (taking one cow as one cattle unit, one buffalo as two cows, one bullock as one cow and one goat as half cow unit).

(iii) Fodder productivity from different landuse categories, i.e., forests, grassland and cultivated land was estimated by harvest method. The harvested biomass was brought to the

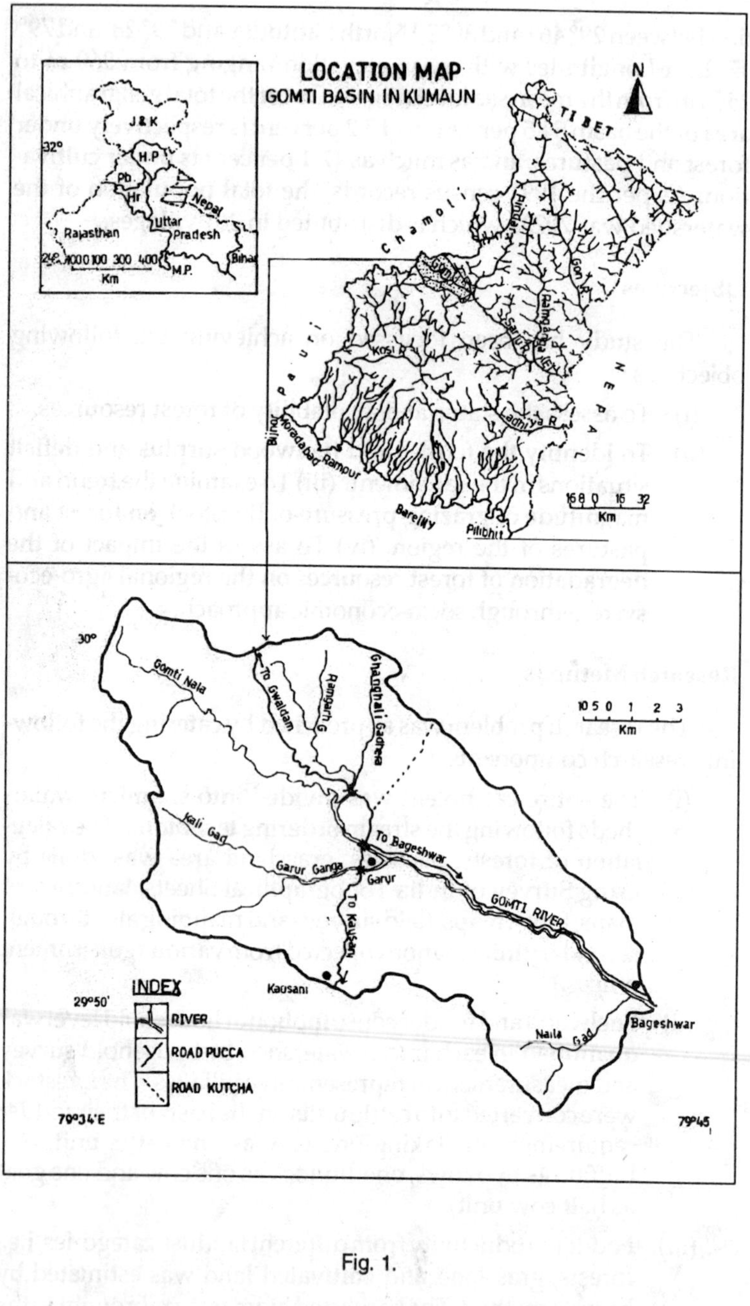
LOCATION MAP
GOMTI BASIN IN KUMAUN
N
TIBET
Chamoli
J&K
H.P
Pb.
Hr
Nepal
Uttar Pradesh
Rajasthan
Bihar
M.P.
32°
Km
Pauri
Kosi R
Ladhiya R
Bijnor
Moradabad
Rampur
Bareilly
Pilibhit
NEPAL
16 8 0 16 32
Km
30°
Gomti Nala
To Gwaldam
Ramgarh G
Ghanghali Gadhera
10 5 0 1 2 3
Km
Kali Gad
To Bageshwar
Garur Ganga
Garur
To Kausani
GOMTI RIVER
Kausani
INDEX
RIVER
ROAD PUCCA
ROAD KUTCHA
29°50'
Bageshwar
Naila Gad
79°34'E
79°45'

Fig. 1.

laboratory and its above ground material were oven dried and weighed till a constant value of its weight was obtained. But for the calculation of leaf fodder production from arable land, the norm of 0.48 tonnes dry matter per year/ha from cultivated land established by Singh et. al. (1988) was used.

(iv) In order to assesss the availability of fuelwood, the total forest area available as resource base was converted into forest stock taking 11.5 tonnes ha/year production of twigs and branches (Singh et. al. 1988).

(v) The impact of forest degradation and depletion on local agro-ecosystem was assessed entirely through socio-economic approach. A comprehensive survey of 10 stratified villages was conducted in the catchment through specifically designed schedules. The information thus generated on important aspects such as temporal variations in the supply of biomass energy to cultivated land, irrigated cultivation, erosion and soil fertility, agricultural productivity, etc. were analysed and interpreted with the help of personal field observations and experience.

**Forest Resources and their Management**

The forested landscape of 197.8 sq km of the basin (54.5 per cent of the total geographical area of the Gomti watershed) covers different type of natural vegetation (trees, shrubs and herbs). About 52.5 per cent of the forested area is under the administration of State Forest Department (i.e. Reserved Forests), and 40.5 per cent are mainly new forests areas which have been developed in village community lands and managed by Civil an Soyam wing of Forest Department in collaboration with village level institutions. Remaining 7 per cent of the forested land includes forests on community and private land (managed and controlled by village Panchayats, and individuals respectively). The felling of green trees in all management types is restricted for few species and is under control of State Forest Department. About 90 per cent forest cover of the watershed area is dominated by Chir pine (*Pinus roxburghii*) extensively utilized for timber, resin and fuel. The various outputs of reserved forests are managed by Government agencies.

In contrast to common pasture land (freely accessible for grazing animal throughout the year), privately owned grassland are under systematic management for optimum production i.e. ban on grazing during fodder growth season (July to September) and harvesting in September to October than storage for less fodder availability period (winter and summer).

**Fodder Availability**

Fodder to the animals is made available mainly from three sources i.e. forest fodder, grasslands and the crop residues. The reserved forests, proportion of Civil forests which remain closed for harvesting during monsoon, individually owned grasslands and cultivated land were taken into account for the estimation of forest and grasslands potentials in terms of fodder production. The analysis reveales that the per hectare production of grass fodder from forest and grassland was 1.23 tonnes per ha, and it was 1.5 tonnes per ha from cultivated land. While for the central Himalayan region Singh et. al. (1988) have estimated 1.6 tonnes average crop residue per year from each ha of farming, which is quite similar to the results obtained through the method used here.

A fodder deficit ranging between 1455.1 and 18312.6 tonnes/ year was observed in different microwatersheds however, the extent of deficit was more in case of the increase in the number of cattle units. The estimates of resource potential in terms of fodder production and its actual requirement bring out the fact that the watershed has great scarcity of fodder, i.e. 82998.8 tonnes/year, (Table 1 & Fig. 2). This can be attributed to the following :

(i) Increase in cattle population and non-availability of proportionate forest areas.

(ii) Almost all the forest areas under all categories of forest are dominated by Chir-pine, which has no fodder utility, this is inspite of the fact that physio-climatic conditions of the watershed, in many areas, are quite suitable for Oak and some other species which have great utility for fuel as well as fodder.

(iii) The wasteland of the region is not being managed properly for fodder production.

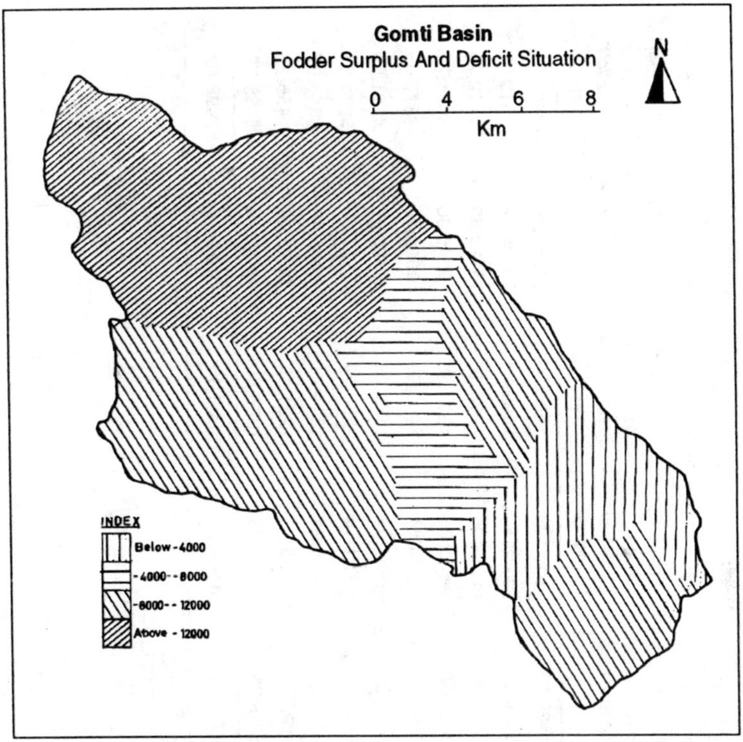


## Fuelwood Availability

The analysis of fuelwood requirement revealed that during summer months it ranges between 10.17 and 17.50 kg per family, per day and in winter the fuelwood demands was approximately between 17.12 and 22.86 kg per day per household, which brings a daily use of 1.58 kg fuelwood per person per day in the studied region. In some villages of Garhwal Himalaya daily per capita use of fuelwood has been estimated to be 1.8 to 2.3 kg per day in summer and 3.6 kg per day during winters (Monech, 1989). Tata Energy Research Institute (TERI), New Delhi, has considered 1.9 kg/day/person fuelwood for the Himalayan region. The micro-watershed-wise estimates of fuelwood production and requirement, and the surplus and deficit situations have been shown in Table 1 & Fig. 3. The overall fuel shortage of -8143.2 tonnes was computed for the entire drainage area. The maximum shortage (-2073.8 tonnes/year) of fuelwood was estimated for Ghanghali micro-watershed, whereas, a surplus of 392.9 tonnes was calculated for Chhira Raula microwatershed (Table 1).

**Table 1 :** Fuel and Fodder availability and requirement in Gomti Basin

| *Micro Watersheds* | *Area Km²* | *Fuelwood Availability Tonnes/year* | *Fuel Wood Requirement Tonnes/Year* | *Fuel wood Surplus/ Deficit Tonnes/year* | *Fodder Availability Tonnes/year* | *Fodder Requirement Tonnes/year* | *Fodder Surplus/ Deficit Tonnes/year* |
|---|---|---|---|---|---|---|---|
| Gomti Nala | 78.57 | 20043.24 | 21129.63 | -1086.39 | 1609.59 | 14051.20 | -12441.61 |
| Ghanghali | 42.55 | 38310.83 | 40384.67 | -2073.84 | 2092.54 | 20405.12 | -18312.58 |
| Garur Ganga | 62.53 | 51361.84 | 52515.90 | -1154.06 | 1708.16 | 13120.89 | -11412.73 |
| Bheta Gad | 22.28 | 11801.67 | 12445.53 | -643.86 | 1234.18 | 9545.90 | -8311.72 |
| Dukara Gadhera | 28.85 | 7208.33 | 7492.77 | -284.44 | 1104.56 | 9982.19 | -8877.63 |
| Pachgari Gadhera | 23.80 | 22469.32 | 2401.24 | -1547.92 | 1159.08 | 5420.67 | -4261.59 |
| Baneshwar | 21.60 | 16213.01 | 16734.15 | -521.14 | 2256.29 | 8127.60 | -5871.31 |
| Kali Rau | 25.34 | 13799.05 | 14017.29 | -218.24 | 1314.21 | 4630.18 | -3315.97 |
| Chhira Raula | 21.98 | 4570.21 | 4177.35 | 392-86 | 1221.15 | 2676.28 | -1455.13 |
| Naila Gad | 35.80 | 41750.00 | 42756.25 | -1006.25 | 1826.60 | 10565.15 | -8738.55 |
| Gomti Basin | 362.80 | 227527.50 | 23670.78 | -8143.28 | 15526.36 | 98525.18 | -82998.82 |

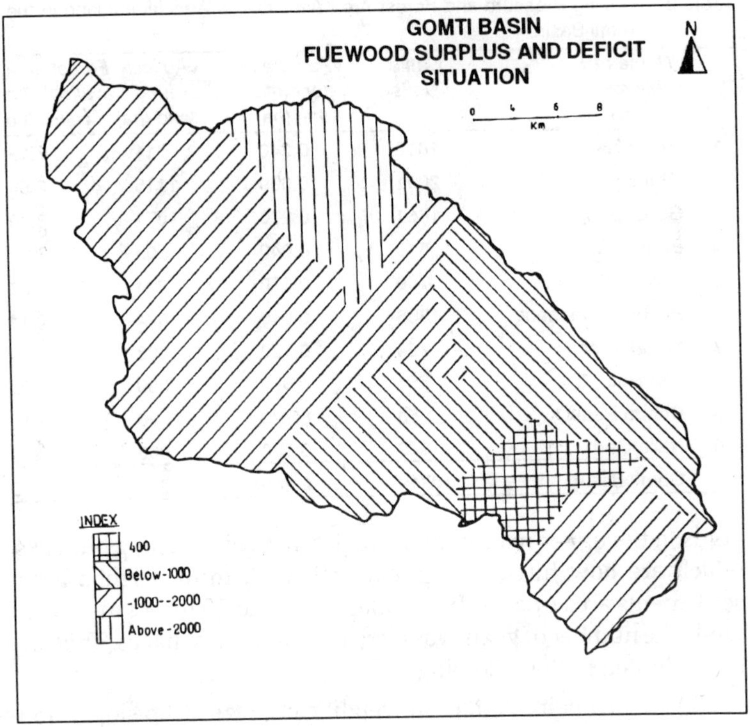


Fig 3

## Grazing Pressure

There were 10719 cattle in the watershed in 1991, Ghanghali microwatershed emerged as the village having the highest cattle population (2091), and Chhira Raula microwatershed has the lowest cattle units (58) in the region.

There was on an average 0.24 ha grazing land per cow unit available in the watershed area (Table 2), whereas, the ecologically permissible capacity of the pasture in the region is considered to be 3.5 ha per cattle (Singh et. al. 1984). The pressure of grazing is the highest in Ghanghali microwatershed (0.009 ha per cattle) and the lowest in Chhira Raula microwatershed (8.79 ha per cattle) Table 2 & Fig. 4. With the increase in cattle population there is an increase in the grazing pressure.

The great scarcity of grazing area has led many of the house-

**Table 2 :** Grazing Pressure and Forest Area Available to Agricultural land in the Gomti Basin

| | *Name of Micro-basin* | *Cattle Units* | *Pastures per cattle unit (ha)* | *Agricul-tural land (ha)* | *Forest area available/ha cultivated* |
|---|---|---|---|---|---|
| 1. | Gomti Nala | 1818 | 0.340 | 711 | 7.18 |
| 2. | Ghanghali | 2091 | 0.009 | 1115 | 2.63 |
| 3. | Garur Ganga | 1661 | 0.270 | 1071 | 3.19 |
| 4. | Bheta Gad | 819 | 0.040 | 503 | 3.00 |
| 5. | Dukara Gadhera | 740 | 0.570 | 294 | 3.37 |
| 6. | Pachgari Gadhera | 1388 | 0.040 | 624 | 2.14 |
| 7. | Baneshwar | 724 | 0.120 | 608 | 1.59 |
| 8. | Kali Rau | 551 | 0.570 | 390 | 2.20 |
| 9. | Chhira Raula | 58 | 8.790 | 98 | 3.15 |
| 10. | Naila Gad | 869 | 0.120 | 803 | 2.93 |
| | Gomti Basin | 10719 | 0.240 | 6217 | 3.18 |

holds in the upper reaches of the region to avoid keeping bullocks which are now hired for agricultural work from neighbouring settlements area or even from villages located 10 to 24 kms away, while the number of goats was increasing in some villages. This can be attributed to the following :

(a) Goats being lighter in weight can graze steep slopes and other such areas which are completely inaccessible to cows and bullocks.

(b) They do not need any additional fodder and maintenance after grazing.

(c) They give very high economic return.

### Forests and Farmland

Table 2 also reveals that the forest land available to per ha of agricultural land is 3.18 ha in the watershed which is higher than the average situation (1.58 ha) reported for Central Himalaya (Singh et. al. 1984, 1984b). This shows the degree and magnitude of pressure on forest area. The agricultural pressure on forests ranges from 1.59 to 7.18 ha in different microwatersheds of the region. The pressure was very accute in Baneshwar microwatershed (1.59 ha) owing to almost complete absence of forest land, and it is also very high in Ghanghali, Pachgari, Kali Rau and Naila Gad microwatersheds where the availability of forest area ranges between 2.14 and

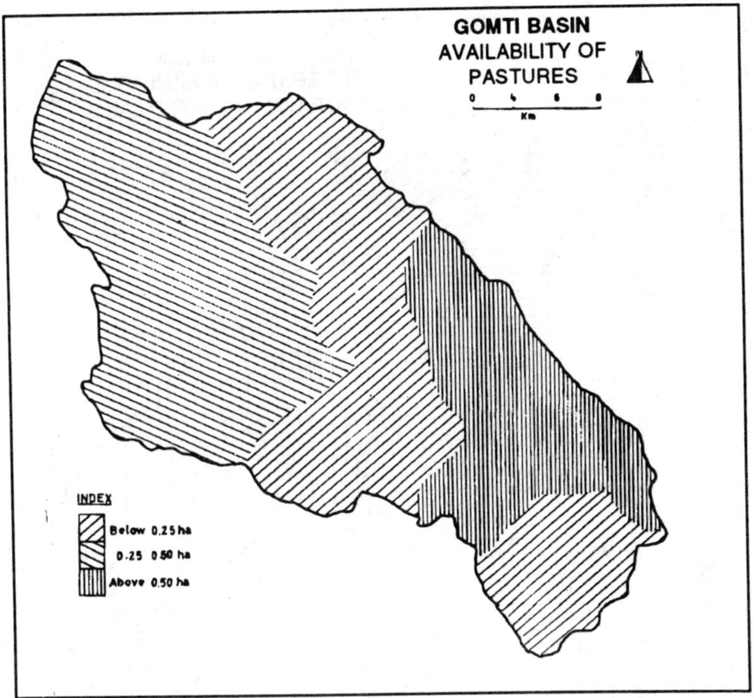


Fig. 4

2.93 ha for each hactre of farmland (Fig. 5). Whereas, the other microwatersheds have considerable proportion of reserved and civil forests within their boundaries, the availability of support area in terms of forest is little higher in these areas compared to other microbasins.

Ecologically, 5 to 12 ha of well stocked forest area is needed to meet the energy requirement of 1 ha of arable land in Himalayan argroecological system (Singh, 1984b). In the central Himalayan region of U.P., the availability of forests area per ha of cultivated land is 1.58 ha (Singh et. al. 1984a, 1984b). Whereas, in some part of Garhwal Himalayan it has been estimated to be 7.3 ha, for per ha of farming (Monech, 1989).

## Impact on Agroecosystem

### Energy Supply Declined

The flow of organic matter and plant nutrients from support

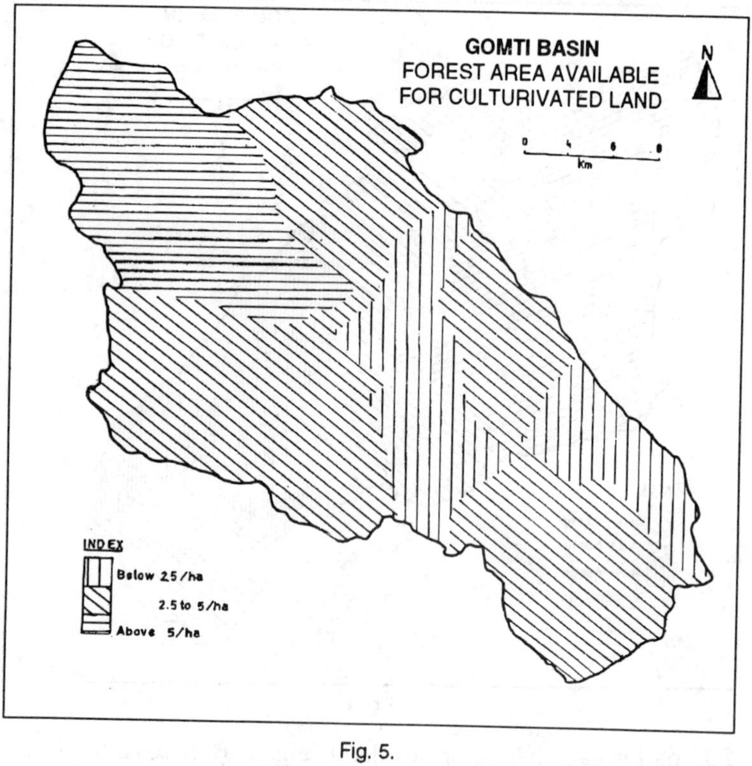


Fig. 5.

area to cropping is the most significant ecological process operating in the hill agricultural system. It was investigated that the average supply of biomass energy has declined by 7.49 per cent during 1980-1990 despite increasing cattle population in many parts of the watershed (Table 3).

## Loss of Irrigation Potential

In the studied region it was investigated that about 30-35 per cent water sources in the sample villages have gone dry or become seasonal during the last 20-25 years. This phenomenon of considerable reduction in the ground water recharge has posed a serious threat to the regional agricultural economy. The watershed had lost about 10 per cent of its irrigation potential during the last 10 years mainly owing to reduced ground water recharge. A recent study revealed that the Himalayan rivers will carry 20-30 per cent less water in coming years (Centre for Science and Environment, 1982).

**Table 3 :** Supply of Manure, Irrigation Potential and Agricultural Productivity in Gomti basin during 1980-90

| | *Name of the Sample Villages* | *Decrease In supply of leaf Manure* | *Loss of Irrigation Potential* | *Decline in Agricultural Productivity* |
|---|---|---|---|---|
| | | *(%)* | *(%)* | *(%)* |
| 1. | Kunjhali (Gomti Nala) | 5.05 | 14.98 | 9.56 |
| 2. | Titoliya (Ghanghali) | 3.25 | 5.28 | 5.18 |
| 3. | Chamoli (Garur Ganga) | 23.11 | 11.10 | 4.07 |
| 4. | Nakuri Talla (Bheta Gad) | 9.98 | 12.16 | 6.65 |
| 5. | Mate (Dukara Gadhera) | 8.51 | 9.87 | 5.03 |
| 6. | Rampur Talla (Pachgari Gadhera) | 12.56 | 10.00 | 4.16 |
| 7. | Baneshwar (Baneshwar) | 3.05 | 14.08 | 9.15 |
| 8. | Kunar (Kali Rau) | 2.18 | 13.10 | 11.12 |
| 9. | Burhgarh (Chhira Raula) | 3.18 | 9.11 | 13.76 |
| 10. | Bijori (Naila Gad) | 4.01 | 5.23 | 9.98 |
| | Total | 7.49 | 10.49 | 7.87 |

## Agricultural Productivity Declined

The productivity and efficiency of agricultural system have declined to a considerable extent during the last 10-20 years, in spite of continuous expansion of farming on hill slopes and increased cropping intensity. A comparative analysis of cropping intensity and productivity of agriculture in the region in 1980 and 1990 brings out the fact clearly that both the variables have inverse relationship between them. The higher the intensity of cropping the lower the agronomic production in the region. The entire watershed had recorded an average increase of about 2 per cent in cropping intensity, during 1980-1990. Whereas, the production of agriculture dropped 7.5 per cent during the same period (Table 3).

This is explained by the fact that the higher cropping intensity requires increased supply of water, soil nutrients and biomass energy, and consequently put increased stress on the support area which has already been passing through a rapid degradational process. Therefore, the agronomic yield is very low and declining progressively.

After measuring the degree of association among different parameters the co-efficients of co-relation were worked out which revealed that the variables of forest area available per unit of cultivated land, supply of biomass energy and percentage of irrigated land had a very high degree of positive relationship with agronomic production. Whereas, the agricultural productivity in the watershed was found to have an inverse association with the variable of cropping intensity (Table 4).

**Table 4 :** Magnitude of Relationship Among Different variable in Gomti Watershed

| *R Value* | *Forest area available per unit of farming* | *Supply of Biomass energy* | *Percentage of Irrigated* | *Cropping Intensity* |
|---|---|---|---|---|
| Agricultural Productivity | 0.72 | 0.75 | 0.76 | -0.74 |

## Conclusion

On the basis of above discussion it may be concluded that the resource dependency pressure of villages on the forests of the catchment is very accute and almost all the microbasins of the region are now faced with great scarcity of fodder and fuelwood. Besides, the grazing pressure of livestock on forests and pastures of the watershed has already approached the critical ecological limit. Consequently the regional forest ecosystem is under great stress. The progressively degrading forests have not only destabilized the fragile geo-ecological balance, through promoting slope failures, accelerating soil erosion and reducing the water generating capacity of land to springs and stream in the region, but have also caused a very profound adverse impact on the productivity and efficiency of the regional agricultural economy and put it in a very deplorable condition. The reduced supply of organic manure loss of irrigation potential and the accelerating pace of soil erosion and soil nutrient loss attributed mainly to forest degradation and depletion, have declined the agronomic production to a considerable extent despite progressively increasing cropping intensity in some areas of the region.

The region, therefore, deserves the specific and adequate attention for protecting the environment and sustainable development of natural resources. In view of this, as a first step, it is necessary to evolve a natural resources information system at village level through detailed analysis, appraisal and mapping of environmental resources particularly with the application of new and emerging areas of science and technology and then to formulate socially acceptable and economically feasible and ecologically sustainable strategies for the development of natural resources.

## References

Centre for Science and Environment, 1982. *The State of India's Environment*, New Delhi.

Monech, M., 1989. Forest Degradation and the structure of Biomass Utilization in a Himalayan Foothill village. *Environmental Conservation*, Vol. 16 (2) : 137-146.

Singh, J. S., S. P. Singh and J. Ram, 1988. Fodder and Fuelwood Resources of Central Himalaya : Problems and Solutiomns. *Report Submitted to Planning Commission*, Govt. of India, New Delhi, 148.

Singh, Jasbir and S. S. Dhillon, 1984. *Agricultural Geography.* Tata McGraw Hill, New Delhi.

Singh, J. S., Y.S. Rawat and O. P. Chaturvedi, 1984. Replacement of Oak Forests with Pine in the Himalaya Affects the Nitrogen Cycle. Nature, London, 311 : 54-6.

Singh, J. S., Uma Pandey and A. K. Tewari, 1984. Man and Forests : A Centre Himalayan Case Study. Ambio, 13 (3): 80.7.

# Chapter 12

# Ethonobiology of Kumaun Himalayas

*P.C. Pande*

*Pramila Joshi*

*G.C. Joshi*

Kumaun Himalaya, an eastward offshoot of the Western Himalayas, is situated between 28°43'-30°49'N lat. and 78°44'-81°5' E long. and comprises three districts namely, Almora, Nainital and Pithoragarh.

A number of workers have explored the utility of Kumaun flora (Atkinson, 1882; Shah & Joshi, 1971; Pangtey, 1980; Pangtey *et al.*, 1982; Pande & Pangtey, 1987; Shah & Jain, 1988; Pande *et al.*, 1989) but by now no account is available as far as the use of animals, birds, reptiles and insects etc. in folk medicine and magico-religious belief is concerned.

The present article is based on either on the information gathered from medicinemen or from the elderly people of this region. The following list gives their Kumauni names with English names in parenthesis and their probable zoological equivalents alongwith their utility. For ready reference, the Kumauni names are arranged alphabetically under each class.

**Phyllum : Mollusca**

**Class : Gastropoda**

1. KACHHUWA-HAR/KAURI (SEAP) : THE SHELL OF MOLLUSCS.

(i) Seap obtained from sea are made into 'Tabej' of the children. It is believed to be useful to ward off evil spirits. It is also said to keep them away from bad eyes.

**Phyllum : Annelida**
**Class : Oligocheta**

1. GIDAUL/KAICHUWA (EARTHWORM) : PHARETIMA POSTHUMA

(i) It is found everywhere in Kumaun, generally in damp and shady places where much humus is accumulated. The women, who after delivery, are unable to feed their children due to poor lactation, are given the soup of earthworm. It is said to improve lactation. However, the practice has become very rare now.

**Phyllum : Arthropoda**
**Class : Arachnida**

1. BICHHU/BICHHI (SCORPION) : PALAMNAEUS Spp.

(i) Bichhu is put in enough quantity of mustard oil (*Brassica campests*) and boiled for some time. Scorpion is then taken out and thrown away. The oil so prepared is applied externally for the treatment of eczema (locally known as MOTI KHUJALI) particularly on neck region.

**Class : Crustacea**

1. GANJYAR/Gyanj/KEKRA (CRAB) : CANCER Spp.

(i) It is found near ponds and river banks. People use to eat it like fish. Its soup is deemed to be very tasty. The skeleton ash of this animal is used as medicine to cure various diseases.

**Class : Insecta**

1. GWALI/GWAIE (PRAYING MANTIS) : MANTIS Sp.

(i) When Gwali sits on any part of the body of a person, it is believed that it measures the length of the person concerned and the new clothes are likely to be stiched for him.

2. JHIMAUR (WASP) : VESPA ORIENTALIS

(i) When Jhimaur applies its sting to human being, it is said to prevent fever, like a vaccine.

(ii) A piece of wasp comb, stratum corneum of Snake (s) a few seeds of Rai (*Brassica juncea* (Linn.) Czern. and black gram (*Phaseolus radiatus* Linn.) are taken enchanted by experts and used for the cure of cows and buffaloes having stopped giving milk.

3. KIRMUL/KIRMAI/ (ANT): DIFFERENT KINDS OF TERMITES (PARTICULARLY DEEMAK) : ODONTOTERMES REDEMANNI

(i) Whenever the ants come out in gregarious form or they carry their eggs from one place to another, it is assumed by the local people of Kumaun that the rain showers would come soon.

4. KURMUL/KURMUN (LARVE OF BEATLE) : ANOMALA DIMIATA, HOLOTRICHIA RETICOLLIS, H. LONGIPENNIS, MISNELA FULGIDIOITTATA.

(i) The skin of Kurmul (larve of beatle) is dried and mixed with water thoroughly and given to children for the cure of round worms or to expel the round worms.

5. MANKH/MANKHEE (HOUSE FLY) : MUSCA INDICA

(i) House fly is killed, a paste is made with gur and crustose lichens (particularly growing on rocks) and applied locally for the cure of ringworms (Locally known as 'DAD').

6. MAUN/MOON/MADHUMAKHEE (HOUSE BEE) : APPUS SPP. (A. INDICA IS A COMMON SPECIES)

(i) It is said that if ghee and honey are mixed in equal amounts (by weight) it becomes poisonous.

(ii) The larve of bee (s) are eaten while extracting honey by the traditional method i.e. the entire combs are removed from the hives.

(iii) Frequent murmuring of honey bee around some person is considered as a good omen for that person.

7. PAS-FAIL KIRH (LADY BIRD BEETLE OR CARPET BEETLE): ANTHRENUS SPP.

As etymology of Pas-fail kirh suggests, the little children in Kumaun region used to play with it to get their result in the examination. A child takes the Pass-fail kirh in his hand and it is believed that if it flies up, the child will get sucess in the examination if it remains (sit) in the hand, he/she will be declared failed.

**Phyllum : Chordata**

**Class : Conducthese (Pisces)**

1. MACHHA/MACHI (FISH) : TOR SP. (MAHASHER); BOTIA SP. (CHITTAL); SCHIZOTHORAX SP. (ASYAI) :

(i) Different types of fish are found in the rivers of Kumaun, out of which 'Asyai' fish is liked more.

(ii) It is believed that on the consumption of upper part (including head) of the fish, eye-sight increases.

(iii) Fish are trapped by fishing net which is made of Al (*Girardinia heterophylla* Deche) or fishing-rod (made up of *Chimonobamusa* spp.) in which athread which is also made of 'Al' is used. Some local inhabitants use a mixture of Khina (*Sapium insigne* Trim.) Rambans (*Agave americana* Linn.) and red soil as bleaching powder to trap the fish. An other method of fishing is 'Goda' (a special process by which scattered water from natural sources or of rivers is given a particular direction to flow through it). The 'Goda' is made up of pseudostem of Kela (*Musa paradisiaca*), and stems of *Pennisetum americanum* (Linn.) Leek. or *Chimonohambusa* spp. or *Zea mays* plants.

**Class : Reptilia**

1. GWARH/GOH (INDIAN MONITOR) : VARANUS MONITOR

(i) The Gwarh is dipped in oil of *Brassica compestris* and boiled. This oil is used for the treatment of rheumatic pains.

2. SYAP/SANP (SNAKE) : VIPER RUSSELLI, PTYAS SP. ANCISTRODON SP.

(i) It is common belief in the local people of Kumaun Himalaya that a person who sees for the first time the fight (actually it is not fighting but mating of a couple of snake) of two snakes is very bad omen for that person.

(ii) A snake when crosses ones way while on journey is considered bad omen and it is thought that the work for which the person is going would not be suceess.

(iii) Stratum corneum (Kainchuli) of the snake when hung on the top of entrance door of the house is considered to protect the house against the ill will of the people and protects the children from the attack of ghosts etc.

(iv) The people generally don't kill the snakes when they enter their premises as a safeguard against the course of Lord Shiva and they try to get them away without killing. When it becomes necessary to kill a snake, cow dung is put in its head (mouth).

(v) Some old snakes are supposed to have 'Mani' on their heads. If a person can acquire this 'Mani' it is believed that he will achieve all the success and riches in life.

(vi) A gentle rubbing of fore-head and eye-lids by the tail of snake is supposed to cure the headache and increases the eye sight.

(vii) The Kainchuli (*Stratum corneum*) of the snake is made into ash and surma is prepared by mixing it with mustered oil (*Brassica compestris*). Application of this surma is said to cure many diseases of the eye.

**Class : AVES***

1. GHINAURI/CHARI/PANCHARI/GEERH/GAURAYA (HOUSE SPARROW) : PASSER DOMESTICUS

(i) The excreta of 'Ghinauri' is applied externally when septic wounds appear on nose or ears when using nose-studs and ear-studs. It is also applied externally on mumps.

2. GHUGHAT : GALLUS SP.

(i) The flesh of Ghughat is rarely eaten by Kumaunies.

3. KABUTAR (PIGEON) : COLUMBA LIVIA

(i) The flesh of Kabutar is supposed to be curative against rheumatic pain and paralysis.

(ii) Domestication of Kabutar is supposed to be preventive against infectious diseases.

---

* The heads of birds are not eaten by Kumaunies.

4. KAUA/KAU (CROW) : CORVUS SPLENDENS

(i) Cocks crow in the morning near ones house supposed to indicate that some guest would come to his house that day.

(ii) When excreta of crow falls on the head of any person, it is considered as a very bad omen particularly for his/her health.

5. MALYA (SNOW PIGEON) : CLUMBA SP.

(i) Its flesh is also eaten by the Kumaunies.

6. MOR (PEA COCK) : PARO CRISTATUS

(i) The feathers of Pea-cock are kept in house, worship places and are considered to be a good omen.

(ii) The claws (hind limb) of Pea cock are rubbed with water and the paste used for ear-troubles/particularly sepsis in the ear (Otorhoea)

(iii) Small piece from the inferior umblicus and rachis of the feather are used as ear-studs and nose-studs.

7. MURGI/KUKURH (HEN) : GALLUS FERRUGINEUS

(i) The pealed boiled hen's egg is kept over night covered with salt and eaten in morning immediately after rising from the bed for a few days. It is said to cure jaundice.

8. SINTAI (HILL MAINI) : ACRIDOTHERES TRISTIS

(i) The blood of this bird is rubbed on spots on having tyroma and is said to cure it.

9. TITUR/TITAR : (A) KALA TITAR (BLACK PATRIDGE) : FRANCOLINUS FRANCOLINUSASIAE, (B) SAFED TITAR (GREY PATRIDGE) : FRANCOLINUS PONDICERIANUS INTERPOSTICUS

(i) Cooked Titar is supposed to be a good medicine for rheumatism and goust (meat) for the treatment of anaemia.

(ii) It is considered as a very clever bird and the children are blessed by the older people to be clever like a Titar.

10. TITYA/TITIHAR (LAPWIG REDWATTLED) : VANELLUS INDICUS

(i) The egg yolk of this bird is supposed to cure typhoid fever when frequently applied on forehead. It is said that in olden days the eggs were preserved by Kumaunies.

The eggs were preserved by Vaidyas for very long period in cow dung.

**Class : MAMALIA**

1. BAG/GAGH (LEOPARD) : PANTHRA PARDUS
   - (i) The fat from this animal is used for massage for the cure of bodyache and rheumatic pains.
   - (ii) The nails from the paw of *Panthra* are worn as garland by the children to protect them from evil spirits for to ward off evil spirits.
   - (iii) The shoulder bone (humerus) of this animal when kept in possession is said to fulfil all desire.
2. BAIL/BALD (OX) : BABULUS SP.
   - (i) The skin of the ox is used as a top for the trumphet.
3. BAKAR/BAKARI (GOAT) : CARPA FALCONERI
   - (i) The milk of the goat is drunk and considered as a tonic.
   - (ii) The flesh is eaten.
   - (iii) The dung is grind and applied on fresh wounds to cure them.
   - (iv) The skin is used for making the tops of the trumphet, tabala etc. and the tops of dumbell are made from the portions of stomach wall.
4. BHALU/BHAALU (BEAR) : SELENARCTES THEBETANUS
   - (i) Massage with the fat derived from the body of a bear is said to cure chronic arthritis.
   - (ii) Gall blader (teetee) is used for the cure of malaria and a number of other diseases. For preservation, the gall bladder is removed and filled with the grains of rice and then air dried. The whole turns into yellow powder after drying and preserved for future use.
5. BHAINS (SHE BUFFALO) : BOS BABALUS
   - (i) The allergy of the skin (Bhainsi Pitta : When their skin develop large red spots with itching) is cured by taking curd made from buffalo-milk and also applying fresh buffalo dung on the skin and then bathing with water having buffalo dung dissolved in it.

(ii) The skin of buffalo is used to make the top of the trumphet e.g. Dhol, Nagara etc.

(iii) Ash made from dried buffalo dung is used as an insceticidal treatment of seedlings.

(iv) The skin of the calves of buffaloes is made like a pillow cover tied on loose side and filled with air. These special floats are used for crossing rivers having no bridges and especially by the parties transporting the timber through river (locally known as 'Bahaan'). One or more floats are tied on the ventral side and then these people swim easily in water and give route to the logs which have been side tracked from the main water current.

(v) The fungus* growing on the decaying horns of Buffaloes is said to cure eczema (especially of neck portion, locally known as 'Moti Khujali') when applied externally.

6. BIRALU/BIRAU/BILLY (CAT) : FALIS DOMESTICUS.

(i) Skin erruptions on the face are said to cure when licking by cat.

(ii) Keeping of placenta of the cat in the house is considered good getting richness etc.

(iii) While going far an important work, crossing of the way ahead by a cat is considered very bad and it is said that the work might not be fulfilled.

7. CHUTHRAUL,CHUTRAUL : HERPESTES SP.

(i) The bones of this animal are rubbed with water and the paste so made is applied for the cure of abscess (locally known as 'Nasoor').

8. GAYA/GORU (COW) : BOS INDICUS

(i) Ash of cow dung is applied on grains as pesticide, it is also used on seedling.

(ii) Ghee from cow's milk is used for massage in rheumatism.

(iii) Butter from cow's milk is applied on the head for the cure of excessive heat.

(iv) Cow dung and urine is used in many puja ceremonies and are considered very pious by most of the Kumaunies.

---

* The fungus or the material growing on it, not worked out.

(v) Cow's urine is taken for the treatment of allergy and many other ailments.

9. GHORH/GWARH (HORSE) : EQUUS CABALUS

(i) The hair of the tail of this animal are used to root out the warts.

(ii) Horse dung is dissolved in water, boiled and filtered in course cloth. The filterate is used for the cure of acute stomache.

(iii) Left horse-shoe is nailed on the front door of the house. It is supposed to keep away the evil spirits from that house.

10. GHURAH (GOPAL) : NEMORHAEDUS GORAL

(i) The animal is hunted for its flesh.

11. JATIYA/BHAINSA (HE-BUFFALO) : BOS BABALUS

(i) Jatiya is sacrificed in some temples like-Nanda Devi (at Almora), Kalika temple (at Gongolihat) etc. on Nanda-Astami at Almora and Aswin-astami at Gangolihat. At Gangolihat Kalika Mandir some people dip their feet in the blood at the place of sacrifice and it is said to cure 'Masur-Khum' (small painfull spots on the lower side of the feet.)

12. JARHYA/BARHASINGA (SAMBHAR) : CERVUS UNICOLAR ; C. DUNVANCELI

(i) The horns of this animal are rubbed with water and the paste is applied on the forehead to remove headache.

13. KANKARH/KANKAR : MUNTIACUS MUNTIJAC.

(i) It is a common hunt game animal of this region.

14. KHARGOSH/SHASH (HARE) : LEPUS RUFICAUDTSUS

(i) It is also a common game animal of this region.

(ii) It's flesh is supposed to cure asthma.

(iii) The spine like outgrowth (modified part of the skin) of this animal (locally known as 'Shaul-Pirul') used in thread ceremony.

(iv) The spines are kept into the houses of other people and said to make them quarrel among themselves.

(v) The flesh is considered good for people suffering rheumatism.

15. KASTURI MIRG (MUSK DEAR) : MOSCHUS MOSCHIFERUS

(i) Musk is removed from the navel of Musk dear and is used for the treatment of a number of ailments.

16. KUKUR/KUTTA (DOG) : CANIS SPP.

(i) The kukur is considered to be a most faithful domestic animal. It is believed to be carrier (vahan) of 'Bhairab', a local ('Devata') of the region and therefore no one dares to step acrossing it.

17. MUS/CHUHA (RAT) : RATTUS RATTUS

(i) The flesh of rats is used for the cure of post natal troubles of the women.

(ii) It is also supposed to cure the prolapsus of annus.

(iii) The flesh of this animal also cures chronic alimentary disorder particularly continuous loose motion.

(iv) It is said to improve the general debility and ill health and also increases the eye sight.

(v) The excreta of the rats is mixed and ground with the seeds of Kakari (*Cucumis melo* var. *utitissimum*). The paste is then applied on the naval region of the person to cure them from the trouble of retention of urine.

18. SUWAR (PIG, WILD BOAR) : SUS SCROFA

(i) The flesh of this animal is given to the people suffering from asthma and it is said to keep it under control.

(ii) Ear-bone of the pig is rubbed with water on the stone and then used for the cure of earache.

(iii) The skin of the naval region is removed and dried. When there is asthma, this is also rubbed with water and the paste applied to the throat externally.

(iv) Gall-bladder of the pigs is preserved and given to the patient. When body temperature falls very low (say below 96° or below this). The Gall-bladder powder is dissolved in water and given to patient. It is said to raise temperature. The gall-bladder is preserved in the same manner as prescribed for 'Bhalu-Teetee' (see under the heads Bhalu).

19. SYAL/SIYAR (FOX) : CANIS QUREUS/PULPES BENGALENSIS

(i) The skin is used for the tops of trumphets of small sizes.

20. UDBILAO/PAROD ((OTTER) : LUTRA LUTRA

(i) It is considered to see Ubdilao as a good omen.

**Acknowledgements**

The authors are grateful to numerous local people and medicinemen of this region. We are also thankful to Dr. L.S.Bisht (Almora), Dr. (Miss) N. Prabha (Pithoragarh) for providing useful informations and to Dr. (Mrs.) Ila (Almora), Dr. S.S. Pathani (Almora) and Dr. J.C. Pant (Almora) for providing zoological name of the animals etc.

## References

Atkinson, E.T., 1882. *The Himalayan District of North West Provinces of India*, 3 vols. Cosmo Publication, New Delhi (Reprint, 1973).

Pande, P.C., G.C. Joshi and M.M. Kandpal, 1989. Ethnobotany of Kumaun Himalaya. *In* N.K. Sah, *et al.* (eds.) *Himalaya : Environment, Resources and Development*. Shree Almora Book Depot, Almora : 285-289.

Pangtey, Y.P., 1980. Some wild edible fruit plants of Kumaun hills. *In* J.S. Singh *et al.* (eds.) *Science and Rural Development in Mountains*. Gyanodaya Prakashan, Nainital : 350-363.

Pande, P.C. and Y.P.S. Pangtey, 1987. Studies on ethnobotany–I : On some less known edible and economic ferns of Kumaun region of Western Himalaya. *J. Econ. Tax. Bot. 11* (1) : 81-85.

Pangtey, Y.P.S., G.S. Rawat and B.S. Kalakoti, 1982. Unusual and Supplementary wild food plants of Kumaun. *Himl. Res. & Dev. 1* (1) : 35-40.

Shah, N.C. and S.K. Jain, 1988. Ethno-medico-botany of the Kumaun Himalaya, India. *Social Pharmacology*, 2 (4) : 359-380.

Shah, N.C. and M.C. Joshi, 1971. An ethnobotanical study of the Kumaun region of India. *Econ. Bot., 25* (4) : 414–424.

# Chapter 13

# Pediatiric Ethnobotany of Kumaun Himalaya

*Vineeta Pande*

*Neeta Pande*

The term ethnobotany was first coined by Harshberger in 1895. It broadly means all aspects of direct or indirect relationships between man and plants. The concrete relationship includes mainly aspects of medicine, house-building, agriculture operations, fibre etc. Collection of medicinal herbs in Kumaun Himalaya is traditional from time-immemorial. In 'Artharveda' it is mentioned that 'Kirata girls' used to dig out herbs in the mountains (Shah, 1981).

The Kumaun Himalaya has a rich culture and heritage and a history of over thousand years, still the main treatment for children disease in rural and partly in urban areas start with prescription given by elderly ladies in the household. However, this inherent cultural knowledge is being gradually lost :

(a) due to urbanisation and easy availability of medical help

(b) breakdown of joint family system.

An attempt is being made to preserve the medicinal folklore for children as practised by ladies of the household commonly termed as 'Grand Ma's' remedies. A perusal of available literature (Jain, 1987) indicates that ethnobotany is a rapidly expanding science and the literature on the subject is piling up at a very rapid pace. Unfortunately, the pediatric ethnobotany has remained untouched so far. This gap in the developing subject like ethnobotany and our personal experience as the natives of Kumaun, where 'Grand Ma's' remedies are very common in almost all the communities inspired us to work over the problem.

**Plants and Their Pharmacognistic Uses**

1. *Allium sativum* Linn. (Lehsan) Liliaceae.

Garlic is largely used as a medicine. It is given to the child suffering from cold and cough, vomiting and stomach disorder. It is prescribed that a garlic lobe is burnt on a candle flame until it turns black and no more smoke arise anymore. The powder is then fed with mother's milk two or four times a day.

2. *Azadirachta indica* Juss. (Neem) meliaceae.

The leaves of the tree have antiseptic value. It is prescribed that the leaves should be boiled until black and then a paste is prepared. This paste is applied on the wounds. It is found that it gives immediate relief. This medication is found very effective even in the case of very old wounds.

The 'neem' leaves, boiled with water are also recommended for giving bath to the infants and children.

3. *Brassica compestris* Linn. (Sarson) Brassicaceae.

It is prescribed that the seeds are burnt in olive oil and the preparation is used for the message of the infants and children. The raw oil extracted from the seeds is also considered to be an effective message lotion. It is found effective in the growth of the children. This makes the child strong, stout and fair.

4. *Carum copticum* Benth & Hook (Ajwain) Apiaceae.

It is found effective in burns and give immediate relief. The seeds of 'ajwain' are burned until they turn black. A paste prepared with oil and applied over the burns.

It is also used for the constipation cases. 'Ajwain' boiled in water with 'batasha' (a typical preparation of boiled sugar) is given to the ailing child. This preparation is given twice a day. To avoid cold and cough 'Ajwain' with raw salt is tied around the neck of the infant or child.

5. *Citrus aurantifolia* Swingle. (Kagzi Nimbu) Rutaceae.

'Kagzi Nimbu' with water when given to the child during loose motions helps to overcome dehydration. One glass of boiled water (boiled and cooled) and one 'kagzi nimbu' with pinch of salt (kala namak) is fed to the child frequently during loose motions to overcome dehydration. This given several times a day, one or two spoons at a time, gives immediate relief.

6. *Cocos nucifera* Linn. (Nariyal or Giri) Arecaceae.

It is an effective 'Grand Ma's' prescription for burns. Coconut oil when applied on burns five or six times a day does not allow blisters to appear. Coconut when given to the child to chew helps the teeth to cameout easily and with less pain.

7. *Cuminum cyminum* Linn. (Zira) Apiaceae.

The fruits are generally used for diarrhoea and snake and spider bites. The powdered 'zira' is mixed with dry ginger (saunth) and water and then the paste is applied over the bite. Immediate relief can be felt.

8. *Curcuma longa* Linn. (Haldi) Zingiberaceae.

'Haldi' is used for cough or for internal injuries. One fourth of a spoon of 'haldi' (or depending on the age of the child) when mixed in a glass of milk and fed to the child the relief is immediate. It is also given with milk during internal injuries. This prevents the blood clotting and acts like an antiseptic.

9. *Elettara caradamomum* Maton. (Bari Elayachi) Zingiberaceae.

'Bari Elayachi' has its own medicinal value. It is mostly used to make the milk lighter and easily digestible. One 'elayachi' is boiled with milk and fed to the child. The milk being lighter is easily digested. It also helps in gas problem.

10. *Ferula assafoetida* Spreng. (Hing) Apiaceae.

'Hing' is mainly used for stomach problems :

(a) to stop motions immediately 'hing' is used with red oxide. Red oxide (suhaga) is heated till it swells and becomes white. It is made into a powder and fed to the child with 'hing' and 'Kala namak' two or three times a day.

(b) 'hing' is more effective if the pain is due to gas problem. A pinch of 'hing' when mixed in a tea spoon of water or mother's milk is applied around the naval of the child.

11. *Piper longum* (Peepali) Moraceae.

'Peepali' fruits are very effective in cold and cough. It is prescribed that 'peepali' fruits are boiled with milk and fed to the infant or child during cold and cough. In one litre of milk two or four fruits are boiled and fed to the child five or six times a day.

12. *Foeniculum vulgare* Mill. (Saunf) Apiaceae.

'Saunf' makes the milk light and easily digestable. Before giving the milk to the infant half a spoon of 'saunf' in one cup of milk is boiled. This makes the milk digestable and the child does not face gas problems and constipation.

Soak a spoon full of 'saunf' in half a cup of water. The 'saunf' is removed and the water is mixed to the milk. When this milk is fed to the child, it gives relief in gas trouble.

13. *Musa paradisica* Linn. (Kela) Musaceae.

'Kela' can be given to the child above six months. It has a large medicinal value. When the child is of six months or more and is ready to take solid food, he is fed with 'kela' and 'elayachi' (small). The 'kela' is mashed and 'elayachi' as a paste is added to it and is given to the child. This helps the child to overcome mucus and constipation and also helps in gaining weight.

14. *Ocimum sanctum* Linn. (Tulsi) Labiatae.

Seeds of 'tulsi' and honey can be given to an infant suffering from cold and cough. If seeds are not available leaves can also be used. The leaves are crushed and the juice is used for infants.

The juice of the leaves is extracted and applied into the ear drop by drop (two or three) three or four times a day provides immediate relief to the patient.

15. *Oryza sativa* Linn. (Chaval) (Gramineae) Poaceae.

'Chaval' is very effective during loose motions. The 'chaval' is boiled with water and this water is fed to the child / infant suffering from loose motions and vomiting. The 'chaval' water helps the baby to overcome dehydration. 'Chaval' gruel is also used for disturbed digestion, bowel complaints, diarrhoea and dysentry. It is given in short intervals and in small quantities about two or three tablespoon.

16. *Piper betle* Linn. (Pan) Piperaceae.

'Pan' leaves are effective in cold and cough. A paste of 'pan' with 'ajwain' is poured over a heated bowl. This mixture with honey when cooled is fed to the child, the child feels immediate relief.

17. *Prunus amygdalus* Baill. (Badam) Rosaceae.

Soaked 'badam' are very nutritive for children. The 'badams' are soaked in water (overnight) and a paste is made over a clean stone by rubbing it over. The paste is given to the child or infant in the morning hours. This helps to increase the intelligence, brightens the child, keeps the skin healthy and is very good for the heart. The 'badams' are generally given in odd numbers like one, three, five, seven etc.

It is also used for ear pains. A paste of 'badam' with 'ghee' (prepared from cow's milk) when administrated drop by drop into the patient's ear gives comfort.

18. *Prunus granatum* Linn. (Darim) Punicaceae.

'Darim' is used for cough. It is prescribed that a 'darim' is emptied out and is filled with cow's milk. The covering of 'darim' is covered with mud and heated in the ash of cow dung cakes. It is then given to the child suffering from cough.

19. *Syzygium aromaticum* Linn. (Long) Myrtaceae.

'Long', 'lahasan' and 'ajwain' are fried over a pan and then cooked with 'gur' (a preparation of sugar cane). This mixture is very effective for chronic cold and cough. For the preparation of this mixture two lobe of 'lahasan', one spoon of 'ajwain' and one 'long' is taken and is fed to the child two or three times a day, one fourth of a tea spoon.

20. *Zingiber officinale* Rose. (Adrak) Zingiberaceae.

'Adrak' is used to cure loose motions. To cure loose motions immediately, the stomach is covered with 'urad dal' and the naval is filled with 'adrak' juice. Relief can be felt within an hour.

If the child complains of ear ache due to cold, injury of dirt, 'adrak' can be used to give an immediate relief. 'Adrak' is made into a paste and squeezed through a clean cloth, four or five drops of the juice are put into the ear. The juice is heated until it become lukewarm before using it.

**Conclusion**

The above given analysis of twenty plants referred by local people of Kumaun Himalaya, proves the close association and interaction in between man and plants. However, no scientific and conservative approach has been adopted to evaluate the potential of paediatric ethnobotany. Though, 'Grand Ma's' remedies have vast scope and potential but this knowledge has never found a scientific understanding. Therefore, before formulating and implementing any action program, a data bank should be created in paediatric ethnobotany.

## Chapter 14

# Folklore Insecticidal Plants of Eastern Kumaun (Western Himalaya)

*I.S. Mehta*
*P.C. Pande*
*Pramila Joshi*

Kumaun, a part of Western Himalayas, is very rich in vegetation, and is gifted with a variety of climate (sub tropical, sub temperate, temperate and alpine zone). The eastern part of Kumaun Himalaya is inhabitat mainly by Bhotiya, Raji and other primitive societies.

A number of workers have explored various ethnobotanical potential of this area (Atkinson, 1882; Shah & Joshi, 1970; Gupta, 1980; Pangtey, 1980; Pangtey et al., 1982; Pande & Pangtey, 1987; Pande *et al.*, 1989), Shah & Joshi, 1971 however, no significant contribution has been made in reference to insecticidal plants.

Extensive tours were made and information was collected on insecticidal plants from the abiders of the area. In all, 34 species of angiospermic plants are being used for protecting the agricultural products from pests and insects and for controlling the sanguinivorus insects and worms. The useful plants reported in this paper have been described in alphabetical order.

**Enumeration of Species**

1. *Acorus calamus* Linn. (Araceae) : Vern. BACH.

The powder of rhizome is employed for killing hair lice, bugs, gnat etc. Also reported by Shah & Jain, 1988.

2. *Adhatoda vasica* Nees (Acanthaceae) : Vern. BASAK, BASHING.

Fresh leaves are mixed with manure and used to destroy or repel the pests of crops or cultivated plants.

3. *Aesculus indica* Colebr. ex Cambers (Hippocastanaceae) : Vern. PANGAR

The paste of the fruits is used locally to kill the lice, leaches and other blood sucking worms of domestic animals. It is mainly used to kill the leaches infesting the animal nostrils.

4. *Agave americana* Linn. (Agavaceae) : Vern. RAMBANS.

The juice of leaves is used for killing the worms pests and insects living on the bodies of domestic animals.

5. *Arisaema tortuosum* (Wall) Schott. (Araceae) : Vern. SYAP KA KAKUNI, BANK.

Tuberous roots are used to kill the insects infesting domestic animals but used very rarely. However, the powder of rhizomes mixed with flour is used widely to kill the rats infesting the cultivated fields.

6. *Artemisia vulgaris* Linn. (Asteraceae) : Vern. DATI

The juice of this plant is used to repel the insects. Both fresh and dried leaves mixed with ghee (butter) are used as Dhoop to keep off mosquitoes and other insects.

7. *Begonia picta* Smith (Begoniaceae) : Vern. TAR

The juice of the plant is used to kill leaches infested the nostrils of domestic animals.

8. *Boenninghausenia albiflora* (Hook.) Reocheub. ex Meissn. (Rutaceae) : Vern. PISSU GHAS, UPIYANKANTHI.

The fresh leaves are employed as insecticide and insect repellant. The leaves are also used ot destroy or epel the insects or pests of stored agricultural products.

9. *Betula jacquemontii* Spach (Betulaceae) : Vern. BHOJ PATRA

Bark used as insecticide (cf. Shah & Jain, 1988).

10. *Canabis sativa* Linn. (Cannabaceae) : Vern. BHANGA

The leaves are kept on stored seeds and agricultural products to repel insects.

11. *Centella asiatica* (Linn.) Urban (Apiaceae) : Vern. BRAHMI.

Plants occasionally used to repel the pests infesting human beings.

12. *Chenopodium botrys* Linn. (Chenopocliaceae) :

Fresh plant extract used to kill the worms infesting wounds of domestic animals or used as antiseptic medicine.

13. *Datura stramonium* Linn. (Solanaceae) : Vern. DHATUR.

The powder of the seeds is employed as insecticide.

14. *Delphinium denudatum* Wall. ex Hook. f. & Thoms. (Ranunculaceae) : Vern. KASTURI, MUDILA.

The powder of roots is used to kill the pests infesting human beings eg. lice, bed bugs, worms etc.

15. *Dioscorea deltoidea* Wall. ex Kunth (Dioscoreaceae) : Vern. MANENIYA

The rhizome is used for washing hair for destroying the hair lice.

16. *Diploknema butyraceae* (Roxb.) Lam. (Sapotaceae) : Vern. CHEURA

Paste of the seeds and cake is used to kill the pests of cattle, viz., like, gnat, fleas, leaches etc.

17. *Hedera nepalensis* K. Koch (Aralliaceae) : Vern. JUYAN.

Paste of linas is used to kill the lice of the cattle.

18. *Jasminum humile* Linn. (Oleaceae) : Vern. BAN. JAYAPTRA

Fresh as well as dried leaves are kept with stored cereals as an insecticides.

19. *Juglans regia* Linn. *Juglans regia* Linn. Var. Kumaonica DC (Juglandaceae) : UKHOR.

The entire leaves and powder of leaves are used as insecticides. The leaves are kept in seed store houses or containers (Also noted by Shah & Joshi, 1971). Further, the paste of the rind of unripe fruits is also used to repel leaches.

20. *Litsea umbrosa* Nees (Lauraceae) : Vern. CHIRARA.

The oil extracted from the seeds and oilcake is used to slay the pests infested human beings, cattle and other domestic animals.

21. *Lyonia ovalifolia* (Wall.) Drude (Ericaceae) : Vern. AYAR.

The compost of young leaves and buds is kept in cultivated fields to protect the crops from worms. However, Shah & Joshi

(1971) noted that the cuttings of the tree may be kept in paddy fields to protect them from worms and insects.

22. *Mallotus philippinensis* Muell (Euphorbiaceae) : Vern. RUEN, ROLI.

Red pubescent covering of ripe fruits is used as insecticide for killing lice and worms.

23. *Nicotiana tabacum* Linn. (Solanaceae) : Vern. TAMAKU

The juice of the plant is used to avoid leaches.

24. *Piptanthus repalensis* Don (Rapilionaceae)

The paste of the young leaves is used to kill the blood sucking insects of domestic animals.

25. *Plectranthus japonicus* (Burn. f.) Koidz (Lamiaceae) : Vern. MOLCHARA.

The juice of the fresh leaves is used as insecticides.

26. *Polygonum amplexicaule* D. Don (Polygonaceae) : Vern. DURGJARI.

The paste and powder of rhizome is used to repel the household pests.

27. *Roylea cinerea* (D. Don.) Baill (Labiataceae) : Vern. TIT PATI.

The juice of fresh leaves is used in killing the pests.

28. *Sapindus mukorosii* Gaertn. (Sapindaceae) : Vern. REETHA, RITH

The powder of the fruits (pericarp only) is used to kill the pests infesting the human beings (especially lice). It is also used to kill the leaches infesting the nostils of domestic animals.

29. *Sapium insigne* Trim. (Euphorbiaceae) Vern. KHIN.

The juice of the leaves is used to kill the sanguinivorus insects. According to Pande *et al.*, (1989) the seeds used as a vermicide.

30. *Tinosopora cordifolia* (Willd.) Miers (Menispermaceae) : Vern. GURGU.

The paste of the linas (stem) and leaves is applied on the body of domestic animals to repell or kill the sanguivorus insects.

31. *Valeriana hardwickii* Wall. and *V. jatamasi* Jones (Valerianaceae): Vern. SHAMEO.

The dried rootstock and roots mixed with ghee are used as dhoop which repels the pests of the houses.

32. *Vitex negundo* Linn. (Verhinaceae) : Vern. SIMALI.

The leaves are employed to repel the pests of stored agricultural products.

33. *Zaxthoxylum armatum* DC. (Rutaceae) : Vern. TIMUR.

The aromatic stem, bark, leaves, fruits and seeds are utilized to abolish the household nuisance pests of human beings, stored agricultural products and insects of cultivated field.

## Acknowledgement

One of the author (PCP) is thankful to Dr. N.C. Shah, Head of the Botany and Pharmacognosy Division, Central Institute of Medicinal and Aromatic Plants, Lucknow, for guidance and encouragements and to CSIR, New Delhi, for financial assistance.

## References

Atkinson, E. T. (1882). The Himalayan Gazetteer. Vol. 1-3, Cosmo Publication, New Delhi.

Gupta, R. K. (1960). Some useful and medicinal plants of Nainital in the Kumaun Himalayas. *Indian Forester 57* : 307-329.

Pande, P. C., G. C. Joshi & M. M. Kandpal (1989). Ethnobotany of Kumaon. In : The Himalayas : Environment, Resources and Dvelopment. (Eds. : N. K. Sah, S.D. Bhatt & R. K. Pande), Shree Almora Book Dept, Almora. Pp. 285-294.

Pangtey, Y. P. S. (1980). Some wild edible fruits of Kumaon Hills. In : "Science and Rural Development in Mountains" (Eds. J. S. Singh, S. P. Singh & C. Sharstri), Gyanodaya Prakashan, Nainital. Pp. 350-363.

Pangtey, Y. P. S., G. S. Rawat & B. S. Kalakoti (1982). Unusual and supplimentary wild food plants of Kumaun. *Himal. Res. & Dev. 1* (1) : 35-40.

Shah, N. C. & S. K. Jain (1988). Ethno-medico Botany of the Kumaon Himalaya, India, *Social Pharmacol.*, 2 : 359-380.

Shah, N. C. & M. C. Joshi (1971). An ethnobotanical study of Kumaon Himalaya, India, *Eco. Bot.*, 25 : 414-422.

# Chapter 15

# Water Resources, their Depletion and Conservation in the Dehra Dun Valley

*M.S. Anantharaman*

*R.K. Sehgal*

Thus, the valley with an average annual rainfall of about 200 cm has a rich potential of water as a major resource. But the water resource of the valley is depleted and lost by the way of heavy surface runoff due to high gradient, infiltration due to the presence of porous subsurface and structurally weak planes, solufluction, landslides, unplanned quarrying, deforestation, gully formation and overgrazing.

The suggested conservational measures include construction of major and minor canals, wheelshifters, pumping sets, water mills, form reservoirs at suitable suggested sites and development of spa-resorts at various spring zones in the valley.

## Geo-Identity

The Dehra Dun valley, the largest of all the Duns in the Himalaya, is a tectonic synclinal trough lying between 29°55'-30°30'N : 77°35'-78°20' E in the Garhwal Himalaya and is enclosed by Lesser Himalaya, Siwalik range, holy rivers Ganga and Yamuna in the North, South, Southeast and Northwest respectively.

The valley with an area of 2002 sq.km. ranges between 300 to 2500 meters above mean sea level in height. The parallelogram shaped valley has maximum length of 70 kms. (NW–SE) and breadth of 35 kms. (N-S).

Geologically, the valley is composed of Jaunsar and Mussoorie groups of rocks in the north. Siwalik rocks in the south and Dun gravels in the centre. The Lesser Himalaya rocks and the Siwalik rocks are separated by main boundary fault (Anantharaman, Saxena & Pandey, 1982).

Climatically the valley possesses subtropical to sub temperate climate with maximum temperature of 42°C during summers, minimum temperature of 4°C during winter and mean annual rainfall of 200 cms (Anantharaman, 1985a).

**Drainage**

The Dehra Dun valley consists of two intra-river basins namely the Suswa river basin forming the Eastern Dun and the Asan river basin forming the Western Dun (Fig. 1). The Eastern and Western Duns are demarcated by the Dehra-Asarori water divide (Anantharaman, 1985b). The two rivers are contributed by number of major and minor perennial and seasonal streams arising from Lesser Himalaya and Siwalik drainage lines (Fig. 1a).

In the Western Dun, Asan river receives the waters of Jiwangarh Rao, Sitla, Swarna, Kaligad, Nalota, Tons, Bhitarli, Kairkurli, Nun, Nimmi, Arnigad and Chorkhala which rise from north-western Lesser Himalayan drainage line and Dhaurwala Rao, Dharawat Rao, Kalawala rao and Ramgarh Rao II which rise from South-western Siwalik drainage line (Saxena & Anantharaman, 1983).

In the Eastern Dun, Suswa river receives the waters of Bindal, Baldi, Bandal, Song, Jakhan, Rispana, Lalpani and Chandrabhaga which rise from north eastern Lesser Himalayan drainage line and Fanduwala Rao, Sukh Rao, Ramgarh Rao I, Kans Rao, Bariwara Rao, Chorpani Rao and Motichur Rao which rises from South Eastern Siwalik drainage line (Saxena & Anantharaman, 1983).

The two rivers Suswa and Asan ultimately confluence with Ganga and Yamuna respectively. In the centre of the valley lies the Dun aquifer zone is bounded by two major spring lines along the two said drainage lines (Saxena, Anantharaman & Pandey, 1979).

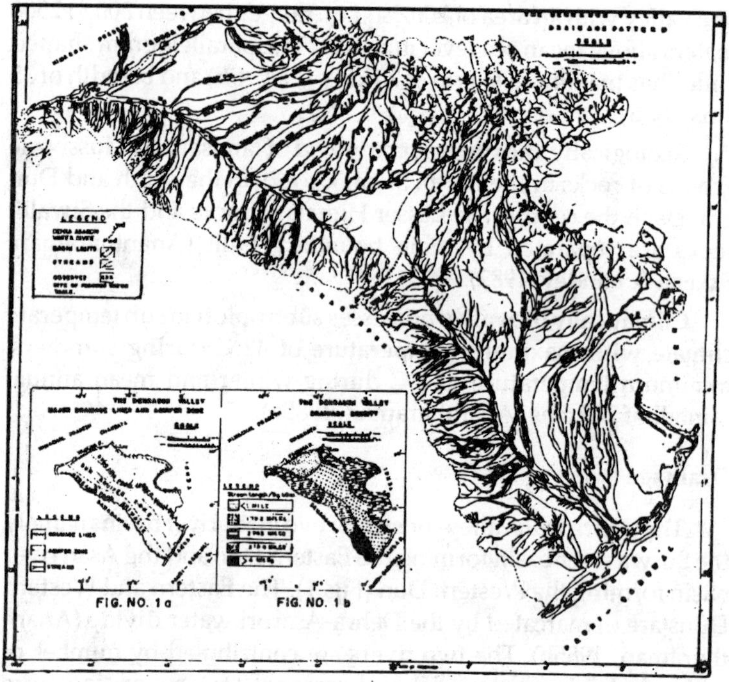


Fig. 1

The valley is infested with number of springs lying between 300 meters and 2500 metres (Fig. 2). On the basis of their location and relationship between altitude, all the springs are grouped into six major springs zones (Saxena Anantharaman & Pandey, 1979) namely–(1) Chandrabhaga-Rishikesh spring zone (450 m, 5 springs), (2) Dun aquifer zone (450 m–900 m, 14 springs), (3) Swarna-Ghuchupani-Bhogpur-Ogal spring zone (900 m–1350 m, 46 springs), (4) Bandal-Baldi-Sahastradhara spring zone (1350 m–1800 m, 20 springs), (5) Aglar-Khattapani–Bidhalana spring zone (1800 m, 33 springs) and (6) Ghhattudhar Murray spring zone (1800 m, 13 springs). The waters of springs at Sahastradhara, Majhara, Sera, Chipaldi and Nirgadu (at Rishikesh) contain fair amount of dissolved sulphur (Anantharaman & Bahukandi, 1984).

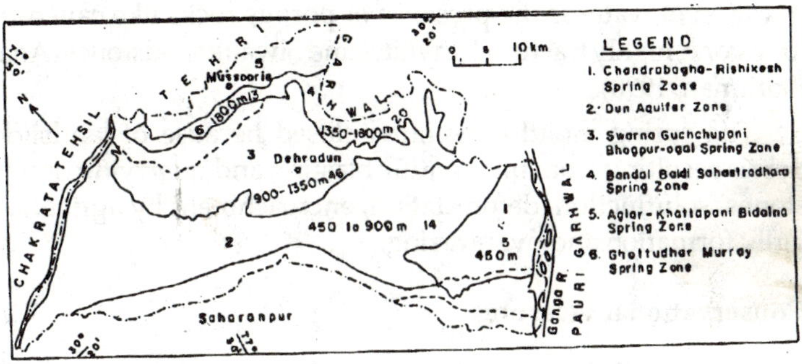


Fig. 2 : Spring Zones in Dehra Dun Valley

## Water Depletion

The Dehra Dun valley is richly endowed with vast reserves of water resources (Fig. 1b) which have not been fully utilized rationally. It is therefore, highly desirable that the reserves of the water resources be conserved from prevailing water resource depletion factors and then be developed for the sake of economic development of valley. It is further studied to diagnose the intensity of depletion and to pinpoint the area for ecological treatment therein.

Maximum runoff of the well defined drainage in the valley occurs during rainy season. The water loss percentage between June and October is 27.1 per cent in fodder grassland, 71.1 per cent in bare followed ground, 59.6 per cent in bare-ploughed ground and 21.2 per cent in natural grass, while between November and March, the water loss percentage in the areas mentioned above 11.4, 58.7, 45.6 and 4.1 per cent respectively (Anantharaman *et. al.* 1984).

Excluding the centre part of the valley where the slope angle ranges between 0 and 2°, in other parts, the slope angles range between 8° and 16° in terraces, 16° and 32° in the Southern Siwalik structural and denudational hills and between 32° and 64° in the northern colluvial hills tracts and Lesser Himalayan structural hills, (Anantharaman, 1985). Because of such high gradient and high velocity of the waterflow the rate of runoff is very high. In the Lesser Himalayan region because of the presence of Main Boundary Fault, Krol-Thrust, minor faults, fracture zones and neotectonic activities the water sinks underground. Another reason for infiltration of water is the presence of porous rocks like cavernous limestone, highly fractured phyllites and Siwalik sandstones (Anantharaman, 1990).

The spring mouths are often closed because of landslides, debris resulting from unscientific blasting and quarrying of limestones, solufluction, deforestation, encroachment by agriculture, gully formation and overgrazing.

**Conservational Measures**

Conservation of water resources deals with the protection of water resources from depletion factors and ecological measures have been discussed to treat the resource ailments. The present haphazard utilisation of supply of water accentuated by development of forest, mineral and agro-based industries as also power generation, demands maximum ration utilization of available hydraulic resources. A thorough investigation of the area reveals that out of the total rainfall, about 50 per cent is spent in the evapotranspiration and infiltration processes, about 25 per cent flows as surface runoff every year which is used in agriculture and for domestic purposes. But 25 per cent rainwater goes unused. This 25 per cent of water of the valley can be channelized through minor canals to fulfil the need of irrigation.

It is purposed that suitable minor canals, wheel shifters and pumping sets are set up at localities like Sripur, Maldevta, Malsi, Baniawala along Song Kalanga Canal, Narwada, Lachchiwala, Ramgarh, Raiwala, Shyampur along Suswa river, Nainiwala, Tunwala, Burronwala, Ghari, Kaowli, Bhurpur, Shabhawala and Judli along Asan river and Bizapur Canal and Kairkurli, Chandrauti and Kotra along Tone river (Fig. 3).

Besides these, there is one more resource like the spring line which runs parallel to the Dun aquifer zone and along the Lesser Himalayan and Siwalik drainage lines. The spring line is associated with oozing water incessantly which is locally called as 'Ogal'. The Ogals are formed in great number along the junction of Dun aquifer zone and uplands of Lesser Himalaya and Siwalik. These Ogals can be converted into small reservoirs and form ponds to store the water for irrigational purpose. This proposal can be successfully implemented at Bhupalpani, Chorpani, Nalapani,

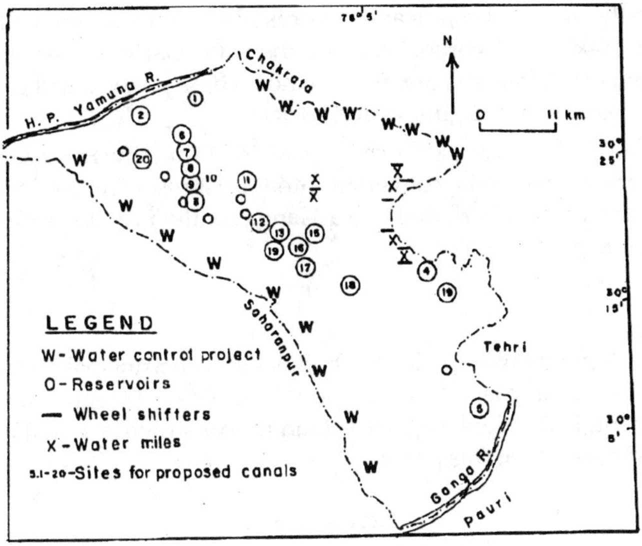


Fig. 3. : Water resources utilisation in Dehra Dun Valley

Karwpani, Khattapani, Guchchupani, Ranipokhari, Judli, Chandberbani, Ogalwala, Bhogpur and Dohari Ogals.

From these reservoirs water can be lifted by pumping sets to irrigate the upper level terraces of the valley.

Small dams can be constructed in the upper reaches of the Song river near Sripur village. It can be worked out that the gradient of the Song valley ranges from 6 to 90 metres/1.6 kms. with the discharge of water available small power generators to operate pumping sets can be set up.

As regards the use of waters of the springs at Sahastradhara, Sera, Majhara, Chipaldi and Nirgudu, these spring waters are similar in composition to the Sulphur spring waters of Bedmeander springs of Germany, white springs of U.S.A. etc. (Saxena, Anantharaman & Pandey, 1979). These waters can be utilized for medical purposes as these water can cure skin ailments. These springs sites should be developed "Spa-resorts" (Health centres) properly on the lines of such centres in Germany, Switzerland, U.S.A., U.K. and U.S.S.R. Further bottling of the spring waters as mineral tonic for general health will give boost to the economy of the valley.

Water and torrent control projects should be implemented

along the seasonal rivers and streams like Nalota, Swarna, Sitla, Bindal, Jakhan, Sukhrao, i.e. along the water parting lines of the Lesser Himalayan and Siwalik uplands. These projects will check the expansion of the gulleys and ravines.

The Robber's cave at Guchchupani, the Tapkeshwar gorge and the Tons gorge can be converted into tourist resort centres. Similarly, at Ganga Lahari, the Suswa-Ganga confluence can be developed into tourist centre.

## Acknowledgements

Grateful thanks are due to Dr. B.B. Raizada principal, for providing necessary facilities, and Dr. P.B. Saxena of Geography Department, D.B.S. College, Dehra Dun for constructive remarks in the preparation of this paper.

## References

Anantharaman, M.S. 1985, (a). Major Physiographic Regions in Dehradun Valley (Garhwal Himalaya). *Publ. C.A.S. in Geol.*, Punjab. Univ., Chandigarh (N.S.), Vol. 1 : 155-165.

Anantharaman, M.S., 1985, (b). A study of Drainage Patterns in Dehra Dun Valley (Garhwal Himalaya). *Curr. Trend Geol.*, Vol. 7-8, Today & Tomorrow Pub. New Delhi : 613-624.

Anantharaman M.S., 1990. A Geomorphic study of the Dehra Dun Valley (Garhwal Himalaya). In *Himalaya : Environment, Resources* and *Development* (eds.) N.K. Sah *et al.*. Gyanodhya Pub. Nainital, U.P. : 10-20.

Anantharaman, M.S. and P.C. Bahukhandi, 1984. A study of Mineral Resources of Mussoorie Syncline, Garhwal Lesser Himalaya, U.P. *Sed. Geol. Him.* (ed.) Srivastava. Today & Tomorrow Pub., New Delhi : 191-196.

Anantharaman, M.S., P.B. Saxena, and B.K. Pandey, 1982. A study of Soil Morphogenesis, its depletion and Conservational Measures in the Dehradun Valley (Garhwal Himalaya). *Him. Geol.*, Vol. 12 : 225-235.

Anantharaman, M.S., P.B. Saxena, & B.K. Pandey, 1984. The role of Sedimentary and Morphogenetic Processes in Soil Formation, its Depletion and Conservation–A case study from the Dehradun Valley (Garhwal Himalaya). In *Sed. Geol. Him.*, (ed.) Srivastava. Today and Tomorrow Pub., New Delhi : 143-154.

Saxena, P.B. and M.S. Anantharaman, 1983. A study of Drainage Pattern in Dehra Dun Valley (Garhwal Himalaya). *Him. Geol.*, 9th Sem., Vol., 13, C.A.S. (Geol.) Punj. Univ. : 162-169.

Saxena, P.B., M.S. Anantharaman, and B.K. Pandey, 1979. A Resource Appraisal of Springs of Krols in Dehradun Valley (Garhwal Himalaya). *Him. Geol.*, Vol 9, (II) : 845-853.

# Chapter 16

# Water Quality Problems in the Urban Areas of Kumaun Himalaya

*R.K. Pande*

*N.S. Bhandari*

Almost all the urban areas of the Kumaun Himalayas are facing acute potable water supply problem, in terms of quantity as well as quality, since time immemorial. Despite of government planning no perfect solution for this problem has yet been noticed. This problem has taken a shape of public unrest and slowly changing the minds of people against the establishment. Furthermore, recent studies on water quality classifications revealed the fact that the waters of river Gomti at Bageshwar and springs of Almora Town are not suitable for human consumption Table 1 & 2. Apart from the potable water crises and quality considerations in Kumaun Himalayas, the importance of water resources in eco-developmental formulations have prompted the author to review the water quality problems within the environmental framework.

Although the total amount of water on earth is generally assumed to have remained virtually constant, the rapid growth of population, together with the extension of irrigated agriculture and industrial development, are stressing the quantity and quality aspects of the natural system. Because of the increasing problems, man has begun to realize that he can no longer follow a 'use and discard' philosophy—either with water resources or any other natural resources. As a result, the need of a consistent policy of rational management of water resources has become evident. Water is an essential material for human life and the presence of a reliable and safe water supply is a vital factor in the development of a community. However, as well as its life-giving properties water also has enormous potential for spreading ill health and death.

Numerous projects have been funded by the government and its agencies to study how to conserve and preserve different elements of Himalayan Ecosystem. Besides this, a number of NGOs are also working in the same direction. Water, an important component of environment has been included with respect to its relation with ecosystem as a whole These studies have surfaced water quality considerations, in general of some water bodies, rivers in particular. Others, springs and pons which are in frequent use and even a boon for local inhabitants where there is water scarcity, are the examples of total negligence. Inadequate and ill considered developmental plans of urbanization in the region have made the people indifferent towards their traditional water resources. Result is the poor condition of most of these resources. Moreover here it is pertinent to say that environmental elements of Himalayas have been the subject of debates, discussions and social movements in and outside the region. But in these activities water, as potable water has got on or little space/share. In nutshell, much have been said and done with respect to the other aspects of environment of Himalayas but nothing or little came to light on potable water quality and its health impacts. It is therefore imperative to go for an exhaustive study of this very aspect of the environmental component.

The concern over deterioration of water quality due to pollution has attracted wider attention from all spheres of mankind. The pollution of water resources is increasing continuously and time is not too far when water may be inadequate for normal living conditions. The important contributors for deteriorations of water quality are domestic waste, sewage discharge, industrial and agricultural waste. These contain organic, inorganic and biological pollutants. The major water uses to which it can be put are potable water supply, power, food protection, irrigation, fishing, industrial water supply, recreation etc. All these uses are directly or indirectly related to mankind. Quality considerations for these uses are different. However, in general, all have common quality parameters (physico-chemical and biological) with different acceptable range.

**Table 1 :** Water quality index (WQI's) for Rome Indian rivers

| *River* | *WQI* |
|---|---|
| Brahmputra[1] | 81.6 |
| Jhelum[1] | 92.0 |
| Chenab[1] | 218.0 |
| Kosi[2] | 30.2 |
| Gaula[2] | 12.4 |
| Panai[3] | 33.2 |
| Gomti (at Baijnath)[3] | 30.9 |
| Gomti (at Bageshwar)[3] | 185.2 |
| Ramganga[3] | 18.4 |
| Sarju[3] | 25.7 |

1. Tewari & Ali (1987)
2. Pathak & Bhatt (1991)
3. CSIR Project (1990)

Freshing water bodies (springs, streams and ponds) of the Himalayas, which ensure the permanent supply of water for domestic as well as other uses, are day by day reaching towards the alarming conditions. The geometrically increasing population has started telling upon the quantity and quality of water resources. The day by day increasing demand by increasing population in urban areas has already generated such a tremendous stress on the water resources that in next five or ten years and region is going to face a terrific problem of qualify water supply. This estimation is not an exaggerating of the problem as it is evident from the recent resentment from the people appeared in the form of news paper writings, demonstrations, representations etc. This problem will have a long term effect on the economy, culture and social values of this requiem. Ill conceived construction activities, unplanned urbanization, lack of proper sewage disposal system, lack of environmental consciousness, absence of any treatment plant in entire Kumaun and lack of water testing laboratory in the region together has added much to this problem. In a case study of Almora and Nainital towns some interesting results have been registered Table 2 & 3. It is noticed that though the physico-chemical impurities are within acceptable limits today but they are just ready to jump over the safe zone and then the situation will be beyond control. Both the towns have also registered biological impurities which are found directly associated with the water related diseases. It is also interesting to note that the water quality index (WQI) has already reached to a danger point and if the future stress on water resources will not be checked the consequences will be disastrous.

**Table 2 :** Water quality index & tape and spring waters in Almora town of Central Himalayas

| *Sample Type/Season* | *Tape water* | *Spring water* |
|---|---|---|
| 1. Summer | 23.91 | 26.64 |
| 2. Monsoon | 54.79 | 34.94 |
| 3. Winter | 42.05 | 26.11 |
| 4. Average | 49.04 | 27.19 |

Pande et. al., (1992)

**Table 3 :** Physio-chemical properties of Naina Lake

| | *Parameters* | *Range* |
|---|---|---|
| 1. | Dissolved organic material (ppm)[1] | 24.80-60.00 |
| 2. | Bottom organic material (g/100g)[1] | 3.60-11.26 |
| 3. | Suspended solid material (mg/100 ml)[1] | 28.20-182.60 |
| 4. | Chloride (ppm)[1] | 23.20-34.20 |
| 5. | DO (ppm)[1] | 3.10-10.00 |
| 6. | $NO^3$-N (1)[2] | 4.99-76.25 |
| 7. | $NO^4$–N(mg/l)[2] | 119.00 |
| 8. | Free $CO^2$ (mg/l)[2] | 22.00 |
| 9. | Alkalinity (ppm)[1] | 263-363 |
| 10. | Ph[1] | 7.77-9.30 |
| 11. | Temp (<198>C)[1] | 8.72-22.20 |
| 12. | $NH_4$-N (ppm)[1] | 0.04-0.028 |
| 13. | Ammonia Concentration (mg/l)[1] | 1.01 |

1. Das, et. al. (1981)
2. Pant, et. al. (1981)

A considerable amount of study has been done by various scholars on the problems related to the water quality in environmental perspective. In the recent years many scholars and institutions in different parts of India have produced remarkable models, thoughts, ideas and case studies on water and water related problems. Varshney (1985) edited some seventeen meaningful articles on water pollution and management. Trivedi (1988) has done a lot of work on water pollution and related phenomenon and has developed a model of water quality. In 1989 Sinha et al. produced a very significant work on water pollution. Tayal et al. (1989) worked on the status of major rivers in Uttar Pradesh. Tripathi and Sikandar (1989) reported some interesting findings in their work on ecology of lakes and pounds in India. Tewari et al. (1988) developed water quality index for Indian rivers.

Himalayan waters are not the exceptions. Many workers from different fields have done a lot of work of lotic and lentil water of Kumaun and Garhwal Himalayas. Kumaun University (Almora and Nainital Campuses), Garhwal University, G.B. Pant Institute of Himalayan Environment and Development, Vivekanand Lab for Hill Agriculture (ICAR), Defence Research Lab etc. have generated a lot of date on water and water ecosystems. Ahmad (1989), Bhatt et al. (1990), Bhandari et al. (1991), Gupta (1990) and Pande et al. (1992) have produced some remarkable work on water. However, most of the work reported about suffered from two basic requirements : (1) they are of more scientific and academic value without any implementation scope and (2) they are based on ecosystem studies and do not cover the potable water problems of Kumaun Himalayas.

In the light of day by day deteriorating conditions of water due to increasing population pressure, construction activities, ignorance of the people towards their environmental rights, lack of environmental consciousness, unplanned urbanization etc. the present situation demands for a systematic study of potable water quality in the hill towns of Kumaun with the following objectives:

to assess the physico-chemical and biological considerations of the water, used for domestic purposes ;

detection and determination of trace metals and their correlation with environment;

determination of water quality index (WQI), an important index of assessing the quality of water ;

to allocate points in respect to the poor sanitation arrangements;

efforts to make people understand about the water, they use through the data of investigation on water body of their interest ;

organise field demonstrations, preparation of water quality, guide book for the common people and educating the local inhabitants about their environmental rights and duties.

to make water quality classification of springs (Naula), in particularly, based on DO and BOD concentrations.

preparation of water quality maps talking into account of pollution caused by sewage overflows or 'accidental' pollution incidents.

to evaluate qualitatively the impact of extermination of forests and construction activities in the vicinity of potable water supply source.

Environmental impact on drinking water quality parameters is to be studied by collecting samples from the source of the main water supply system and from the consumers end. Besides, these sampling sites, sampling spots at springs, in and around of towns frequently utilised by the people should also be included in the investigation.

The authors are of the view that mere collection and compilation of data on potable water quality does not serve any purpose until the efforts are not made to make the consumers aware from the situation existing and of their rights and duties in making environment clean. Therefore it should be planned that the conclusions drawn from the data collected should be supplied to the consumers are knowledge package for discussion.

## References

Ahamd, A., 1989. Eutrophication with special reference to Himalayas and strategies for its rehabilitation. In *Management of Aquatic Ecosystem*. Narendra Publishing Housing, New Delhi.

Bhandari, N.S. and R.K. Pande, 1991. Solute dynamics of river Sarju in the Central Himalayas, India. In *Ecology of Mountain Waters*, Ashish Publishing House, New Delhi.

Bhatt, S.D. and J.K. Pathak, 1990. Assessment of water quality and aspects of pollution in a stretch of river Gomti (Kumaun Himalayas). *I.J. Environ. Biol.* 13 (1).

Bhatt, S.D. and R.K. Pande, J.K. Pathak and D.S. Rawat, 1990. Relationship among hydro-soil, water chemistry and biota with special reference to fish production and their development rhythms in the Sarju-Ramganga (E) water basins of the Kumaun Himalaya. *Final Technical Report, CSIR.* Govt. of India, New Delhi.

Bhatt, S.D. and R.K. Pande, 1991. *Ecology of Mountain Waters*. Ashish Publishing House, New Delhi.

Bhatt, S.D. and J.K. Pathak, 1992. *Water Quality of Himalayan Drainage Basins*. Almora Book Depot, Almora.

Gautam, A., 1991, *Ecology and Pollution of Himalayan Waters*. Ashish Publishing House, New Delhi.

Gupta B.P., 1990. Natural perennial water sources of Almora town. *Uttrakhand-4*.

Khulbe R.D., 1989. *Perspectives in Aquatic Biology*. Papyrus Publishing *House*, New Delhi.

Pande, R.K., 1988. Seasonal rhythms in water quality of Nana Kosi—a spring fed river of Kumaun Himalayas. In *Ecology and Pollution on Indian Rivers*. Ashish Publishing House, New Delhi.

Pande, R.K., 1989. Evaluation of environmental stress on Himalayan towns : the experience of Nainital. *Pollution Research* 8(3).

Pande, R.K., 1992. Quality of urban water supply in Central Himalayas, India. In *Conserving Wildlife in India*. Vinod Book Depot. Jammu Tawi.

Sah, N.K., S.D. Bhatt and R.K. Pande, 1991. *Himalayas : Environment, Resources and Development*. Shree Almora Book Depot, Almora.

Sinha, A.K., R. Boojh and P.N. Whwanathan, 1989. *Water Pollution*. Gyanodaya Prakashan, Nainital.

Singh, S. and G.N. Misra, 1989. Status of major rivers in Uttar Pradesh. In *Water Pollution*. Gyanodaya Prakashan, Nainital.

Tayal, S. and A.K. Sinha, 1989. Physico-chemical and biological monitoring techniques of water analysis with reference to pollution. In *Water Pollution*. Gyanodaya Prakashan, Nainital.

Tewari, T.N. and M. Ali, 1988. Water quality index of Indian rivers. In *Ecology and Pollution of Indian Rivers*. Ashish Publishing House, New Delhi.

Tripathi, B.D. and Mohd. Sikandar, 1989. Ecology of lakes and ponds in India. In *Water Pollution*. Gyanodaya Prakashan, Nainital.

Trivedi, R.K., 1988. *Ecology and Pollution of Indian Rivers*. Ashish Publishing House, New Delhi.

# Chapter 17

# Demographic Profile of Uttrakhand

*Rajnish Pande*
*P.C. Pande*
*R.K. Pande*

The capacity of a country/region for economic growth and development primarily depends upon the availability of human, physical and financial resources. Of these three resource factors, the human resource factor appears to be the most crucial and strategic. A country/region may possess abundant and inexhaustible natural and physical resources and the necessary capital equipment but it cannot make rapid strides towards economic and social advancement unless and until there are men who can mobilise, organise and harness nature's bountiful resources for the production of goods and services. Thus, the most important task which all the developing countries/regions of the world are confronting today is the building up of human capital.

In most developing countries/region, rapid population growth has constituted a formidable obstacle in the formation of human capital, in the raising of productivity, incomes and the levels of living of the people. Besides this, rapid population growth has tended to cancel out the margin of economic advances, and any significant improvements in the conditions of life and work. Moreover it has also necessitated the diversion of the limited physical and financial resources from the production to consumption channels, leaving narrower resource base for future economic development. Thus, it has become imperative for the developing countries/

regions to analyse and examine the magnitude of the problem posed by the rapid population growth particularly from the view point of current and future employment situation and development and utilisation of human resources.

It is in this light, the present study attempts to highlight and analyse the main characteristics of population growth in 'Uttrakhand, U.P.' on the basis of secondary data collected from Census Handbooks. In order to obtain a clear picture of population structure of the region, some important variables like population distribution, population growth rates, population density, sex ratio and literacy are considered in the study.

**Study Area**

Uttrakhand region extends between 70°34'27"E and 81°01'22" E longitude and between 28°43'24" N and 31°27'5" N latitude in Uttar Pradesh. It consists of nine hill districts of Uttar Pradesh, viz. Uttarkashi, Chamoli, Tehri Garhwal, Dehradun, Garhwal, Pithoragarh, Almora, Nainital and Udam Singh Nagar. Of these nine districts, the first five districts are in Garhwal Commissionary while the remain four are in Kumaon Commissionary. Since Udam Singh Nagar district has been formed only in 1994, it has been included in Nainital district for the present study as prior to formation it was a part and parcel of Nainital district.

The region is bounded by Himachal Pradesh in the west. It is separated from Tibet in the north by a natural waterdivide while in the east the river Kali forms its boundary with Nepal. The southern boundary of Uttrakhand touches the plain districts of Pilibhit, Bareilly, Rampur, Moradabad, Bijnor, Hardwar and Saharanpur. The region covers an area of 51122 sq. km. with a maximum north-south length of about 235 kms. Its altitudinal range varies between 204 and 7436 metres above MSL. The northern and eastern boundaries of the region are natural as well as international. Therefore, the region is politically sensitive and is of great strategic significance.

**Population Distribution**

Table 1 reveals that the total population of Uttrakhand was 5,874,353 in 1991 which was estimated to be 4.22 per cent of the Uttar Pradesh's Population. Districtwise, population in Uttrakhand,

Nainital recorded the highest population (1,557,415) and ranked 45th in the state whereas Uttarkashi district with a population of 237,772 had the lowest population in the region and ranked 63rd in the State. Moreover, it may be pointed out here that Nainital and Dehradun districts account for nearly 44 per cent of the total population of Uttrakhand. This situation is explained by the fact that of all the districts of Uttrakhand, Nainital and Dehradun are the most developed ones in all respect–agriculturally, industrially, commercially, educationally etc. Further, more than four-fifth of the total area of these two districts comprises of Tarai, Bhabhar and duns which have several urban settlements, trade centres, industrial estates and is agriculturally highly developed.

**Table 1 :** Population Distribution

| *District* | *Total Population* | | | |
|---|---|---|---|---|
| | *1981* | | *1991* | |
| Uttarkashi | 190,948 | (0.17) | 237,772 | (0.17) |
| Chamoli | 373,605 | (0.34) | 441,667 | (0.32) |
| Tehri Garhwal | 497,710 | (0.45) | 575,352 | (0.41) |
| Dehradun | 761,668 | (0.69) | 1,014,700 | (0.73) |
| Garhwal | 628,618 | (0.57) | 666,165 | (0.48) |
| Pithoragarh | 489,267 | (0.44) | 557,148 | (0.40) |
| Almora | 757,373 | (0.68) | 824,134 | (0.59) |
| Nainital | 1,136,523 | (1.03) | 1,557,415 | (1.12) |
| Uttrakhand | 4,835,712 | (4.37) | 5,874,353 | (4.22) |
| Uttar Pradesh | 110,862,512 | (100.00) | 138,760,417 | (100.00) |

*Note* : Figures in parenthesis indicate percentage

## Growth of Population

Out of eight districts of Uttrakhand five districts, namely Almora, Pithoragarh, Chamoli, Garhwal and Tehri-Garhwal register below 20 per cent growth in 1981-91 (Fig. 1). Uttarkashi records 25 to 27.49 per cent decade growth. However, Nainital and Dehradun are districts where the decadal growth (1981-91) is registered to be above 30 per cent.

As far as per cent increase in population (1981-91) is concerned Uttarkashi falls in 22.50 to 24.99 group, while Almora, Pithoragarh, Garhwal, Chamoli, Tehri-Garhwal records below 20 per cent

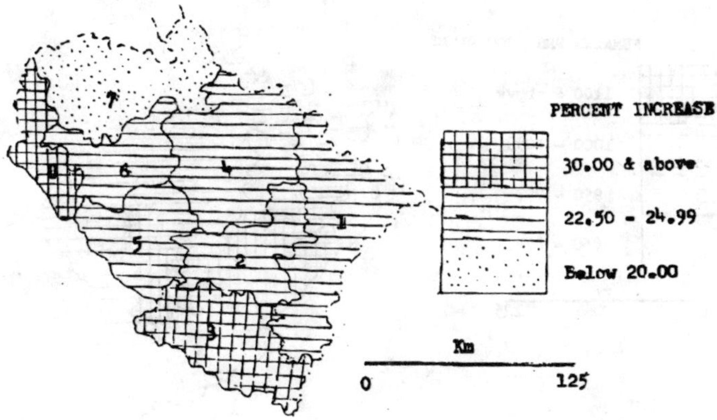


Fig. 1. Growth of population in Uttrakhand (1981-91) (1-Pithoragarh, 2-Almora, 3-Nainital, 4-Chamoli, 5-Pauri, 6-Tehri, 7-Uttarkashi, 8-Dehradun).

increase (Fig. 2). Nainital and Dehradun register above 30 per cent increase.

Table 2 reveals the percentage decadal variation in population of eight districts of Uttrakhand.

**Table 2 : Percentage Decadal Variation in Population of Uttrakhand**

| *Year* | *UK* | *CH* | *TG* | *DDN* | *G* | *PTH* | *ALM* | *NTL* | *UP* |
|---|---|---|---|---|---|---|---|---|---|
| 1901-11 | +11.88 | +11.69 | +11.88 | +15.25 | +11.69 | +15.88 | +15.88 | –0.15 | –0.97 |
| 1911-12 | + 5.85 | + 1.16 | + 5.85 | + 3.59 | + 1.16 | + 0.90 | + 0.90 | –14.42 | –3.08 |
| 1921-31 | + 9.79 | +10.04 | + 9.78 | + 8.48 | +10.04 | + 9.99 | + 9.99 | + 0.15 | +6.66 |
| 1931-41 | +13.67 | +12.78 | +13.67 | +15.63 | +12.78 | +17.83 | +17.83 | + 5.26 | +13.57 |
| 1941-51 | + 3.69 | + 6.23 | + 3.72 | +36.08 | + 6.23 | +12.46 | +12.46 | +14.92 | +11.82 |
| 1951-61 | +15.82 | +17.09 | +13.53 | +18.61 | +13.86 | +19.88 | +13.80 | +73.10 | +16.66 |
| 1961-71 | +20.33 | +15.71 | +14.28 | +34.57 | +14.57 | +20.64 | +17.32 | +37.57 | +19.78 |
| 1971-81 | +29.19 | +24.36 | +25.25 | +31.93 | +15.31 | +17.85 | +16.77 | +43.85 | +25.49 |
| 1981-91 | +24.52 | +18.22 | +15.60 | +33.22 | + 5.97 | +13.87 | + 8.81 | +37.03 | +25.16 |

*Note :* UK = Uttarkashi; CH = Chamoli; TG + Tehri Garhwal, DDN = Dehradun ; G = Garhwal; PTH = Pithoragarh; ALM = Almora; NTL = Nainital & UP = Uttar Pradesh

Here it may be observed that in every case there has been an increase, the only exception being the decades of 1901-11 and 1911-21 for Nainital district where the population decreased by 0.15 per cent and 14.42 per cent respectively. This decrease in population was due to epidemic. In the population of Uttrakhand as a whole,

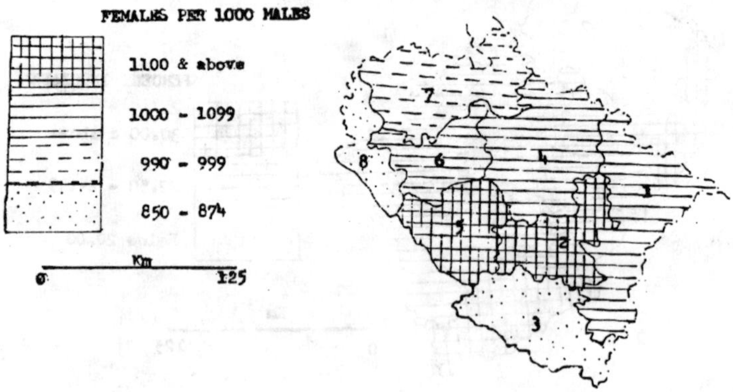


Fig. 2. : Ser ratio (1991) in Uttrakhand (1-Pithoragarh, 2-Almora, 3-Nainital, 4-Chamoli, 5-Pauri, 6-Tehri, 7-Uttarkashi, 8-Dehradun)

the largest increase of 73.10 per cent is observed in Nainital district during the decade 1951-61. This gigantic increase in explained by the fact that the reclamation of Tarai Bhabar areas of Nainital district was undertaken during this period. Moreover, the uprooted persons from west Punjab (now Pakistan) were resettled in this region.

Status of Uttrakhand districts among ten slowest growing districts of U.P. during the period 1901-91 is revealed in Table 3. Here it is interesting to observe that during the decade 1901-11 none of the districts of Uttrakhand figured among the first ten slowest growing districts of the State. During the decades 1911-21 and 1921-31, Nainital ranked first among the slowest growing districts of the State. During the decade, 1981-91 we notice that five districts of the region are among the ten slowest growing districts of Uttar Pradesh. Garhwal district ranked first while Almora ranked second, Pithoragarh ranked fourth, Tehri Garhwal ranked fifth and Chamoli ranked sixth.

Table 4 highlights the status of Uttrakhand districts among ten fastest growing districts of Uttar Pradesh during 1901-1991. It is interesting to note that during the decade 1901-11, five out of the eight districts of the region where among the first five fastest growing districts of the State. Almora ranked first followed by

**Table 3 :** Status of Uttrakhand Districts among Ten Slowest Growing Districts of U.P. (1901-91)

| Rank | *1901 –11* | *1911 –21* | *1921 –31* | *1931 –41* | *1941 –51* | *1951 –61* | *1961 –71* | *1971 –81* | *1981 –91* |
|---|---|---|---|---|---|---|---|---|---|
| 1 | – | NTL | NTL | | – | – | – | G | G |
| 2 | – | – | – | – | – | – | – | ALM | ALM |
| 3 | – | – | – | – | – | – | – | PTH | – |
| 4 | – | – | – | NTL | TG | – | – | – | PTH |
| 5 | – | – | – | – | UK | – | TG | – | TG |
| 6 | – | – | – | – | G | – | – | – | CH |
| 7 | – | – | – | – | CH | – | G | – | – |
| 8 | – | – | – | – | – | – | – | – | – |
| 9 | – | – | – | – | – | – | CH | – | – |
| 10 | – | – | – | – | – | – | – | – | – |

Note : UK = Uttarkashi; CH = Chamoli; TG = Tehri Garhwal ; G = Garhwal, PTH = Pithoragarh; ALM = Almora ; and NTL = Nainital.

**Table 4 :** Status of Uttrakhand Districts among Ten Fastest Growing Districts of U.P. (1901-91)

| Rank | *1901 –11* | *1911 –21* | *1921 –31* | *1931 –41* | *1941 –51* | *1951 –61* | *1961 –71* | *1971 –81* | *1981 –91* |
|---|---|---|---|---|---|---|---|---|---|
| 1 | ALM | UK | – | – | DDN | NTL | NTL | NTL | – |
| 2 | DDN | TG | – | – | – | – | DDN | – | – |
| 3 | UK | – | – | – | – | – | – | – | NTL |
| 4 | TG | – | – | – | – | – | – | – | – |
| 5 | G | DDN | – | – | – | – | – | DDN | DDN |
| 6 | – | – | – | – | – | – | – | – | – |
| 7 | – | – | G | – | – | – | – | – | – |
| 8 | – | – | ALM | – | – | – | – | – | – |
| 9 | – | – | – | – | – | – | – | – | – |
| 10 | – | – | UK | – | -- | – | – | – | – |

*Note :* UK = Uttarkashi ; TG = Tehri Garhwal ; DDN = Dehradun ; G = Garhwal; ALM = Almora; NTL = Nainital.

Dehradun, Uttarkashi, Tehri Garhwal and Garhwal, while during 1981-91 decade we notice that only two districts of the region i.e., Nainital and Dehradun figure among the ten fastest growing districts of the state. Nainital ranked third whereas Dehradun ranked fifth.

It is interesting to note that immediately after independence i.e., during the decades 1951-61, 1961-71 and 1971-81, Nainital district ranked first among the fastest growing districts of the State. This position of Nainital district may be explained by the fact, as mentioned earlier, that the reclamation work of Tarai Bhabar areas of the district was done during this period. Besides this, the displaced persons from Pakistan after partition were resettled in the Tarai Bhabar areas of the district.

**Population Density**

Population density of Uttrakhand is shown in Figure 3 and table 5. It is clear from the Table 5 that Dehradun is the most densely populated district in Uttrakhand. The number of persons per square kilometer of area in 1991 was reported to be 329 in Dehradun district, followed by Nainital (229). Uttarkashi is the most thinly populated district of the region recording only 30 persons per square kilometer.

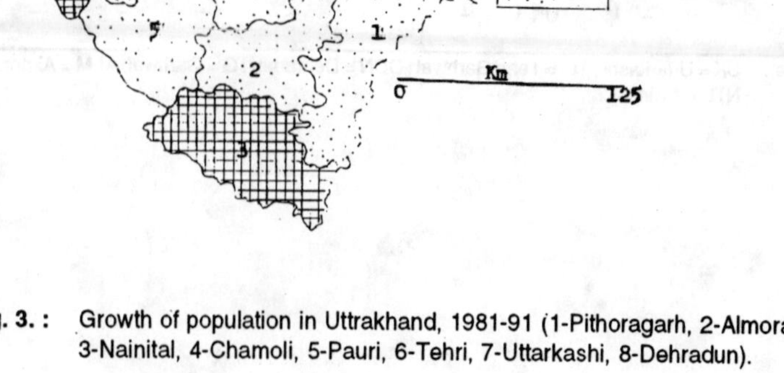


**Fig. 3. :** Growth of population in Uttrakhand, 1981-91 (1-Pithoragarh, 2-Almora, 3-Nainital, 4-Chamoli, 5-Pauri, 6-Tehri, 7-Uttarkashi, 8-Dehradun).

**Table 5 :** Population Density

| *District* | *Population Density* | |
|---|---|---|
| | *1981* | *1991* |
| Uttarkashi | 24 | 30 |
| Chamoli | 41 | 48 |
| Tehri Garhwal | 113 | 130 |
| Dehradun | 247 | 329 |
| Garhwal | 116 | 123 |
| Pithoragarh | 55 | 63 |
| Almora | 141 | 153 |
| Nainital | 168 | 229 |
| Uttrakhand | 94 | 115 |

## Sex Ratio

Sex ratio is an important aspect of population structure and is a measure used to indicate the balance between the two sexes. Figure 4 and Table 6 reveals the number of females per 1000 males at each district of Uttrakhand in 1981 and 1991. It may be observed that on the whole the number of females has increased in each of the 8 districts of the region during the decade 1981-91. However, very significant observation is that in 5 out of 8 districts of the region the sex ratio is in favour of females with Garhwal district leading the tally. This is explained by the fact that the economic and employment situations in the hill areas even today are such that the able-bodied males generally try to seek employment in other regions of the country to the best of their capacity. This male dominated migration from the region necessitates the female members to be left behind to look after the small patch of land and animals. Moreover, the life in industrial areas being more expensive particularly in terms of housing, the very purpose of migration would be defeated if the whole family is shifted. An interesting outcome of this phenomena is the money order economy in the region.

## Literacy

Literacy occupies an important position in the economic development of any country / region and educational progress provides

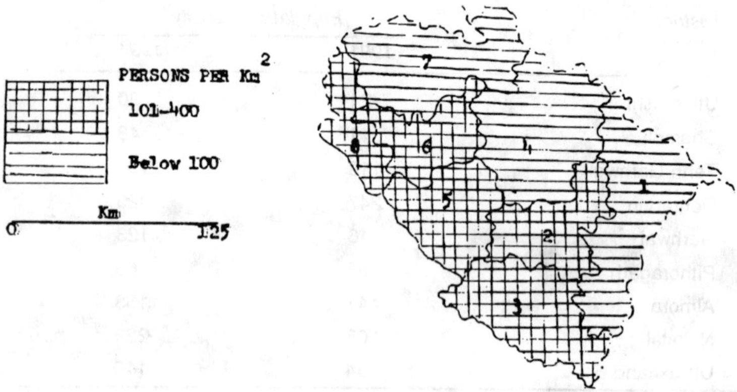


**Fig. 4. :** Density of population, 1991 (1-Pithoragarh, 2-Almora, 3-Nainital, 4-Chamoli, 5-Pauri, 6-Tehri, 7-Uttarkashi, 8-Dehradun)

**Table 6 :** Sex Ratio

| *District* | *Sex Ratio (Female per 1000 males)* | |
|---|---|---|
| | *1981* | *1991* |
| Uttarkashi | 881 | 928 |
| Chamoli | 1042 | 1059 |
| Tehri Garhwal | 1088 | 1073 |
| Dehradun | 811 | 851 |
| Garhwal | 1092 | 11112 |
| Pithoragarh | 1014 | 1031 |
| Almora | 1081 | 1106 |
| Nainital | 841 | 874 |
| Uttrakhand | 959 | 976 |

the main hopes for any society. The proportion of literates in the population is thus an important indicator of the quality of population. Considering the physical handicaps of Uttrakhand region, the present level of education in the area are satisfactory.

Table 7 gives the number of literates in both sexes separately in each of the districts of Uttrakhand. It may be observed from the table that 6.06 per cent of the total literates in the State reside in Uttrakhand. Analysing the data districtwise, we notice that in

Uttrakhand region the highest number of literates are in Nainital (707,415), followed by Dehradun (594,124). Uttarkashi has the lowest number of literates (90,588) in the region.

**Table 7 :** Literacy Sex-wise : 1991

| *District* | *Literate Population* | | |
|---|---|---|---|
| | *Persons* | *Males* | *Females* |
| Uttarkashi | 90,588 (0.19) | 68,697 (0.21) | 21,891 (0.16) |
| Chamoli | 214,578 (0.46) | 136,848 (0.41) | 77,730 (0.57) |
| Tehri Garhwal | 224,846 (0.48) | 159,512 (0.48) | 65,334 (0.48) |
| Dehradun | 594,124 (1.27) | 361,920 (1.09) | 232,204 (1.71) |
| Garhwal | 357,999 (0.76) | 210,716 (0.63) | 147,283 (1.08) |
| Pithoragarh | 264,591 (0.56) | 174,120 (0.52) | 90,471 (0.67) |
| Almora | 391,100 (0.83) | 246,477 (0.74) | 144,623 (1.06) |
| Nainital | 707,415 (1.51) | 452,117 (1.36) | 255,298 (1.88) |
| Uttrakhand | 2,845,241 (6.06) | 1,810,407 (5.44) | 1,034,834 (7.61) |
| Uttar Pradesh | 46,871,095 (100.00) | 33,268,503 (100.00) | 13,602,592 (100.00) |

*Note :* Figures in parenthesis indicate percentage.

It is interesting to note that number of male literates in every district of the region is more than that of females. This is mainly due to age old social customs prevailing in the region particularly in rural area. Despite the much needed awareness about the benefits of education, there is still a general feeling that females' domain is home and fields. Recently, it has been observed that the emphasis on female education in rural areas has also increased and thus the future picture looks much brighter.

**Conclusions**

In all the Himalayan states/regions of Indian, where census was conducted in 1991, Uttrakhand region of Uttar Pradesh reported highest population which was estimated to be 0.70 per cent of the country's population. It was followed by Himachal Pradesh with 0.61 per cent of country's population. During the decade 1981-91 Uttrakhand witnessed population growth rate of 22.03 per cent. In 1991, the region had a density of 115 persons per square km. while the sex ratio was 976. 48.51 per cent of the total population in the region were literate in 1991. Only 2 per cent of the population in Uttrakhand belongs to backward classes whereas in Uttar Pradesh they are 52 per cent. Regarding the level of development of the region in comparison to other regions of the country, it may be pointed out that the region is fairly backward in most of the sectors, as only 13.12 per cent of total geographical area is being used for agricultural purpose. The overall situation of industrialization is also not good, thereby having only marginal employment potential whereas the literacy rate is better than other regions of the country. All this results in the wide ranging frustration among the masses, particularly among the educated youth.

The foregoing discussion of some of the demographic and economic characteristics of Uttrakhand, provide some broad guidelines for the formulation and implementation of a comprehensive, integrated and coordinated policy, designed to secure economic development and a more rational and efficient utilisation of the human resources. Here it may be pointed out that the main basis of planning for Uttrakhand should be economic development rather than social justice as has been mentioned earlier that only 2 per cent of the total population of Uttrakhand is backward. In formulating the policy for the region, future demographic growth and the likely changes in size, composition and distribution of population should be considered so that the problems of the human resources development planning in the region are solved. Besides this, the strategy should be such that more employment opportunities be created and rapid economic development of the region may be achieved.

## References

Census of India, 1981, Series-22, Uttar Pradesh.

Census of India, 1991, Series-25, Uttar Pradesh.

Correa, H. (1963). The Economics of Human Resources, Amsterdam.

Swarup, R. (1991), Agricultural Economy of Himalayan Region, Gyanodaya Prakashan, Nainital.

Zoya, Hasan (1989). The State, Political Processes and Identity. Sage Publications, New Delhi.

# Chapter 18

# U.P.F.C. and Industrial Development of Garhwal Region (Appraisal and Suggestion)

*R.C. Dangwal*

*A.K. Sarkar*

*K.S. Negi*

Garhwal is situated between 29°45′N to 37°27′N latitudes and 77°45′E to 80°7′E longitudes. The region comprises the five districts of Chamoli, Dehradun, Pauri Garhwal, Tehri Garhwal and Uttarkashi and has an area of about 11325 square miles. Dehradun district falls partly in hill and partly in plains. The region lies in the northern part of Uttar Pradesh and rises from a height of 300 m to more than 6000 m above sea level. Garhwal is conspicuously different from other parts of the State due to its topography, soils, climate and vegetation. The area gets a relatively higher rainfall in U.P. and many of its peaks are permanently covered with snow. The population of this region is about 24.53 lakhs according to the census of 1981 (J.D.I. undated).

The Garhwal region has a very week industrial basis. This is the result of various factors such as distance from rail head, distance from established market, lack of local market, inadequate transport facilities, high cost of transportation, lack of local entrepreneurship, locational disadvantage and migration of skilled man power from the region to out-side (plains). Another factor restricting the development of industries is the strict control on exploitation of local forest resources by the government, to protect the reserve forest. As a result, despite extensive forest cover, forest base industries has not been able to come up.

To quote the example of Switzerland will not be out of place here. According to the Encyclopaedia Britannica, Switzerland-a land largely mountainous with damp climate and few materials was at one time, endowed by nature with no more than the basis of meagre pastoral economy. Modern Switzerland, however, has been able to base its industry on a highly skilled urban force, having 92 per cent population employed in non-agricultural pursuits. The ratio of the farming population fell from nine-tenth to about one-tenth of the total between 1800 and the mid-twentieth century (Prasad, 1981). There is no reason why Garhwal region cannot achieve similar results if proper planning is adopted and the enterprising spirits of the inhabitants is developed.

The region has the potential for the development of small scale industries, industries based on forest products, tourism, hotel and electronic industries. The factors which are directly affecting the development of the industries in the region are lack of enterpreneurial abilities and various types of finances. An institution was urgently required which can provide not only the finance to the industries but may also provide technical assistance in the formation and development of the industries. U.P.F.C. being the important financing institution in the region has been taken for the study purpose. An attempt has been made here to assess and evaluate the performance of U.P.F.C. in the industrial development of the region.

The principal objective of the corporation is the industrial development of the U.P. State by providing essential financial assistance to small and medium scale industrial units. The Corporation fulfills these objectives by granting loans to various industrial units of the state under its various schemes : Main Loan Scheme, Composite Loan Scheme, Handloom Weavers Scheme, SRTO Scheme, Rehabilitation Scheme for sick units, Promotional Scheme of Central and State Governments, Special Share Capital Scheme etc.

The U.P.F.C. grants long term loans upto Rs 60 lakhs to companies and registered cooperative societies; large project upto Rs 3 Crore are jointly financed with the PICUP/Scheduled Bank, bridging loan upto 90 per cent of term loan sanctioned are granted and medical practitioners are assisted by x-ray clinic scheme. The new tiny and small scale units are assisted for fixed capital or working capital by single window scheme by IDBI. Financial assistance is provided to industrial units that are set-up or would be set-up in the State. The units may be constituted as Public Ltd., Private Ltd. companies, registered cooperative societies, partnership or sole proprietorship concern.

### U.P.F.C. and Garhwal Region

#### A. Schemewise Analysis of Assistance

(1) Corporation Main Loan Scheme : The maximum number of units that have been sanctioned maximum amount of loans are in Dehradun district. Under the scheme the amount sanctioned to 708 units is Rs 4856.04 lakhs and disbursement amount is Rs 3949.67 lakhs to 631 units since the inception of the U.P.F.C. to 31-3-1989. The total loans sanctioned and disbursed under the scheme to Dehradun district is 80.32 per cent and 81.45 per cent respectively of the total loans sanctioned and disbursed to Garhwal region. The total number of units that have been sanctioned and disbursed loans under the scheme in Dehradun district are 79.10 per cent and 78.48 per cent respectively. In this regards Dehradun district is followed by Tehri, Pauri, Uttarkashi and Chamoli district.

The reason of such distribution of loans to various districts of Garhwal region in comparison to Dehradun district, is that other districts are industrially backward. This is due to the fact that Chamoli, Pauri, Tehri and Uttarkashi districts are located in remote areas of hill which have little scope for industrial development.

The total loans sanctioned and disbursed under the scheme in U.P. is Rs 90176.04 lakhs and Rs 68742.89 lakhs respectively to 15438 units and 14553 units respectively since its inception to the end of March 1989. In this, the share of 895 and 804 units of Garhwal region is 6.71 per cent and 7.05 per cent respectively (Table 1, 2 and 3).

(2) Composite Loan Scheme : Under the scheme, the effective

**Table 1 :** U.P.F.C. Scheme and Districtwise Loans Sanctioned (Effective) and Disbursed as on 31.3.1989 (Garhwal Region)

(Rs. in Lakhs)

| District | Main Loan Scheme | | | | Compsite Loan | | | | Handloom Weavers | | | | Small (Road.) Transport | | | | Total Loan | | | |
|---|---|---|---|---|---|---|---|---|---|---|---|---|---|---|---|---|---|---|---|---|
| | S | | D | | S | | D | | S | | D | | S | | D | | S | | D | |
| | No. | Amt. | No. | Amt. | No. | Amt. | No. | Amt. | No. | Amt. | No. | Amt. | No. | Amt. | No. | Amt. | No. | Amt. | No. | Amt. |
| 1. Chamoli | 19 | 62.39 | 18 | 34.60 | 13 | 2.76 | 12 | 1.78 | – | – | – | – | 6 | 14.75 | 5 | 11.12 | 38 | 79.90 | 35 | 47.50 |
| 2. Dehradun | 708 | 4856.04 | 631 | 3949.67 | 239 | 48.38 | 237 | 47.09 | – | – | – | – | 15 | 29.41 | 13 | 27.78 | 962 | 4933.83 | 882 | 4024.54 |
| 3. Pauri Garhwal | 80 | 396.54 | 68 | 296.11 | 26 | 5.43 | 24 | 4.14 | – | – | – | – | 9 | 14.17 | 4 | 4.89 | 115 | 416.14 | 96 | 305.14 |
| 4. Tehri Garhwal | 708 | 648.01 | 70 | 515.72 | 12 | 2.50 | 12 | 2.50 | – | – | – | – | 19 | 48.39 | 16 | 39.45 | 101 | 698.90 | 98 | 557.67 |
| 5. Uttarkashi | 18 | 83.20 | 17 | 53.58 | 4 | 0.76 | 4 | 0.76 | – | – | – | – | – | – | – | – | 22 | 83.96 | 21 | 54.34 |
| Total of Garhwal Region | 895 | 6048.16 | 804 | 4849.68 | 294 | 59.83 | 289 | 56.7 | – | – | – | – | 49 | 106.72 | 38 | 83.24 | 1238 | 6212.73 | 1132 | 4989.19 |

*Source :* Data compiled from U.P.F.C. Regional office at Dehradun.

S = Sanctioned

D = Disbursed

**Table 2 :** U.P.F.C. Percentage of Scheme and Districtwise Loans Sanctioned (Effective) and Disbursed as on 31.3.1989 (Garhwal Region)

| District | (%) Main Loan Scheme | | | | (%) Composite Loan | | | | (%) Handloom Weavers | | | | (%) Small (Road) Transport | | | | (%) Total Loan | | | |
|---|---|---|---|---|---|---|---|---|---|---|---|---|---|---|---|---|---|---|---|---|
| | S | | D | | S | | D | | S | | D | | S | | D | | S | | D | |
| | No. | Amt. | No. | Amt. | No. | Amt. | No. | Amt. | No. | Amt. | No. | Amt. | No. | Amt. | No. | Amt. | No. | Amt. | No. | Amt. |
| 1. Chamoli | 2.13 | 1.03 | 2.24 | 0.71 | 4.42 | 4.62 | 4.15 | 3.36 | – | – | – | – | 12.25 | 13.82 | 13.16 | 13.36 | 3.07 | 1.29 | 3.09 | 0.95 |
| 2. Dehradun | 79.10 | 80.32 | 78.48 | 81.45 | 81.29 | 80.86 | 82.02 | 83.69 | – | – | – | – | 30.62 | 27.55 | 34.21 | 33.37 | 77.71 | 79.41 | 77.92 | 80.67 |
| 3. Pauri Garhwal | 8.93 | 6.56 | 8.46 | 6.10 | 8.85 | 9.07 | 8.30 | 7.36 | – | – | – | – | 18.36 | 13.28 | 10.52 | 5.88 | 9.29 | 6.70 | 8.48 | 6.12 |
| 4. Tehri Garhwal | 7.83 | 10.72 | 8.70 | 10.64 | 4.08 | 4.18 | 4.15 | 4.44 | – | – | – | – | 38.77 | 45.35 | 42.11 | 47.39 | 8.15 | 11.25 | 8.66 | 11.18 |
| 5. Uttarkashi | 2.01 | 1.37 | 2.12 | 1.10 | 1.36 | 1.27 | 1.38 | 1.35 | – | – | – | – | – | – | – | – | 1.78 | 1.35 | 1.85 | 1.08 |
| Total | 100% | 100% | 100% | 100% | 100% | 100% | 100% | 100% | – | – | – | – | 100% | 100% | 100% | 100% | 100% | 100% | 100% | 100% |

**Table 3 :** U.P.F.C. Percentage of Schemewise Loans Sanctioned (Effective) and Disbursed in U.P. to Garhwal Region as on 31.3.1989

| *Schemes* | *U.P.* | | | | *Garhwal Region* | | | | *%age of (3) to (2)* | | | |
|---|---|---|---|---|---|---|---|---|---|---|---|---|
| | *Sanctioned* | | *Disbursed* | | *Sanctioned* | | *Disbursed* | | *Sanctioned* | | *Disbursed* | |
| | *No.* | *Amt.* | *No.* | *Amt.* | *No.* | *Amt.* | *No.* | *Amt.* | *No.* | *Amt.* | *No.* | *Amt.* |
| *1* | *2* | | | | *3* | | | | *4* | | | |
| 1. Corporation Main Loan Scheme. | 15438 | 90176.04 | 14553 | 68742.89 | 895 | 6048.16 | 804 | 4849.68 | 5.80 | 6.71 | 5.52 | 7.05 |
| 2. Composite Loan Scheme | 7467 | 1186.42 | 7315 | 1101.04 | 294 | 59.83 | 289 | 56.27 | 3.94 | 5.04 | 3.95 | 5.11 |
| 3. Handloom Weavers Scheme | 6185 | 345.11 | 5813 | 328.12 | – | – | – | – | – | – | – | – |
| 4. Small Road Transport Operators Scheme | 389 | 613.82 | 336 | 529.95 | 49 | 106.72 | 38 | 83.24 | 12.60 | 17.39 | 11.31 | 15.71 |
| Total Loan | 29479 | 92321.39 | 28017 | 70702.00 | 1238 | 6212.73 | 1132 | 4889.19 | 4.20 | 6.73 | 3.57 | 7.06 |

*Source :* Data compiled from U.P.F.C. Annual Report, 1988-89.

sanctioned and disbursed loans by the Corporation since the implementation of the scheme upto 31st March 1989 is displayed in table 1 and 2. The total assistance sanctioned and disbursed under the scheme aggregated Rs 59.83 lakhs to 294 units and Rs 56.27 lakhs to 289 units respectively. The Table 2 clarifies that the highest percentage of assistance sanctioned to Garhwal region goes to Dehradun district (80.86 per cent), followed by Pauri (9.07 per cent), Chamoli (4.62 per cent), Tehri (4.18 per cent) and Uttarkashi district (1.27 per cent). So far as disbursement is concerned, the highest percentage of assistance disbursed to Dehradun district is (83.69 per cent) followed by Pauri, Tehri, Chamoli and Uttarkashi districts.

In U.P. under the scheme, the total financial assistance sanctioned and disbursed by the U.P.F.C. as on 31.3.1989 is Rs 1186.42 lakhs to 7467 units and Rs 1101.04 lakhs to 7315 units. The Garhwal region shares 5.04 per cent and 5.11 per cent of the total assistance sanctioned and disbursed to 3.94 per cent and 3.95 per cent units in U.P. (Table 3).

(3) Handloom Weavers Scheme : In the Handloom weavers scheme no sanctions or disbursements of the assistance have been made in all the five districts of the Garhwal region. This is due to the fact that handloom weaving is not popular in the region and due to the poor availability of raw material.

(4) Small Road Transport Operators Scheme : Under the scheme, which is being operated in hill areas and for the scheduled caste and scheduled tribe persons, the Corporation sanctioned and disbursed loans in Garhwal region to the tune of Rs 106.72 lakhs and 83.24 lakhs respectively as on 31st March 1989. Out of this the maximum assistance has been sanctioned to Tehri district (Rs 48.39 lakhs to 19 units). The percentage of total assistance sanctioned and disbursed under the scheme in the Garhwal region to Tehri district is 45.35 per cent and 47.39 per cent respectively. Tehri district is followed by Dehradun, Chamoli and Pauri districts.

The Corporation granted and disbursed total loans of Rs 613.82 lakhs and Rs 529.95 lakhs respectively to 389 and 336 operators respectively since the implementation of the scheme to 31-3-89 in U.P. The percentage of total loans sanctioned and disbursed under the scheme in U.P. to Garhwal region is 17.39 per cent and 15.71 per cent respectively to 12.60 per cent and 11.31 per cent of operators respectively.

**B. Districtwise Analysis of Assistance of Small Scale Sector**

The SFCs were set up in different states to provide the financial assistance to small and medium sector enterprises. The major function of U.P.F.C. is to extend term loan assistance for setting up industrial units in the small and medium sector. The share of small scale industries excluding SRTOs is given in Table 4 and 5. Upto the end of March 1989 the Corporation sanctioned 84.19 per cent (Rs 5230.37 lakhs) of its total assistance in Garhwal region to small scale industries. This assistance was availed by 1153 small scale units, which constituted 93.05 per cent of the total number of industrial units. In U.P., 82.20 (Rs 75884.91 lakhs) assistance is provided to 96.20 per cent (28359 units) of the total units (Table 4).

**Table 4. :** U.P.F.C. Districtwise (Effective) Sanctioned as on 31.3.1989 in the Garhwal Region

| *Districts* | *SSI Amt.* | *SRTO Amt.* | *Others Amt.* | *Total Amt.* | *%age of (2) to (5)* |
|---|---|---|---|---|---|
| *1* | *2* | *3* | *4* | *5* | *6* |
| 1. Chamoli | 65.15 | 14.75 | — | 79.90 | 81.54 |
| 2. Dehradun | 4254.13 | 29.41 | 650.29 | 4933.89 | 86.22 |
| 3. Pauri Garhwal | 316.97 | 14.17 | 85.00 | 416.14 | 76.17 |
| 4. Tehri Garhwal | 510.16 | 48.39 | 140.35 | 698.90 | 72.99 |
| 5. Uttarkashi | 83.96 | — | — | 83.96 | 100.00 |
| Total of Garhwal Region | 5230.37 (1153) | 106.72 (49) | 875.64 (36) | 6212.73 (1238) | 84.19 (93.05) |
| Total of U.P. | 75884.91 (28359) | 613.82 (389) | 15822.66 (731) | 92321.39 (29479) | 82.20 (96.20) |

*Source :* Data compiled from U.P.F.C. Regional office at Dehradun and Annual Report of U.P.F.C. during the year 1988-89, p. 52.

*Note :* Figures in brackets indicate number of units.

It is clear from the Table 5 that out of the total assistance sanctioned by the Corporation to small scale industries in Garhwal region, the highest percentage of assistance goes to Dehradun district 81.34 per cent (Rs 4254.13 lakhs) followed by Tehri (9.75 per cent), Pauri (6.05 per cent), Uttarkashi (1.61 per cent) and Chamoli district (1.25 per cent).

With the view to have deep insight into the performance of the institute, some other aspects of the financing activities of the institute have also been analysed here.

**Table 5:** U.P.F.C. Percentage of Districtwise (Effective) Sanctioned as on 31.3.1989 in the Garhwal Region

| | *Districts* | *SSI* | *SRTO* | *Others* | *Total* |
|---|---|---|---|---|---|
| | | *% of Amt.* | *% of Amt.* | *% of Amt.* | *% of Amt.* |
| 1. | Chamoli | 1.25 | 13.82 | — | 1.29 |
| 2. | Dehradun | 81.34 | 27.55 | 74.26 | 79.41 |
| 3. | Pauri Garhwal | 6.05 | 13.28 | 9.71 | 6.70 |
| 4. | Tehri Garhwal | 9.75 | 45.35 | 16.03 | 11.25 |
| 5. | Uttarkashi | 1.61 | — | — | 1.35 |
| Total in Garhwal Region | | 100 | 100 | 100 | 100 |
| %age of Total Loan given in U.P. to Garhwal Region | | 6.89 | 17.39 | 5.53 | 6.73 |

## C. Assistance as General Loan and Assistance to Hotel Industry

The financial assistance granted and disbursed by the U.P.F.C. under general loans and hotel industry have been shown in Table 6, 7 and 8. Taking the number of units into consideration in the five districts combined, out of the total assistance sanctioned to small scale sector by the Corporation, 82.86 per cent units have been sanctioned general loans, 13.06 per cent units got hotel industry loan and remaining 4.08 per cent units received transport industry. However, when the amount loaned is taken into consideration, it is found that out of the total assistance sanctioned to small scale sector, 66.83 per cent has gone to general loans, 31.17 per cent to hotel industry and 2.00 per cent to the transport industry. Similar trend is found in case of loans disbursements under general loans, hotel industry and transport industry as on 31-3-1989 in the region.

**Table 6 :** U.P.F.C. Financial Assistance Sanctioned and Disbursed (Since Inception to 31st March 1989) under Corporation Loan Schemes (Garhwal Region)

(Rs. in Lakhs)

| *Particulars* | *Sanctions (Effective)* | | | | | | *Disbursement* | | | | | |
|---|---|---|---|---|---|---|---|---|---|---|---|---|
| | *Small* | | *Others* | | *Total* | | *Small* | | *Others* | | *Total* | |
| | *No.* | *Amt.* | *No.* | *Amt.* | *No.* | *Amt.* | *No.* | *Amt.* | *No.* | *Amt.* | *No.* | *Amt.* |
| **Loans :** | | | | | | | | | | | | |
| **Corporation Loans :** | | | | | | | | | | | | |
| (i) General Loans (Other than for Hotel and Transport Industry) | 996 | 3567.03 | 36 | 875.64 | 1032 | 4442.67 | 925 | 3005.72 | 31 | 712.88 | 956 | 3718.60 |
| (ii) Hotel Industry | 157 | 1663.34 | — | — | 157 | 1663.34 | 137 | 1187.35 | — | — | 137 | 1187.35 |
| (iiii) Transport Industry | 49 | 106.72 | — | — | 49 | 106.72 | 39 | 83.24 | — | — | 39 | 83.24 |
| Total | 1202 | 53337.09 | 36 | 875.64 | 1238 | 6212.73 | 1101 | 4276.31 | 31 | 712.88 | 1132 | 4989.19 |

*Source* : Data Compiled from U.P.F.C. Regional office at Dehradun.

**Table 7 :** U.P.F.C. Amount of Total Loan Sanctioned and Disbursed to General Loan, Hotel Industry and Transport Industry as on 31.3.1989 in Uttar Pradesh and Garhwal Region

| Particulars | | Uttar Pradesh (%) | | | | | | Garhwal Region (%) | | | | | |
|---|---|---|---|---|---|---|---|---|---|---|---|---|---|
| | | Sanctioned (Effective) | | | Disbursed | | | Sanctioned (Effective) | | | Disbursed | | |
| | | Small | Others | Total | Small | Others | Total | Small | Others | Total | Small | Others | Total |
| **Loan :** | | | | | | | | | | | | | |
| **Corporation Loans :** | | | | | | | | | | | | | |
| (i) General Loans (other than for Hotel and Transport Industry) | No. | 97.28 | 97.26 | 97.28 | 97.57 | 97.56 | 97.57 | 82.86 | 100 | 83.36 | 84.02 | 100 | 84.45 |
| | Amt. | 94.70 | 97.21 | 95.13 | 94.86 | 97.30 | 95.25 | 66.83 | 100 | 71.51 | 70.28 | 100 | 74.53 |
| (ii) Hotel Industry | No. | 1.32 | 2.74 | 1.36 | 1.20 | 2.44 | 1.23 | 13.06 | — | 12.68 | 12.44 | — | 12.10 |
| | Amt. | 4.50 | 2.79 | 4.21 | 4.25 | 2.70 | 4.00 | 31.17 | — | 26.77 | 27.77 | — | 23.80 |
| (iii) Transport Industry | No. | 1.40 | — | 1.36 | 1.23 | — | 1.20 | 4.08 | — | 3.96 | 3.54 | — | 3.45 |
| | Amt. | 0.80 | — | 0.66 | 0.89 | — | 0.75 | 2.00 | — | 1.72 | 1.95 | — | 1.67 |
| Total | No. | 100 | 100 | 100 | 100 | 100 | 100 | 100 | 100 | 100 | 100 | 100 | 100 |
| | Amt. | 100 | 100 | 100 | 100 | 100 | 100 | 100 | 100 | 100 | 100 | 100 | 100 |

**Table 8 :** U.P.F.C. Percentage Amount of Total Loan given in Uttar Pradesh to Garhwal Region under General Loans, Hotel Industry and Transport Industry as on 31.3.1989.

| *Particulars* | *Sanctions (Effective) (%)* | | | | | | *Disbursement (%)* | | | | | |
|---|---|---|---|---|---|---|---|---|---|---|---|---|
| | *Small* | | *Others* | | *Total* | | *Small* | | *Others* | | *Total* | |
| | *No.* | *Amt.* | *No.* | *Amt.* | *No.* | *Amt.* | *No.* | *Amt.* | *No.* | *Amt.* | *No.* | *Amt.* |
| **Loans :** | | | | | | | | | | | | |
| **Corporation Loans :** | | | | | | | | | | | | |
| (i) General Loans (Other than for Hotel and Transport Industry) | 3.69 | 4.92 | 5.06 | 5.69 | 3.73 | 5.06 | 3.46 | 5.34 | 4.85 | 6.47 | 3.50 | 5.52 |
| (ii) Hotel Indusry | 42.90 | 48.29 | — | — | 40.67 | 42.81 | 41.64 | 47.04 | — | — | 39.71 | 41.95 |
| (iii) Transport Industry | 12.60 | 17.39 | — | — | 12.60 | 17.39 | 11.61 | 15.71 | — | — | 11.61 | 15.71 |
| Total | 4.33 | 6.98 | 4.92 | 5.53 | 4.35 | 6.73 | 4.02 | 7.20 | 4.73 | 6.29 | 4.04 | 7.06 |

**D. Industrywise Analysis of Assistance**

The financial assistance sanctioned by the U.P.F.C. in Garhwal region covered a wide range of industries. The small scale units have obtained maximum amount of sanctions and disbursements (Rs 5337.09 lakhs and Rs 4276.31 lakhs respectively) as compared to the other units. This could, perhaps, be attributed to the fact that in comparison to the small scale industries the other ones have faced more difficulties in their dealings with the Corporation.

The hotel industry has claimed maximum share (27.29 per cent) of the U.P.F.C. sanctioned assistance, followed by miscellaneous manufacturing (15.17 per cent), textiles (14.23 per cent), chemical products (12.48 per cent) and metal products (9.05 per cent). These five industries together account for 78.22 per cent of assistance. The industrywise disbursement of assistance to Garhwal region, shows almost the similar trend.

The above analysis reveals that the largest share of the assistance sanctioned and disbursed by the Corporation except SRTOs has gone to Dehradun district which, incidentally, is the most developed of the five districts. It seems, therefore, that the Corporation's objective of promoting industrial development especially in the less developed areas has not yet been fulfilled. Under SRTO scheme the highest percentage of financial assistance sanctioned and disbursed was to Tehri district. It is important to note that the Corporation is main loan scheme have made significant contribution in the industrial development of Garhwal region. Under the scheme maximum loans have been sanctioned and disbursed.

One of the principal function of U.P.F.C. is to promote small scale industries, especially in the specified backward districts of the State. It has been observed that the share of small scale sector in terms of total number of beneficiaries as well as the total volume of assistance granted by the U.P.F.C. has been greater than that the medium sector. It claimed about 84.19 per cent of the total assistance sanctioned by the Corporation in Garhwal region as on 31-3-1989.

It is interesting to note that the hotel industry has been given special attention by the U.P.F.C. in the Garhwal region. Out of the total loans sanctioned and disbursed to hotel industry in U.P., Garhwal region claimed 48.29 per cent and 47.04 per cent respectively.

The yearwise lending shows that the Corporation has not performed satisfactory till the year 1978-79 in Garhwal region. Since its inception (1954) upto 31st March 1979 only Rs 312.13 lakhs to 100 units could be sanctioned. The work of the Corporation picked up real pace only since the year 1979. It has been observed that the Corporation's assistance to Garhwal region has increased progressively from Rs 89.84 lakhs in 1978-79 to Rs 1869.64 lakhs in 1985-86. But it declined from Rs 1661.64 lakhs in 1986-87 to Rs 1152.26 lakhs in 1988-89 (Dangwal 1990).

**Suggestions**

The conclusion drawn on the basis of the above analysis points to some fundamental defects in the planning for the industrial development of the region. Some of the suggestions for the industrial development of the region to the right direction are as follows:

1. The U.P.F.C. should give wide publicity of its different financing schemes and the facilities it provides to entrepreneurs. The procedure of financing should be simple, easy and should not be time consuming.
2. Special arrangements should be made to provide proper training to the entrepreneurs and they must be provided further assistance in the formation and development of the industries.
3. Efforts should be made to establish electronics and telecommunication industries. The industries based on electronics and related field have good feasibility in the region.

   Specialised industries should be established at the foothills of the region which must be connected with the main line of the transportation. Different ancillary industries should be scattered in the remote areas where potentialities for the development of the industries exist. This will certainly help in the industrial development of the region in the real sense.
4. Apart from establishing electronic and tele-communication industries, tourism and hotel industries have good potential. Some tourist centres should be spotted and developed.

## References

Dangwal, R.C. 1990. *Contribution of U.P. Financial Corporation in the field of U.P.'s Industrial Development, with Special Reference to Garhwal Region* (Unpublished D.Phil. Thesis), Garhwal University, Srinagar.

Prasad, Ram 1981. Industrial potentiality of U.P. Himalaya. *Science and Rural Development in Mountains* (ed.) J.S. Singh, S.P. Singh and C. Shastri, Gyanodaya Prakashan, Haldwani (U.P.).

U.P.F.C., 1988-89. Annual Report, Kanpur.

# Chapter 19

# Some Aspects of Growth and Management of the Human Resource in the U.P. Himalaya*

*Raj Laxmi Singh*

*O.P. Singh*

The U.P. Himalaya is a well recognized regional entity comprised practically of the eight mountainous districts of north-western Uttar Pradesh forming a part of the Himalaya-Nainital, Almora and Pithoragarh, of Kumaun division (and region) and Dehra Dun, Garhwal (Pauri Garhwal), Tehri Garhwal, Uttar Kashi and Chamoli, of Garhwal division (and region). Having about 17.4 per cent area (51,122 sq. km.) and 4.3 per cent population (48,35,712 persons in 1981) of the state, the region has a diversified natural setting with the well known tripartite geotectonic divisions of the Himalayas besides the Trans-Himalaya and the adjoining Tarai-Bhabar sub-region.

Demographically too, the region is highly interesting having a variety of trends and patterns. Population density is still low (94.6 per sq. km. in 1981) and concentration of population is more naturally towards south which is more favourable from the point of view of relief, climate and other resultant and related aspects.

---

* The present paper is partly based on a paper presented in the International Interdisciplinary Seminar on 'Population Growth : Issues and Response Patterns with Special Reference to Developing Countries', held in the Department of Geography, Banaras Hindu University, Varanasi from Feb. 26 to March 1, 1989.

The detailed final figures of the 1991 census are still not available to the author. The present paper therefore analyses population growth in the 1901-81 period and discusses some problems and traits of the human resource and offers certain suggestions for its management. The term *'human resource'* includes human population with its various traits and elements and aspects directly related to these. The need of analysing the human resource patterns and trends and relevance of attempts to manage them are hardly a matter of debate.

## Overall Growth of Population (1901-81)

The overall growth of population in the region is quite similar to that of the country, with certain exceptions. The total as well as districtwise population growth of the region has been presented in Table 1 alongwith respective figures of the state and the country given for the sake of comparison.

Population growth in the region, as is evident from Table 1, has generally been higher than in the state-always higher except during 1951-61 decade (in which too, it was not much lower). The lowest growth of 2.5 per cent has been recorded in 1911-21 period, but this too was not negative as in case of the state and the country.

In the post-Independence period (1951-81), the growth figures correspond more to those of the state rather than to those of the country. During the last decade (1971-81), the growth is the highest among the census decades (26.52 per cent). The growth rate has also increased considerably since 1951, as is evident from percentage figures.

The region has never shown negative growth. Even during 1901-11 period when the state's population slightly declined, the region has registered a good percentage of growth. All this may be attributed mainly to the geographic environment of the region because of which the impact of famines and epidemics was minimized. Moreover, the mountainous barriers have functioned to check the spread of epidemics and other negative factors interregionally as well as intraregionally.

## Interval Variations in Growth

The internal growth variations of population too are great and quite interesting. The prominent features are :

**Table 1 :** Population growth in the U.P. Himalaya, 1901-81 (in per cent)

| *S.No.* | *District/Region* | *Total population 1981* | *Population Growth* | | | | | | | | |
|---|---|---|---|---|---|---|---|---|---|---|---|
| | | | *1901-11* | *1911-21* | *1921-31* | *1931-41* | *1941-51* | *1951-61* | *1961-71* | *1901-1971* | *1971-1981* |
| 1. | Dehra Dun | 761,668 | 15.3 | 3.6 | 8.5 | 15.6 | 36.1 | 18.6 | 34.6 | 225.0 | 31.93 |
| 2. | Uttarkashi | 190,948 | 11.9 | 5.8 | 9.8 | 13.7 | 3.7 | 15.9 | 20.3 | 114.0 | 29.19 |
| 3. | Chamoli | 364,346 | 11.7 | 1.2 | 10.0 | 12.8 | 6.2 | 16.7 | 15.6 | 101.0 | 24.53 |
| 4. | Garhwal | 637.877 | 11.7 | 1.2 | 10.0 | 12.5 | 6.2 | 14.1 | 14.7 | 95.0 | 15.34 |
| 5. | Tehri Garhwal | 497,710 | 11.9 | 5.8 | 9.8 | 13.7 | 3.7 | 13.5 | 14.3 | 99.0 | 25.25 |
| 6. | Pithoragarh | 489,267 | 16.7 | 0.9 | 10.0 | 17.8 | 12.5 | 18.5 | 19.0 | 140.0 | 17.85* |
| 7. | Almora | 757,373 | 15.9 | 0.9 | 10.0 | 17.8 | 12.5 | 15.1 | 18.4 | 132.0 | 16.77* |
| 8. | Nainital | 1136,.523 | 16.4 | 4.9 | 9.8 | 19.5 | 19.9 | 22.3 | 29.6 | 205.0 | 43.85 |
| | U.P. Himalaya | 4835,712 | 13.8 | 2.5 | 9.8 | 15.4 | 12.5 | 16.2 | 20.4 | 131.0 | 26.52 |
| | U.P. | 110,862,013 | – 1.0 | – 3.1 | 6.7 | 13.6 | 11.8 | 16.7 | 19.9 | 81.7 | 25.49 |
| | India | 685,184,692 | 5.7 | – 0.3 | 11.0 | 14.2 | 13.3 | 21.5 | 24.8 | 129.6 | 25.045 |

* Champawat tahsil, which formed a part of Almora district in 1971 and is now a part of Pithoragarh district (since 1981), has been excluded from Almora for finding out the growth percentage and included in Pithoragarh for the same.

(1) Kumaun districts and Dehra Dun have observed in general much higher growth than the remaining parts upto 1951, except during 1911-31 period.

(2) Nainital and Dehra Dun districts continue to record highest growth in the largest part of the period with some exception being again in the 1911-31 period.

(3) During the crucial period of 1911-21 decade, Tehri Garhwal State having present Uttarkashi and Tehri Garhwal districts as its constituents showed the highest growth because of the negative impact of wars etc. being the least. Almora and Pithoragrah registered the ever lowest growth of just 0.9 per cent.

(4) During the other two early decades (1901-11 and 1921-31) and also during 1951-61, the growth was more or less uniform throughout the region.

(5) In the partition decade of 1941-51, the growth was exceptionally high in Dehra Dun (36.1 per cent) followed by Nainital (19.9 per cent) and the remaining two Kumaun districts. Dehra Dun and Nainital both have received refugees' influx from Pakistan in this period.

(6) During the last two post-independence decades (1961-81), Uttarkashi and adjoining parts have shown enhanced growth.

(7) The growth rate of Pauri Garhwal is the lowest at present; and in most of other decades up 1971, the two Garhwals have shown the lowest as well as almost a stagnating growth. Growth of Almora has been the lowest among Kumaun districts. Almora and Pithoragarh too have recorded much slower growth during the last decade of 1971-81. The two Garhwals and Almora have outmigration as the most dominant factor to explain such a trend, as stated in the sequel.

**Role of Migration**

Migration in various forms has been the most dominant single factor of population growth in the region. Though social customs of polyandry in Jaunsar-Bawar and polygamy in main Garhwal have also been playing some role in the growth structure of population through influencing natural increase and balance of sex ratios, but migration has been affecting the growth of population directly. Though nearly each district has both the currents of migration–outmigration and immigration, but their dominance and intensities differ. The spatial variability of migration therefore is high, as elucidated below :

(1) Pauri Garhwal, Tehri Garhwal and Almora have shown excessive outmigration–particularly of young males being pushed outside for jobs etc. This fact is confirmed by their extraordinarily high sex ratios in the young age groups–the overall sex ratio in 1981 being respectively 1091, 1088 and 1081, but reported to be as high as 2066, 2390 and 1606 respectively in 20-24 age group in 1971.

(2) Pithoràgarh and Chamoli, with sex ratios of 1041 and 1026 respectively, are districts of moderate outmigration.

(3) The remaining three districts—Dehra Dun, Nainital and Uttarkashi (with sex ratios 811,841 and 881) are districts of dominant immigration. Dehra Dun and Nainital both have been experiencing high immigration–particularly during the partition decade, owing mainly to colonization and repatriation in the economically more prospective areas of Dun, Tarai and Bhabar.

(4) Uttarkashi is a more balanced district having an edge in favour of immigration. Greater growth of population in Uttarkashi and Tehri in recent years is accounted for mainly by immigration in recent years. Construction activities pertaining to Tehri Dam Projects are one important factor in case of Tehri Garhwal.

(5) Outmigration from interior parts of the region has given birth to what is known as the *'money order economy'* very much prevailing in the entire mountains region, but with varying degrees three types of outmigration–Temporary or seasonal, periodic and permanent.

## Growth Types and Associated Factors

Because of the migration being the most dominant factors of growth, the broad growth types coincide more or less with the broad migration categories listed above.

The sub-Himayalan zone experiences both-dominant immigration and high growth. The interior parts–particularly of the Middle Himalaya and more specifically the main Garhwal and Kumaun, are dominated by outmigration and the resultant slow growth. Exceptions and anomalies presented in recent years by Uttarkashi, Tehri Garhwal and Chamoli–all the three having higher growth–can be explained by a number of factors including the enhanced developmental activities. Greater urbanization and economic development of the sub-Himalaya are other important factors of its more rapid population growth in the Great Himalayan region, having naturally the lowest densities of population, intensity of nearly all other demographic traits (e.g., growth, migration and sex ratio) has been moderated (Singh, 1990), making it thus a separate type of region even demographically too.

**Quality of the Human Resource and Its Management**

Quality of life of population in the region is somewhat inferior, as evinced by demographic indicators too. Growth of population has been faster here than in the state and the country. The highly agrarian and underdeveloped economy of the region is not only handicapped and limited in area, but also it faces the problem of ecological viability in case of unscientific cultivation and cattle rearing and in case of sensitive areas. The non-primary sectors of economy including the infrastructure-particularly transport, are not developed sufficiently. Natural resources, except of course a few, are either limited or are economically and ecologically unexploitable. The region therefore, despite having low arithmetic density, has got high pressure on its resource base.

Natural hazards of various types and intensities coupled with environmental austerity particularly in areas of high relief and inhospitable climatic conditions enhance the difficulties further. Lack of fuel and water for actual use, outflow of soil, water and work force from the region–resulting mainly from the processes like deforestation and overgrazing and scarcity of food and employment, malanutrition and under mutrition, lesser longevity of staying males illiteracy of rural females of the level of about 80 per cent, socio-cultural deteriorzation caused mainly by the tourist influx and the assimilation-modernization processes, and the ecology-development dilemma (Ives and Messerli, 1989)–these are the major factors which have made the quality of the human resource quite poor.

Besides these, physiological density of population is quite high in the region. The per capita net sown area, carrying capacity of land, calories of food supply and per capita income-all the naturally therefore considerably low (Singh, 1990). Outmigration of a high percentage of capable males, for availing better opportunities and economic gains outside (upto a level varying from about 4 to 18 per cent of the total population), who send back money to their dependants and semi-dependents, has aggravated the already worsened management of the economy.

For sound management of the human resource, the following measures or actions are suggested :

(1) People's education in the right direction and raising the level of literacy, particularly of the rural female; as a very high positive correlation between knowledge and development is fully well established.

(2) Stopping as well as reverting the outmigration process, as far as possible by developing economically more gainful as well as viable activities and by promoting a system of conserving and integrating all the domestic subsystems–i.e., those of crops, grasslands, forests and water, etc.

(3) Strengthening the institutional framework and infrastructure, particularly to provide sufficient accessibility and linkages and to promote a sound hygiene, so that people's qualities and inherent capabilities may be developed and used as best as possible.

(4) Arresting or at least minimizing the high growth rate of population.

(5) Enforcing methods and laws to stop delinquent and unhealthy activities and practices in order that true development of personality traits may create a favourable situation for emergence of the human resource at its best-to be utilized for all round development and for solving the local, interregional as well as national problems.

If the people of the region are sufficiently educated, enlightened and awakened; if they have been developed to the level of possessing a sound physical, mental and spiritual health and if they come forward themselves subsequently to find solutions to the problems of their own and their region and nation at large, any objective of development an be achieved. Proper and right type of education appears to be the final and the most important single step for the desired development of the human resource which again is the most significant key of all other types of development.

## References

Ives, Jack D. and Messerli, Bruno, 1989. *The Himalayan Dilemma–Reconciling Development and Conservation*, London and New York, Routledge ; reprinted, 1990.

Singh, O.P. 1990. Population Dynamics and Pressure in the Uttar Pradesh Himalaya. In Aijazuddin Ahmad, john I. Clarke.C.B. Shrestha and A. Trilsbach (eds.). *Mountain Population Pressure*, Vikas Publishing House, New Delhi : 192-205.

# Chapter 20

# Women, Fuel and Forest : The Experience of Central Himalaya

*Ravindra K.Pande*

*Anil K. Singh*

Success in Everest expedition by Bachandri Pal is really an event of National importance more especially for the women of hills, but such mirracles could not sole women folk tragedy. It is case of one woman, who took pains to climb up on such a height of more than 29,000 feet, and certainly it can not be applied in the case of all women of Himalaya. However, thin golden victory can not solve the problems of thousands of 'pahari' women.

There is no doubt that there are thousands of hill women travelling daily from kitchen to forest since their early childhood to late old age. They feed their children only after becoming mother but feed their kitchens since the early age of eight or so. Thus, by all means, to talk about the rural Himalayan economy is ultimately to talk about hardworking 'pahari' women. Due to the very high level of anthropomorphization the core areas of kitchen energy are retreating fast but all kitchens remain at the places where they were previously. Thus, the distance between kitchen to forest is increasing day-by-day.

The author has attempted to understand this short but frequent journey by interviewing the women of rural Himalaya. By appearance, most of them were looking innocent, hardworking, little bit straight forward and apart from these they were looking feared and disgusted. Really for them, the sunrise and sunset have no meaning if the kitchens are devoid of firewood. The retreating or shrinking forests have enhanced many times their problems by increasing the distances between kitchen and forest. Such increasing distances are proving as malady by increasing the burdens of domestic life, in so many ways—by maximizing the efforts, time consumption, energy loss and by the minimization of the opportunities and comforts, etc. Indeed such widening distances have wreaked a havoc on women fold by adding various kinds of covert and overt physical pains and mental strains and so on. Who is responsible for this misery? This question can be properly answered 'by the following statement—

1. Socio-cultural setup of Kumaun, which has compelled women to arrange the kitchen fuel and if in case women fail to do (actually they rarely fail), they have to suffer a lot by the various kinds of problems of domestic life.
2. Multiplying we, you and development of the day and in vogue on the cost of forests.
3. Traditional system of this sub-Himayalan region, which has given an overheard and more compact daily routine system to Kumauni women.
4. Population increase which have also disturbed the man-forest ratio due to the high demand for fuel.
5. No other alternative to solve the fuel problems of rural Himalaya.
6. Illiteracy and orthodoxian outlook of the village folks.

Thus the kitchen economics of fuel for Kumauni women gives somewhat different calculations balancing only hard work in their part.

## Discussion

Not randomally but all randomally, 200 women were selected for interview. Out of this chosen number 28 per cent failed to appear before the investigation team due to the religio-ritual traditionality. Thus, the total inference of this article is based on only 144 women and their views. Answering the first question almost all women were of the same opinion–as the family size increasse, the consumption level of fuel wood also increases Table 1. From the statistics presented in Table 1, it is clear that seasonality has a great role to play in the consumption pattern of fuel. The fuel wood consumed in the kitchen is carried out by the women and they usually carry it twice or thrice in a week 92.5 Kg per day is considered to be standard fuel consumption in Central Himalaya). Except 48 women, out of 144, had no supporters to share their burden of management and 96 women were alone in their fate of serving kitchens.

**Table 1 :** Family size Vs fuel consumption

| *Family Size* | *Per day consumption (Kg)* | |
|---|---|---|
| | *Summer* | *Winter* |
| Below 4 | 3.5 | 5.0 |
| 4–8 | 6.5 | 8.0 |
| 8–12 | 8.0 | 10.5 |
| 12–14 | 10.0 | 12.5 |

Almost all women were looking highly depressed from the kitchen's 'Super Micro-Environment' as they were bound to stay 3 to 5 hours in their damp and depressive kitchens having a very small window and a circular shaped hole in the roofs of about 8-12 cm radius. This is the only way from which light can enter the kitchen. During cooking hours they are not allowed to go out owing to the myths of pollutions. Thus, in terms of environmental quality, they suffer the worst. Table 2 depicts time devoted in the kitchen and perception of the kitchen.

**Table 2 :** Time devoted in the kitchen and perception of the kitchen

| | | | | |
|---|---|---|---|---|
| Devoted hours | 1–2 | 21 | 2–3 | 86 |
| in the kitchen | 3–4 | 30 | + 4 | 07 |
| Frequency of | 01 | 10 | 02 | 80 |
| outing from the kitchen | 03 | 21 | + 4 | 10 |
| Feelings | OK | 20 | Not well | 60 |
| | Troublesome | 25 | Very bad | 39 |
| Smoke problems | High | 14 | Moderate | 30 |
| | Low | 60 | No | 40 |

One most important aspect, which was interviewed with great difficulty, was as to how they manage cooking during menstruation period. It was surveyed that during this period they are not allowed to enter the kitchen or any other room. They are asked to remain in a separate room, in most cases the 'ghoth', of the house. According to the customs they are entitled to touch only dry material. However, they are not departed from their regular duty to bring fuel wood from the forests.

Distance and time aspects in the collection of fuel wood are important parameters. Table 3 depicts a very high percentage of women covering a distance 12-28 Km once a week. Age factor does not reflect a major deviation except in the age group below 12. It is worth mentioning that they take little bit less time in coming back to their homes, inspite of having a considerable weight of fuel wood on their heads.

**Table 3 :** Distance, time and age factors related to fuel collection

| Distance covered (Km) | below 12 | 12–20 | 20–28 | 28–32 |
|---|---|---|---|---|
| No of women | 15 | 48 | 69 | 12 |
| Age groups | above 40 | 40–30 | 30–20 | below 20 |
| No of women | 39 | 40 | 50 | 25 |
| Time consumed (per week) | | | | |
| Approaching the place | 3–4 hrs | 5–6 hrs | 6–8 hrs | 8–9 hrs |
| Coming to home | 2–4 hrs | 4–5 hrs | 5–7 hrs | 7–9 hrs |

***Note :*** During the approaching hours, the time of minor haltages is included.

During this 'fuel conspiracy' which is more or less an imposed one, they suffer a lot. Few of the pains they can not express, i.e., infant baby's cry for mother's milk, a highly fevered baby in search of mother's affection or a husband crolling on the bed and so many of this nature but it can be extracted from their talks Table 4. Out of 144 there were 74 women having their dependents ill while 70 were free from such problems and 12 women were in family way. Table 5 provides information regarding their physical injuries during fuel wood collection.

**Table 4 :** Sufferings during fuel collection

| *Women having* | *Above 4 years* | *1–2 years* | *Below 1 year* |
|---|---|---|---|
| babies | 39 | 33 | 18 |
| Women having | son/daughter | husband | others |
| their dependent ill | 24 | 10 | 40 |

**Table 5** : Injuries received

| *Very minor* | | *Minor* | | *Major* | | *No injury* |
|---|---|---|---|---|---|---|
| 53 | | 23 | | 10 | | 58 |
| D | H | D | H | D | H | |
| 52 | 01 | 17 | 06 | 02 | 08 | |

D–Domestic cure and H-Hospital cure.

## Conclusions

On the basis of the above mentioned discussion the following conclusions and suggestions may be derived :

1. The participation of man and woman in almost every sector of village-ecosystem must be equal. For this, a change in socio-cultural attitude in rural society is required which can be developed with the help of education.
2. Assuming that 'there is not forest to begin with' a new model for afforestation is in the need. The demand for fuel wood can be met from village forest or 'panchayati forest'. This will also reduce the distance between forest and kitchen.
3. The alternative sources of energy should reach to villages.

## Chapter 21

# Environmental Study of Farm Fringe : The Experience of U.P. Himalaya

*Devi Datt*

All these factors, external to the man either in the reaction systems or otherwise directly affect the maintenance, growth and reproductive functions, constitute its environment. Now a days, ecology has acquired a new broad term–the environmental science which focusses greater attention on man and his environment which cannot be viewed separately. Thus the environmental problems are global in character and they need a global solution. The solution to the environmental problems also requires a holistic approach- considering all components together. The global environmental problems should really be studied as a whole. This is obviously an impossible task, and thus, these problems usually should select the study units of biosphere of a more convenient size. Considering the view, the present communication deals with the environmental study of farm forest fringe of a part of the Central Himalaya.

The present study is confined to a Himalayan village, i.e., Village Kelani (29° 54′ 40″ N latitude and 79° 11′ 20″ E logitude) which is considered representative village of the Bino basin, a tributary basin of the Ramganga in the Lesser Himalaya. The village is located at the mid-spur at an elevation of 1400 m above sea level. It has a total area of about 221.60 acres and the total population was 242 persons in 1981. The forests cover encloses the village from three sides except in the west.

Mean monthly temperature (Jaurasi Station) ranges between 7° and 26° C which represents the coldest and the hottest months of the year respectively. Total annual rainfall is 955 mm (1981). The maximum rainfall occurs in the months of July and August. In winter (January and February) occasional snowfall is recorded.

The History of the area, coupled with the mode of occupation of the natives is an index of the degree of the ill-effects of human interference in changing the landscape and turning it into bold country (Pandey, 1976). Thus the natives are dependent directly or indirectly on the natural resources of their environment.

An assessment of the present status of farm forest fringe of the village and the degree of impact of human interference are the principal aims of the present study. The author himself was able to observe the dynamics of landuse patterns associated with a number of ecological changes that took place during the last fifteen years. However, efforts were directed to trace out happenings and find out data pertaining to the topic for the last eighty years. All these attributes are analysed for the present study which is of an exploratory character.

**Discussions and Results**

It has been observed that the early settlers of the village had occupied the fertile '*talaon*' land (Dutt, 1983) near and around the settlements. Later on, in due course of time, more and more land in excess of the previously occupied land was used for cultivation as shown by the new census. Further population of the village had increased due to the rapid mode of immigration that stressed upon the agricultural land, and thus the land under cultivation expanded simultaneously from the settlement sites to the fringe site. This process continued for many decades. Due to the extension of agricultural land and cultivable waste land, the forest cover in and around the village was degraded. As a result the stability of the village eco-system began to lose its entity. The data on the magnitude of landuse pattern during the last 80 years are furnished in Table 1.

**Table 1 :** Magnitude of Land utilization during the last 80 yrs.

| Use of land | 1900 Area in acre | 1935 Area in acre | per-centage | 1970 Area in acre | per-centage | 1980 Area in acre | per-centage |
|---|---|---|---|---|---|---|---|
| Cultivated land | NA | 148.25 | 66.9 | 164.75 | 74.35 | 166.65 | 75.20 |
| Cultivable Waste Land | NA | 23.35 | 10.76 | 25.15 | 11.35 | 12.80 | 5.78 |
| Area not available for cultivation | NA | 15.90 | 7.18 | 15.35 | 6.93 | 15.35 | 6.93 |
| Forest land | NA | 24.45 | 11.03 | 4.00 | 1.81 | 14.25 | 16.43 |
| Area under Settlements | NA | 6.15 | 2.78 | 7.35 | 3.32 | 7.55 | 3.41 |
| Other uses | NA | 3.00 | 1.35 | 5.00 | 2.26 | 5.00 | 2.25 |
| | | 221.60 | 100.00 | 221.60 | 100.00 | 221.60 | 100.00 |

*Source :* Revenue Department & Personal surveys

It is evident that prior to 1935 the are under cultivation was only 66.90 per cent and, in turn, it was increased to 74.35 per cent and 75.20 per cent respectively during 1971 and 1981. As such the magnitude of cultivated land was increased 11.13 per cent during a period of approximately 35 years (1935 to 1971). The percentages of cultivable waste land was 10.76, 11.35 and 5.78 respectively in 1935, 1971 and 1981. The rapid extention of agriculture coupled within consequential grazing adversely affected the forest cover, as a result the area of cultivable waste land was increased from 10.72 to 11.26 per cent during the period of 1935-1971. The area of cultivable waste land was decreased to about 49.11 per cent due to the forestation programme on the cultivable waste land during the 1971-81 decade. The area classified as not available for cultivation was 7.18, 6.93 and 6.93 per cent respectively in 1935, 1971 and 1981. This category of land is characterised by various rivulets, steep slopes, footpaths and bare boulders.

The land under forest cover occupied the second place in the study area and contained 11.03 per cent cover in 1935 and it was reduced to about 1.81 per cent in 1971. The data reveals that how the forest cover was decreased by 83.64 per cent within 35 years in the Kelani village. In 1971, the upper part of the village was replanted by the villagers and was kept reserved upto 1981, hence the area of forest land was increased (6.43 per cent).

The rapid growth of population and families is not only confined to food problems but also creates a number of environmental problems. The village Kelani, which was a small hamlet about 100 years ago, is now a sizeable village of the region.

**Table 2 :** Population and livestock growth

| *Year* | *No. of families* | *Growth in per cent* | *Population* | *Growth in per cent* | *Livestock population* | *Growth in per cent* |
|---|---|---|---|---|---|---|
| 1900 | 22 | — | 203 | — | 340 | — |
| 1935 | 26 | 18.18 | 219 | 7.88 | 350 | +7.6 |
| 1951 | 28 | 7.6 | 233 | 6.39 | 342 | -2.29 |
| 1961 | 30 | 7.14 | 245 | 5.15 | 335 | -2.05 |
| 1971 | 35 | 16.67 | 261 | 6.53 | 330 | -1.49 |
| 1981 | 42 | 20.00 | 274 | 4.98 | 320 | -3.03 |
| 1985 | 44 | 4.76 | 281 | 2.55 | 315 | -1.56 |

*Source* : Based on village records and personal surveys.

The village records show that the population and the number of families increased in every 10 years. After 1961 the growth rate increased because of the declined rate of mortality associated with disintegration of the joint family system in the village.

The population growth during the decade of 1971-81, was comparatively lower (4.31 per cent) due to the migration of families to other areas. The combined effect of population growth and disintegration of families is an important and alarming factor in modifying the environmental gradients in and around the settlements.

Table 2 also gives the livestock statistics for the village. In the year of 1900 the total livestock population was 305, it was found increasing (7.69 per cent) in the following 35 years due to traditional farming. However, after 1951 the growth rate were negative. This might be attributed to the attitudinal change in natives to cultivate fields to gain more crop yields, instead of keeping goats and sheeps. This view is supported by and based on the current field survey of 44 families. In the year 1985, the total livestock population was only 315, predominently oxen,. cows and buffaloes and the least of goats and sheep.

## Environmental Observations of Farm Forest Fringe

Land use pattern of the village plays an important role in

defining the farm forest fringe. Various attributed for this fringe are shown in Table 3.

**Table 3 :** Area and growth of fringe duringe last 80 years

*(Area under fringe in Acre)*

| *Year* | *Area of village* | *Area of Reserve forest* | *Total area* | *Growth in per cent* |
|---|---|---|---|---|
| 1900 | 25 | Nil | 25 | — |
| 1935 | 20 | 20 | 40 | 60 |
| 1971 | 22 | 30 | 52 | 30 |
| 1981 | 10 | 32 | 42 | 19.23 |

*Source :* Based in village records and personal surveys.

Table 3 shows that in 1900, the total area of the village, i.e., about 25 acres land, was under the fringe. In this period reserved forest line (tree line) was at the close proximity to the village and there was no reserve forest land under the fringe (Fig. 1). In 1935 the area of the farm forest fringe was expanded by 60 per cent of its previous location that is 40 acres of land. As such the 20 acres fringe was a part of the reserved forest.

During the next 35 years (1935-71), the area of farm forest fringe was increased further (30 per cent), however, the rate of expansion was lower than the previous record. This decreased rate of expansion was associated with the deforestation in the area. In 1970-71, the upper *Danda*[1] of the village was forested by natives which was previously classified as cultivable waste land (Community land). All this resulted in a decrease in the area of the farm forest fringe of estimated 42 acres in 1981. However, the anthropocentric activities were not only limited to the forest, they also affected the cultivated land of the village. With the increase in population and families the area of cultivated land was also increased towards the fringe which previously was under the waste land (pasture). As a result, the reserved forest began dwindling due to heavy biotic stress (over grazing and fuel) upon the forest community. This process still continues in the reserve forest. About 300 animals (cows, oxen, goats and sheep) are grazed at the fringe nearly 4 hours per day and 44 families bring fuel, leaves and fodder from this adjoining

1 Danda : The gentle slopy hill ridges locally known as Danda.

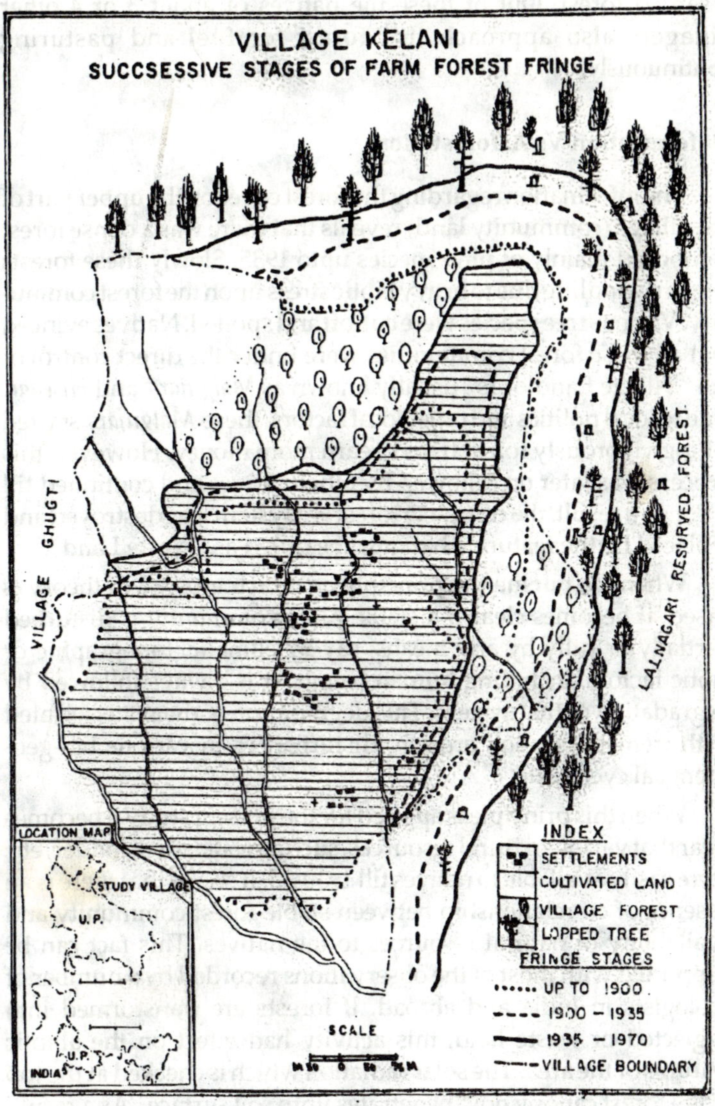
VILLAGE KELANI
SUCCSESSIVE STAGES OF FARM FOREST FRINGE
VILLAGE GHUGTI
LALNAGARI RESURVED FOREST
LOCATION MAP
STUDY VILLAGE
U.P.
U.P.
INDIA
SCALE
INDEX
SETTLEMENTS
CULTIVATED LAND
VILLAGE FOREST
LOPPED TREE
FRINGE STAGES
UP TO 1900
1900 – 1935
1935 – 1970
VILLAGE BOUNDARY

Fig. 1

reserved forest. Out of these the natives of about 3 or 4 other villagers also approach this forest for fuel and pasturing continuously.

**Deforestation Vs Afforestation**

The information regarding the forest cover on the upper part of the village (community land) reveals that there was a dense forest composed mainly of pine species upto 1935. Slowly these forests began dwindling due to heavy biotic stress upon the forest community. Various tree species were cut-off and spoiled. Natives evinced that all these forest communities were under the direct control of the "Village Landlords" usually known as *Malgujars* and *Hisedar.* Due to local politics and other local factors, these *Malgujars* started felling vigorously forest trees to earn more money. However, this process was later on followed by other natives and continued till 1935. As a result, the complete forest ecosystem was destroyed and replaced by the feature what must be called as "Waste Land".

When we pursue the principle on which ecosystem theory is based, it becomes clear if a stable plant community is disturbed partially or fully by any means, say by climatic, topographic or biotic factors, including human beings, it is always followed by degradation of that system. This degradation is always associated with tremendous soil erosion, disturbances in various bio-geo-chemical cycles, etc.

When this principle is applied for the present study it becomes clear that various natural resources, such as water, fuel, fodder, etc; were easily available to natives till 1930-35. It seems that there is an absence of co-relationship between stable forest community and availability of natural resources to the natives. This fact can be supported with most of the observations recorded by a number of ecologists in India and abroad. If forests are transformed into neglected or waste land, this activity had effect on the abiotic features of the area. The solar radiation which is checked at the top of the forest canopy now penetrates upto soil surface. As a result, the rate of avapotranspiration is increased which directly or indirecly effect the water table of the area and thus the compound effect appears in the form of minimizing the water content of springe, rivulets, etc. of the area. Further, when the stable community is destroyed, the water which is available in the form of rainfall is also decreased due to changes in the abiotic component. The above mentioned facts when observed in the present study state that the water content of springs and rivulets started reducing after 1935.

Such a correlation between stable community and availability of water is further supported by the fact that the upper Danda was replanted by various tree species during 1970-71. After approximately 10 years of this plantation when the plant species were acclamated, the water in springs and rivulets started increasing. As a result, now a days natives are able to get water from the springs and rivulets even in the dry season of the year, i.e., the summer season.

Before 1971, all the rivulets which originates from the upper Danda, were over flooded and loaded in the rainy season. These loaded streams were silted and destroyed the *Talaon* (irrigated) fields. After 1971-72, it is observed that the siltation process was reduced. In this way the gulley erosion, land slide and collapse of terraces also stopped in the downward cultivated land.

## Conclusions

The following conclusions may be drawn from the study :

(a) The agricultural land is extending towards the fringe owing to population pressure on agricultural activities.

(b) The earlier cultivable waste land is being converted into cultivated land, forcing pressure on the earlier land-use pattern.

(c) The expansion of agricultural land over cultivable waste land and the consequent replacement of forest fringe into cultivable waste land are detrimental to the reserved forest, which is already under additional pressure of grazing and fuel extraction.

(d) Aforestation in the denuded patches of land has controlled soil erosion, helped in raising the water table, countered decigation of springs and reduced the siltation process in cultivated land.

On the basis of this study and observation, the following recommendations are putforth.

(a) Scientific agricultural methods should be introduced in the village in order to set-up per acre out-put, in order to avoid further encroachment on to the forest fringe.

(b) The present cultivated waste land (rivulets, steep hill tops) should be planted with suitable fuel and fodder tree species (blue gum celtis etc.).

(c) The village forest should be restricted for grazing and fuel / timber extraction for 10 years.

(d) The reserved forest should not be exploited for revenue but should take the pressure of a bonafide controlled grazing and the fuel / timber needs of the village on a rotational forest compartment basis.

## References

Dutta, D., 1983. *Bino Basin : A Study of Landforms and their Utilization*. Unpublished D. Phil Thesis. Garhwal University, Srinagar (Garhwal).

Moddie, A.D., 1981. Himalayan Environment. In Lall, J.S. (ed), *The Himalaya : Aspect of Change*. Oxford University Press, Delhi.

Pandey, S.C., et. al., 1976. *The Environment Cenchrus Grazing lands in Western India. An Ecological Assessment*. Department of Biosciences, Saurashtra University, Rajkot, India, p. 384.

Rieger, H. C., 1981. Man versus Mountain : The Destruction of the Himalayan Ecosystem. In Lall, J. S. (ed), *The Himalaya : Aspect and Change*. Oxford University Press, Delhi.

# Chapter 22

# Agro-ecosystems Approach for Sustainable Production in Himalaya

*B.P. Ghildyal*

Based upon the principles of ecology and population dynamics, the concept of sustainable agriculture propounds that farming practices should be so designed as to maintain an optimal yield indefinitely. The particular biomass that produces this optimal yield defines the land's carrying capacity. Sustainability has long been an important precept for resource management in Agriculture, Animal husbandry and Forestry and through the millenia sustainable husbandary practices have been evolved, in India and especially in Himalayas. But eco-crisis in Himalaya is now well known and threatening the basic life support systems.

Global concern is also now focussed on sustainability due to causes and effects of deforestation, acid rain, eutrophication of water resources, ozone depletion and green house gases. Modern Agriculture is both a contributor to the erosion of sustainability and a victim of other environmental abuses caused by industry, rapid population growth and urbanisation. Within agriculture, we must distinguish the predominantly affluent, urbanized industrial countries from the predominantly poor, agricultural countries, the former are largely in temperate climates and the latter mostly in the tropics where crops, agricultural practices, and farming systems have evolved differently. In poor countries agriculture remains less mechanized and move vulnerable to the risks of uncertain climate. Natural resources largely determine the potential for modification of traditional practices. Compared to heavy dependence on fossil fuel and chemical's inputs by rich industrialized countries' agricultures, agriculture of poor countries generates less chemical and environment pollution.

Even in countries with traditional agriculture, deforestation, desertification, and failure to sustain even low levels of subsistence production is now common. Traditional farming systems that were designed to maintain relatively low population densities from their resource bases are failing to sustain the rapidly increasing human and animal population with those same resources and techniques. There is a gross lack of appropriate technology suited to the resources of the very poor and the more difficult ecological zones because researcher have also failed to build on the best of their traditional practices. Thus far ecology has played only a minor role in increasing this understanding because most agricultural research institutions are exceedingly weak in environmental and social sciences.

Unable to increase productivity from their limited resources, the poor hill farmers are driven to practices that amount to ecological suicide shortening restorative bush fallows in shifting cultivation systems, extending cultivation to forested areas with steep slopes that they need for fuelwood and forage, burning dung for fuel instead of using it to build soil fertility, planting annual crops on erosion–prone slopes and grazing more animals than natural range lands can support. For example, farming in the Uttar Pradesh hills, comprising of the regions of Kumaon and Garhwal is characterised by small and scattered land holdings. Nearly 75 per cent of the land holdings are less than 1 ha. On average, the size of holding is 0.38 ha. The striking fact about the Uttar Pradesh hills is the pressure of people on cultivated land. There are 6.84 people per ha of cultivated land in the U.P. hills. The population of the eight hill districts in Uttar Pradesh increased from 3.822 million in 1971 to 4.836 million in 1981, with a linear growth of 2.56 per cent per annum. Under the existing conditions of high population pressure and adverse topography, farmers are forced to extend cultivation to steep terrain to increase the food supply.

With the increase in the population over the years, forests are felled to accommodate larger cultivated area, and cultivaton extended to marginal and sub-marginal land over steep terrain having a slope of 30 degree or more. This process has aggravated the problem of soil erosion and large scale land slides are common.

Inventory of forest biomass also presents a dismal picture Landsat-1 images were visually interpreted for the assessment of forest cover in the Indian Central Himalaya. Forests have been classified into three status classes, viz., 'good', 'medium', and 'poor'. For each class, biomass ranges were standardized from aerial photointerpreted biomass maps, and biomass for entire area was computed (Kumaun and Garhwal regions had 34.1 per cent, 24.9 per cent in Kumaun, 4.1 per cent in Garhwal, and 4.4 per cent in whole of the Indian Central Himalaya.

Many areas are not suited to the introduction of green revolution, input-intensive technology. Although it has helped in producing more food in some states for increasing urban population–But even their, declining growth rates of productivity since the late 1970s have tempered early euphoria with doubts about sustainability.

New crop varieties with broad based adaptability and standardized packages of green revolution technology are not likely to have such widespread success in the future because the more favourable environments have largely been exploited. The Rice Tungro Virus in Southeast Asia is an example. Ecologists have discovered that the wide-adaptability of the prototypic semi-dwarf wheat and the rice varieties was somewhat illusory, these varieties did indeed spread over a large geographic area, but only in scattered zone with very similar ecological characteristics. Also the loss of genetic diversity increases risks.

Therefore agricultural technolgy must be more closely tailored to specific environments. Failure to understand population resource dynamics and the specific characteristics of local agricultural ecosystems has led to failures in development projects and research programs which are often planned from the top to down and do not adequately accounts for differences between national and local conditions and constraints. Indeed, experiments have shown that environmental differences and genotype-environment interactions influence the success or failure of crop and livestock, breeding programs more than genetic differences alone. Integrating information on environmental conditions and variability in time and space helps the scientists to set research priorities and to identify areas where the new varieties and improved technology can be best adopted. Ecological characterization with the improvements in weather patterns provides early warnings of possible serious crop losses. This will contribute greatly to sustainability, defined by some as 'the ability of a system to maintain its productivity when subjected to environment stress or shock.'

**Goals and Determinations of Sustainability**

A dictionary definition of sustainability—"keeping an effort, going continuously, the ability to last out and keep from falling"–suggests that agriculture would be sustainable if current levels of production could be maintained. However, sustainability should not be treated in static sense, but rather as a dynamic concept that reflects changing needs.

Thus to achieve sustainable agriculture in Himalaya resources must be successfully managed to satisfy changing human needs, to conserve natural resources and to maintain or enhance the quality of the environment. The goal of sustainable agriculture therefore, should be to maintain agricultural production at levels necessary to meet the increasing needs and aspirations of an expanding population without degrading the environment.

A distinction should be made between sustainability and productivity. While greater productivity will be required to achieve sustainability goals, that productivity must be achieved in such a manner as not jeopardize the ability of agriculture to meet the future needs. More explicitly, productivity goals often may be achieved through unsustainable, short-term approaches. Thus, efforts to achieve sustainability must take into account long-term implications and needs.

The sustainability of agricultural production is determined by a complex interaction of biological, physical and socio-economic factors that constitute the basis of all production systems. Although diverse in character, agricultural production systems are inherently neither sustainable not unsustainable. They respond to changes that either enhance or endanger sustainability depending on the nature of the change in relation to the prevailing environment and socio-economic circumstances. If an agricultural system fails to respond to change, however it is unlikely to be sustainable.

The determinants of sustainability might be classified into three primary groupings, biological, physical and socio-economic and legal. The factors that affect sustainability are : Availability and use of plant and animal genetic resources; Crop pests; Animal health and nutrition ; Soil water and nutrients; Air quality; Hazardous chemicals ; Energy resources ; Economic policies ; Infrastructure and markets; Inputs and credit; Institutions of research extension and education and Land tenure and laws and regulations.

**Improving Sustainability with Agro-Ecosystems : An Analysis**

With good land and cheap labour, poor farmers of India do not readily adopt green revolution technology that increases yield per hectare because the inputs are expensive. Only when overpopulation pressure pinches and new land can not be brought in to cultivation easily is there incentive to shift to yield–increasing methods.

Poor and marginal farmers still cling to their traditional cultivation practices (even where rising population pressures–dictate a change) because they lack resources and because, thus far, the new techniques cannot sustain continuous annual crops in the fragile soils of the hill. Targeting research to increase yields in zones with population pressures and to increase stress tolerance in crops for areas where land remains abundant is a sensible strategy. Agro-Ecosystems characterisation may overcome these problems by matching particular agro-ecological approaches to certain ecological and socio-economic constraints, but clearly both factors must be taken into account.

There are also many indications that the rapid rate of growth in food production resulting from the green revolution during the late 1960s' and 1970s' cannot be maintained. Moreover, in both industrialized and developing nations, many natural resources on which agriculture depends have deteriorated (water logging salinity) could make it increasingly difficult to maintain current levels of production, to say nothing of improving productivity to meet growing needs.

The next few decades present a greater challenge to the world's food systems than they may ever face again. The effort needed to increase production in pace with an unprecedented increase in demand, while retaining the essential ecological integrity of food system, is colossal both in its magnitude and complexity. Given the obstacles to be overcome, most of them man-made, it can fail more easily than it can succeed.

In the future ecology, climatology, geography, eco-physiology and physical edaphology and other resource based disciplines must play greater roles in sustainable agricultural research and development. What has been lacking in the past is a conceptual framework to enable them to interact effectively and contribute to the progress toward sustainable agriculture. Agro-ecological characterization provides this framework. The potential value of agro-ecological characterization as an operational tool for planning sustainable agricultural development as a framework for integrating the expertise of different disciplines, is clear from its wide range of uses.

Mapping the current distribution on ecozones and assessing their suitability for specific crops and cultivars are the two important components of planning. Rice, the crop that feeds the largest number of people is grown in at least five kinds of macro-environments or agro-ecosystems, irrigated, rainfed lowland, rainfed upland, swamp, and deep water flooded areas. Seventy five per-cent of the world's rice is currently produced in the more fertile areas by only 25 per cent of the world's rice farmers. Within major rice environments and especially in non-irrigated land, many local environmental–variables affects choice of varieties and production techniques. The challenge is to fit improved production techniques to these local situations.

**Agro-Ecosystem Analysis**

Agro-ecosystems Analysis involves essentially two major steps.

A. Agro-Ecosystems Mapping

B. Systems Diagnosis

**Village Agro-Ecosystems Mapping and Transect Analysis**

Agro-ecological mapping involves several disciplines with members of both the genders taking part. Through a village transect problem, opportunities are identified that future research efforts could address. It is also possible to recognize opportunities for different types of appropriate research and technology to address the problems of different social groups within a village.

Agro-ecosystem mapping involves the following maps.

1. Map of topography and hydrology
2. Map of enterprises
3. Map of social groups
4. Transect of Agro-ecological zones

**Topography and hydrology** : This map is made by eliciting every land type that can distinguish and ascertaining their spatial distribution on a base map of roads, homes, orchards, and other landmarks (temples, water sources etc.,) (Fig. 1).

After establishing land-type boundaries the farmers are then requested to indicate flood and drainage directions and main water sources.

**Enterprises** : This map is made by asking farmers what enterprises they conduct on each land type (Fig. 2).

Major enterprises are typically crops, animals, fodder and fuel plantation, and orchards. Special attention should be given to agricultural and non-agricultural enterprises in the busy housing areas.

One must also be sensitive to gender differentiation in enterprises.

**Social groups** : This map is made by asking farmers to name the castes or social groups living in the village (Fig. 3).

The spatial distribution of houses by caste is then sketched.

Caste access to land areas and other important resources should then be demarcated.

**Transect of village** : This agro-ecological zones transect is constructed from a composite section through every land type (Fig. 4).

For each land type the local name, soil type, crops, trees, livestock, problems and opportunities are listed from information provided by the farmers.

### System Diagnosis and preparation of Systems Diagram

Agro-ecosystems Analysis is done through village mapping and transect analysis, ranking the problems identified and practice

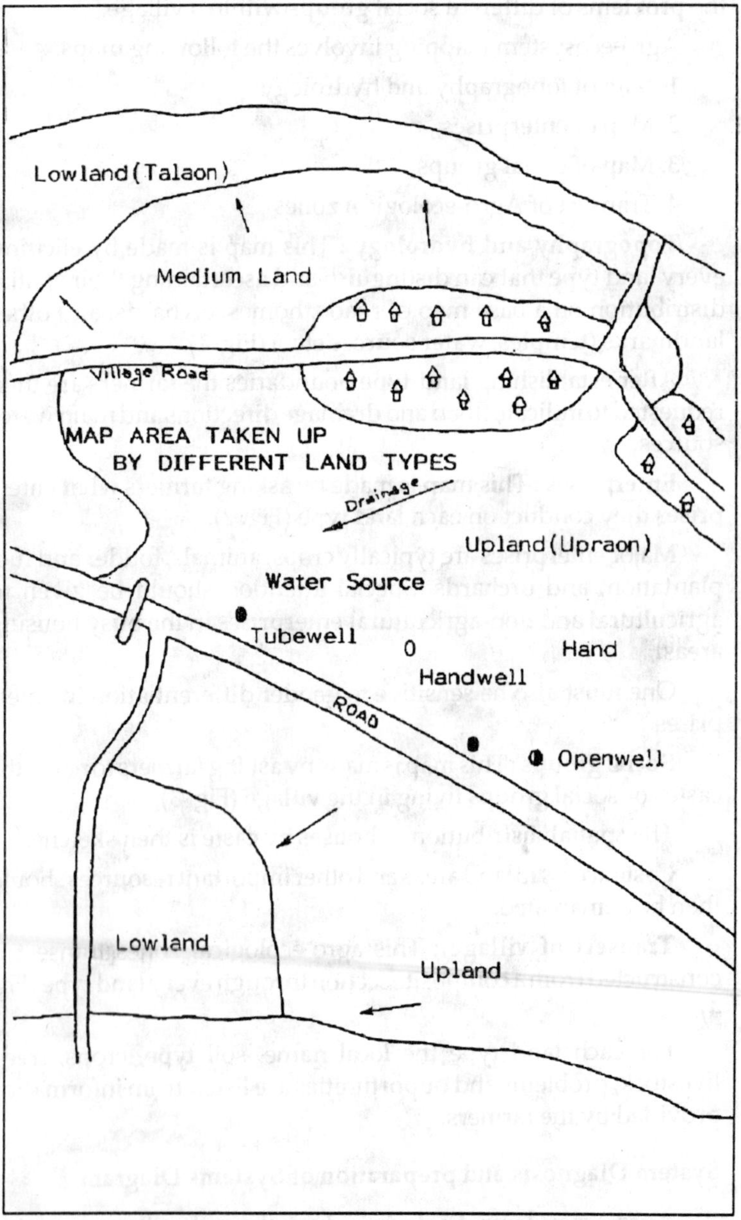


**Fig. 1.: Topography and hydrology map**

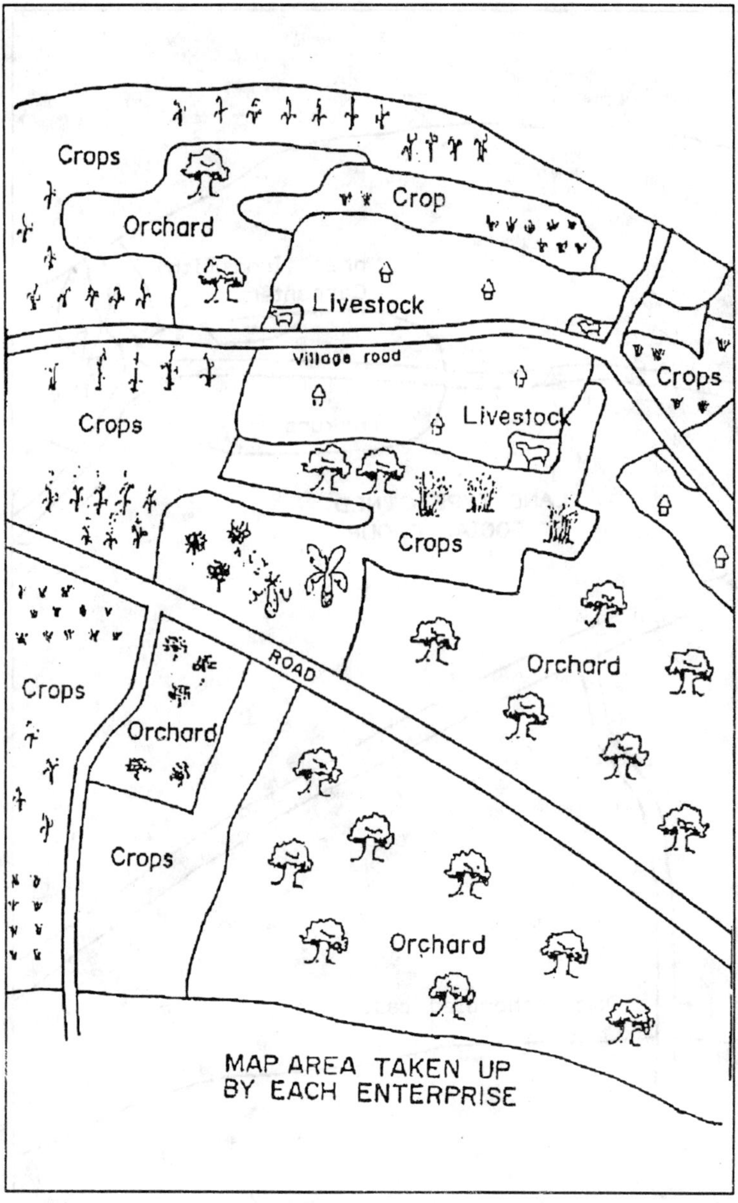


Fig. 2. : Enterprise map

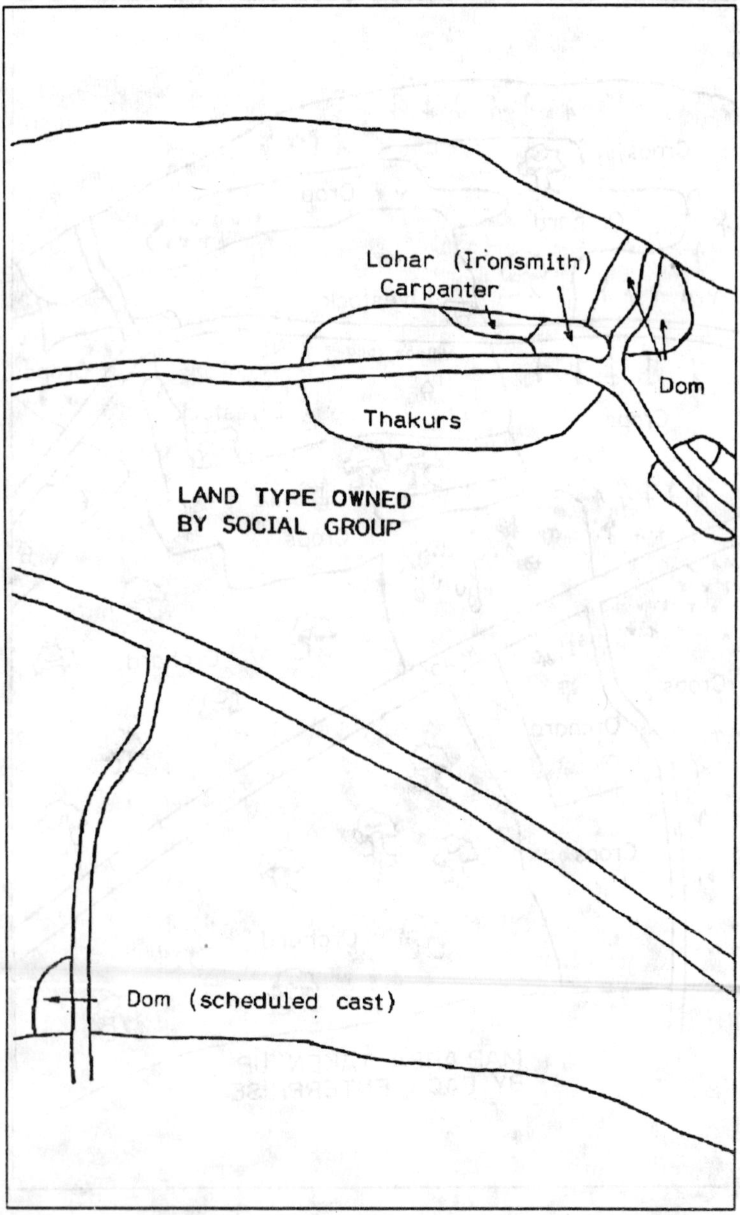


**Fig. 3.** : Social map

| LAND TYPE | UPLAND | MIDLAND | LOWLAND |
|---|---|---|---|
| SOIL | | Silty loam | |
| CROPS | | Millets<br>Pulses | Rice |
| TREES | Mango<br>Citrus | | |
| LIVE-STOCK & FISH | Bullocks<br>Goats<br>Catte | | |
| PROB-LEM | Soil Erosion | | Dainage |
| OPPOR-TUNITIES | Crop storage facilities | High quality seeds | |

**Fig. 4.** : Transect

systems diagnosis and systems diagramming with farmers. This exercise will enable the scientists to identify research and extension opportunities and develop experimental programmes. These terms can be defined as follows.

1. Diagnosis
2. System Diagnosis
3. Systems diagram

1. Diagnosis is a series of consultations and measurement activities to obtain an adequate understanding of farmers' problems so that test solutions can be designed.
2. Systems diagnosis is the process by which farmers draw out or diagram the interactions between biophysical causes and socio-economic constraints to the farmers' problems.
3. A systems diagram is a simplified pictorial representation of how biophysical causes of the farmers' problem are linked together and how the socioeconomic constraints interact and give rise to the farmers' problem.

Agro-ecosystem analysis provides an ecology-based framework for understanding farming systems and identifying problems by land type.

Diagnosis provides sufficient understanding of the farmers' problem and how it interacts with system components so that solutions can be designed.

Problems can be ranked considering the number of farmers affected, the importance of the enterprise involved, and the losses in production and income entailed.

A systems diagram is a tool for analysing relationships among factors affecting a problem, ordering the causes and effects of problems, and providing ideas for intervention points where specific technologies may solve the problem.

The whole process consists of following steps :

**1. Agro-ecosystem analysis**

The problem to be diagnosed is identified through agro-ecosystem analysis. Maps and transects of land types and enterprises are prepared. Problems are listed at the bottom of transect. As there are, of course, many problems a procedure for ranking them should be followed.

**2. Ranking problems**

Problems are ranked using three criteria (Fig. 5)

1. The distribution of the problem.
2. The importance of the particular enterprise to the farming system.
3. The loss of production or income for which the problem is responsible.

| Problem | Distribution of problem | Importance of enterprise | Seriousness of problem | Relative importance of problem |
|---|---|---|---|---|
| Upland | | | | |
| Weed | XXX | XX | XX | 1 |
| Soil erosion | XXX | XX | X | 2 |
| Pest | XX | X | X | 3 |
| | place farmer's score of importance | | | calculate rank |
| Land type 2 | | | | |
| Kidn of problem | | | | |

Fig. 5. : Table of ranked problems

### 3. Informal interview

When conducting interviews the following points should be kept in mind.

1. A target group of farmers should be identified.
2. Where possible, respondents should be encouraged to draw out their problem tree (Fig. 6) on the ground.
3. The order of questions is determined by the flow of conservation and not by the guide topic listings.
4. Questions are best asked in the field because farmers answers are more detailed when they can show what is happening.
5. Questions to guide researchers as they seek a greater understanding of the problem are developed using Rapid Rural Appraisal techniques. Typically, questions cover (a) farm typology (b) descriptions of processes (c) rationale or difficulties and (d) estimations of key economic and biological parameters.

### 4. Systems diagramming

A systems diagram (Fig. 7) is constructed by

1. Placing the farmer's problem in the centre.
2. Assigning each primary biophysical cause of that problem to a box and linking that to the right-hand side of the central box.
3. Assigning each secondary cause a box and linking that to with the appropriate primary cause.
4. Following the same procedure for each socio-economic constraint.
5. Arranging primary and secondary causes and constraints into a circle surrounding the central problem with bio-physical causes on the right-hand side and socio-economic constraints on the left. The size of each segment is determined by the number of responses.

### 5. Using the system diagram

When properly constructed, the outer circle of bio-physical causes presents farmers and researchers with opportunities to

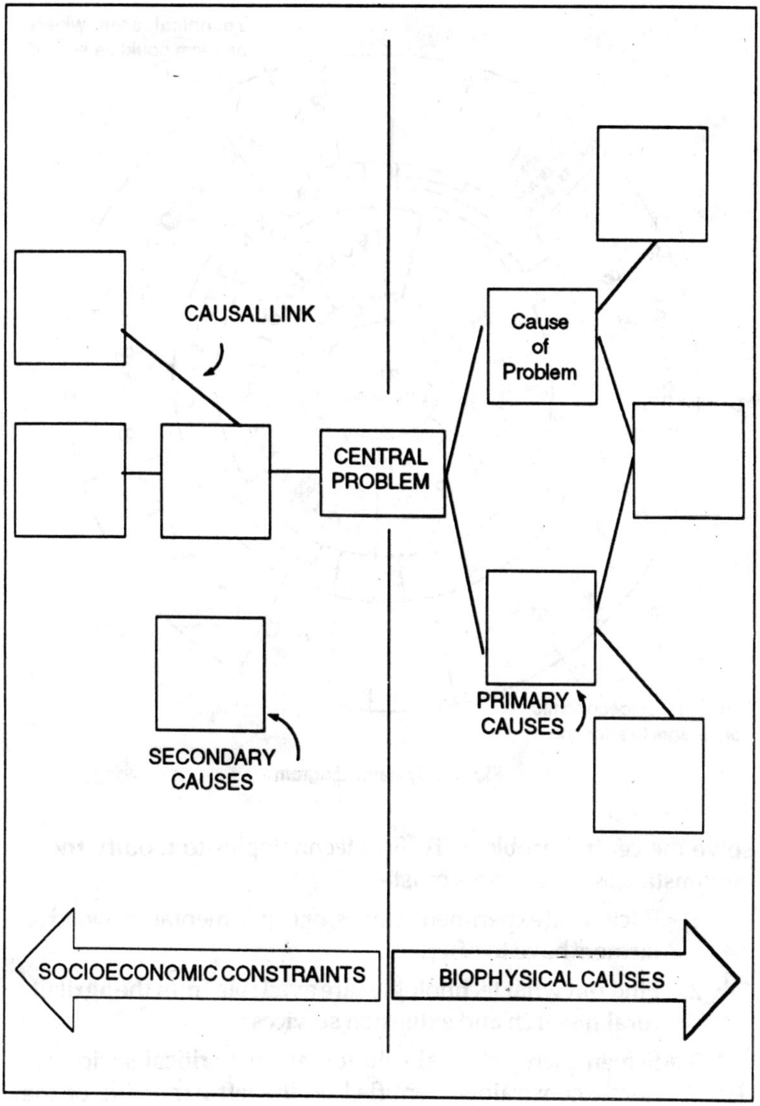


**Fig. 6. : Farmers' problem tree**

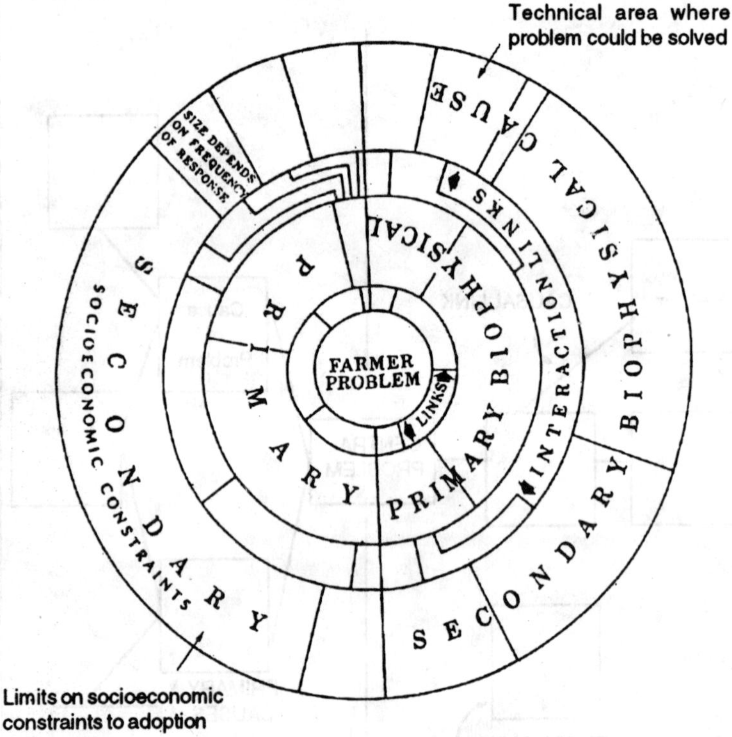


Fig. 7. : Systems diagram

solve the central problem. To find technologies to modify those circumstances, researchers must—

1. Elicit what experiments, ideas, or experimental knowledge farmers have to offer.
2. Find out what technologies are available from the agricultural research and extension services.
3. Screen each potential solution against critical socio-economic constraints identified on the left-hand side of the systems diagram.
4. Hold group meetings to discuss which ideas farmers would like to test on their fields, which leads farmers and researchers straight into experimental design.

# Chapter 23

# Environment and Development with Special Reference to Forest

*H.C. Upadhyay*

The problems related to environment received Government attention in a planned manner with the beginning of the Fourth Five Year Plan which talked of the necessity to introduce environmental dimension into the development planning process. In 1972, a committee on Environmental and Coordination was set up to look into the environmental problems and suggest solution in consultation with the experts and the concerned Ministries / Departments of the Government of India. Another committee was set up in January 1980 for reviewing the existing legislative measures, administrative machinery for ensuring environmental promotion and for recommending ways to strengthen them. On the recommendation of this high powered committee, the Department of Environment was set up by the Government in 1980, to serve as the focal point in the administrative structure of the Central Government for planning, promotion and coordination of environmental programmes. In 1985 a new Ministry of Environment and Forest with Department of Environment, Forests and Wild Life was set up. It serves as the focal point in the administrative structure of the Central Government for planning, promotion and coordination of environmental, forestry and wild life programmes. The Department besides monitoring the protection of environment, also looks after the environmental protection of coastal water, mangroves, coral reefs and wetlands, environmental appraisal of development projects; eco-regeneration and development; assessment, survey and conservation of biotic and abiotic resources, environmental research, education, awareness and information; pollution control and monitoring; checking destruction of forests; restricting division of forest lands to non-forestry uses; developing a rational forest policy, social farm forestry schemes and protecting and preserving wildlife (India, 1987).

Environment has a far reaching influences on the material, culture and general economic life of a community. It promotes the growth of some particular activities while on others, it sets great limitations difficult to be overcome by the community (Swan, 1952). Environment refers all situations, conditions, forces, processes and elements whether natural or artificial which surround, affect or control the biotic world or any of its constituents or subject under consideration (Singh, 1989).

It is true that most of the elements of physical environment are capable of only slight modifications by man. Nevertheless, even this slight control that he has over nature means a great deal towards high progress. For example, man knows very little about the the atmosphere and can control atmosphere conditions even less. But he has studied some of the law that apply to it. Man's material progress is largely determined by his success in adjusting his actions to his physical environment. According to Alfred C. Reed, "Human achievement is consummated by defying and conquering the forces or laws of nature. Invariably it is won by understanding and utilising those law." (Science Monthly, 1927). The natural environment affects large groups of people very directly and in a primarily manner. "Every community, tribe, state, nation and empire on earth is affected by it directly, vigorously and persistently. No major activity of human society is independent of its helps, hindrances, or directives. The natural environment does to human society what the social environment does to the human individuals." (Science Monthly, 1927).

**Protection of Environment**

There are about 30 major enactments related to protection of environment now being administered by the Central and State Governments. The Water (Prevention and Control of Pollution) Act, 1974 and the Air (Prevention and Control of Pollution) Act, 1981, the Factories Act, the Insecticides Act, are some of the prominent ones among these enactments. These acts are implemented through several organizations like the Central and State Pollution Control Boards, Chief Inspectors of Factories and Insecticides Inspectors of Agriculture Department. In spite of several organizations functioning in the field, industrial accidents have led to widespread concern regarding the danger to environment from hazardous substances. The existing laws to cover such hazards do not provide a clear focus of authority to prevent or limit the damages. Taking into account these factors, the Government of India has enacted the Environment (Protection) Act, 1986 to provide a single focus for environmental issues in the country and to plug loopholes in the existing acts. This act has been brought into force with effect from 19 November, 1986.

The important features of this act are : (a) it confers powers on the Central Government to : (i) take all necessary measures for protecting the quality of environment ; (ii) coordinate the actions of State Governments, officers and other authorities under this act or under any other law which is related to the objects of this act; (iii) plan and execute a nation-wide programme for the prevention, control and abatement of environmental pollution; (iv) lay down standards for discharge of environmental pollutions ; (v) empower any person to enter, inspect, take samples and test; (vi) establish or recognise environmental laboratories; (vii) appoint or recognise government analysis; (viii) lay down standards for the quality of environment; (ix) restrict areas in which any industries, operations on processes may not be carried out or shall be subject to certain safeguards; (x) lay down safeguards for prevention of accidents and take remedial measures in case of such accidents; (xi) lay down procedures and safeguards for handling hazardous substances; (xii) constitute an authority or authorities for exercising its powers; (xiii) issue directions to any person, officer or authority including the power to direct closure, prohibition or regulation of any industry, operation or process or process or stoppage or regulation on supply of electricity, water or any other service; (xiv) require any person, officer, state government or authority to furnish any prescribed information; and (xv) delegated powers to any officer of state government or authority; (b) it confers powers on persons to complain to the courts regarding any violation of the provisions of the Act, after a notice of 60 days to the prescribed authorities, (c) the Act makes it obligatory for the person incharge of a place to inform the prescribed authorities regarding any discharge or apprehended discharge of any pollutants in excess of standards by accident and the authorities, on receipt of such information or otherwise, shall cause remedial measures taken to prevent or mitigate the pollution caused by such accidents and the expenses incurred by the authorities in respect to remedial measures or a recoverable with interest from the polluter. (d) the Act prescribed stringent penalties for violation of the provisions of the Act. No distinction is shown between government departments and other companies. (e) The jurisdiction of civil courts is barred under the Act.

## Environment Impact Assessment

The two major criteria—the project should maximise economic returns; and it should be technically feasible—are no longer considered adequate to decide the desirability or even the viability of the project. It is now widely recognised that the development effort may frequently produce not only sought for benefits, but other-often unanticipated-undesirable consequences as well which may nullify the socio-economic benefits for which the project is designed. It is, therefore, considered imperative that the environmental issues are anticipated and incorporated in the development project at the project formulation stage itself for appropriate decisions on the aspects of (i) sitting of the project, (ii) choice of technology and, (iii) choice of preventive and control measures to minimise environmental damage.

## Impact Assessment Procedure

The impact assessment of the development projects is carried out by the Department with the help of inter-Ministerial Appraisal Committees comprising of experts from related disciplines. The feasibility reports along with Environmental Management Plan and or Environmental Impact Statement furnished by the project authorities are scrutinised by these Expert Groups and additional information, as and when required, is collected through field visits of specially constituted Expert Groups. To help the project authorities, the Department of Environment has developed guidelines and questionnaires indicating clearly the environmental issues to be incorporated in the different projects.

Projects approved by the Department of Environment often require implementation of suggested safeguards and mitigative measures for which an effective monitoring mechanism is required. The Department of Environment and its associated agencies like the Botanical Survey of India and the Zoological Survey of India with their network of regional stations provide all necessary help to the project authorities during the project formulation, implementation and monitoring stages.

**Forests**

Forests are renewable resources and have contributed substantially to the economic development of the country by providing goods and service. Forests also have a major role in enhancing the quality of environment.

The country has an area of 747.2 lakh hectares notified as forests. Of which, 397.8 hectares are classified as reserved, 216.5 lakh hectares as protected. Unclassified forests area is spread over 86.7 lakh hectares and 46.3 hectares classified as others. The optimum forest cover proportionate to the geographical area is 60 per cent in the hills and 20 per cent in the plains. As per data obtained from the study of satellite imagery, the extent of forest cover in the country for the period 1981-83 was 648.7 lakh hectares.

**Forest Policy and Law**

India is one of the few countries which has Forest Policy since 1894. The resolution of the Forest Policy stated (i) the sole object to which the management of forests is to be dedicated is to promote the general well being of the country; (ii) the maintenance of adequate forest is dictated primarily for the preservation of the climatic and physical conditions of the country and secondly to fulfil the needs of the people; (iii) permanent cultivation should come before forestry; (iv) satisfaction of needs of the local population at non-competitive rates, if not free, should over ride all considerations of revenues; and (v) after fulfilment of the above conditions the realisation of maximum revenue should be the guiding factor.

The Forest Policy was revised in 1952. It laid down the basic principles for proper management of the forests of the country to derive from them maximum benefits, direct or indirect. These are (i) the classification of forests on a functional basis viz. protection of forests and village forests; (ii) establishment of tree lands whenever possible for the amelioration of the physical and climatic conditions and for promoting the general well being of the people; (iii) ensuring increased supplies of fodder, timber and firewood and to release cattle dung for use as manure; (iv) opposition to indiscriminate extension of agriculture by the excision of forests, as this not only deprives the local population of wood, grass, etc. but also deprives the land of its natural defences against dust, storms, hot winds and erosion; and (v) stipulation that India as a whole should aim at maintaining one third of its total area under forests, the proportion in the hills and plains being 60 per cent and 20 per cent respectively.

Since the enforcement of the National Forest Policy in 1952, developments of far reaching importance have taken place in the economic, political and social fields. There is a much greater realisation today that the tribals and fringe population have to be taken into confidence by making them participate in developmental plans and management decisions to build up a buffer of determined people committed to the cause of forest conservation.

The revised National Forest Policy on the basis of the experience gained in the intervening years, shall have as its basic objective, the maintenance of environmental stability and ecological balance. Revenue earning motive shall be subjected to the primary objective. Recognising the urgent need to halt the dangerous trend where the forests have reached a stage of regression very near, the irreversible threshold, the Department is giving a new orientation to see entire gamut of forests related activities. Some of the important areas for immediate action for effective protection of forests include afforestation and development of wastelands aforestation and replantation in existing forests, strengthening of forest survey, forest settlement, systematic and thorough inspection by forest officials, restriction of grazing, supply of other kinds of fuel, control on economic exploitation of timber, etc.

A number of programmes and projects have been initiated for implementation of the objectives. Afforestation, social forestry and farm forestry are among the significant components of the new 20 Point Programme. An area of 17.62 lakh hectares has been taken up for afforestation during 1986-87.

**Forest Conservation**

Increasing destruction and degradation of forests and tree-lands in the country, especially in the Himalayas and other hill areas, is leading to heavy erosion of top soil, erratic rainfall and recurring floods, besides causing acute shortage of firewood and what is more important, the loss of productivity due to the eroded and degraded land. It is in this perspective that the Forest (Conservation) Act, 1980 was enacted, primarily to check indiscriminate deforestation / diversion of forest lands for non-forestry purposes. The Act provides that no forest land will be dereserved or diverted for non-forest use without approval of the Central Government. Prior to promulgation of this Act, the rate of diversion was 1.5 lakh hectare per annum during the year 1951-80. This rate of diversion has come down to 5,500 hectare per annum during the year after the enactment.

Degradation of forests due to biotic pressure has been another area of concern. Therefore, guidelines have been framed for preparation of working plans and felling in forests. Some of the salient features are : (i) working plans should be up-to-date and stress conservation; (ii) preliminary working plan should have a multi-disciplinary approach; (iii) tribal rights and concessions should be highlighted along with control mechanisms; (iv) grazing situation should be studied in detail and specific prescriptions should cover fodder propagation; (v) shifting cultivation and encroachments need to be controlled; (vi) clear-felling with artificial regeneration should be avoided as far as possible and clear-felling blocks should not exceed 10 hectare in hills and 25 hectares in plains; and (vii) banning all felling above, 1,000 meters altitude for a new year should be considered to allow these areas to recover. Critical areas in hills and catchment, prone to landslips, erosion, etc. should be totally protected and quickly afforested.

Some of the other steps taken for preservation of forests have been as : intensification of action under the Indian Forest Act; states are taking steps to eliminate the agency of contractors in working of forests; smokeless chulha, non-conventional energy sources, such as, biogas, solar cookers, water-heaters, etc., are being popularised, grazing, encroachment, shifting cultivation and the possibility of substitution of wood by other products are being studied to find long term solutions. Townships containing population more than 5,000 in the vicinity of forests have been identified and states and Ministry of Petroleum and Natural Gas has been requested to supply Liquified Petroleum Gas to those areas on a priority basis to bring down the pressure on fuelwood forests.

**Environmental Research Programme**

The main functions of this programme are to develop strategies for sustainable resource management and create facilities and trained manpower to conduct environmental research in the country. Major programmes include (a) integrated Action Oriented Eco-development Research Programmes on Western Ghats, Eastern Ghats, Himalayan Region and River Ganga; and (b) All India Coordinated Research Projects on Ethnobiology, Air Pollution and Plants, Conservation of Endangered Plant Species, Wetlands and Mangroves. Apart from the above, a number of smaller location-specific research projects have been initiated. Two research Committees namely, the Indian National Man and the Biosphere Committee and Environment Research Committee make recommendations on priority areas for environmental research, scrutinise research proposals received from various research institutions, evaluate the progress and recommend suitable means for implementation of research projects.

On the basis of a review research programme undertaken in the past and an assessment of future requirements, a profile identifying the priority areas of research has been drawn up under five fundamental elements viz., land, water, air, space and energy with two basic approaches for study viz., the ecosystem management and the ecological aspects. The ecosystem include the mountains, forests and lands, mangroves, islands and coastal areas, arid and semi-arid zones, rivers and river basins. Similarly the management and technological aspects include monitoring and control, development process systems and instruments, health and toxicology, impact assessment and management. Creation of 'Centres of Excellence' was initiated to strengthen the country's environmental research and training base. Three such centres have been set up so far. The first centre is at Indian Institute of Sciences, Bangalore whose focus is on Western Ghats and air and water pollution. The second centre at Ahmadabad focussed on environmental eduction and is engaged in development of educational material. The third centre is at Indian School of Mines, Dhanbad whose main area of activity is the study of mining environment.

## FOREST RESEARCH

Forestry Research is mainly the responsibility of the Forests Research Institute, Dehradun and its regions stations. The Indian Council of Forestry Research and Education has been set up to look after the forestry research. New sets of research priorities grouped appropriately under basic research, technology and application have been identified. It has also been decided to revamp research set up by upgrading the existing regional research centres into fulfledged Forestry Research Institutes to cater to the needs of different ecogeographic regions of the country.

At present, in addition to the existing research institute at Dehradun, five new research institutes have been set up namely (i) Institute of Arid Zone Forestry Research, Jodhpur, (ii) Institute of Northern Peninsular Deciduous Forests, Jabalpur, (iii) Institute of North Eastern Moist and Wet Evergreen Forest, Jorhat; (iv) Institute of Wood Science and Technology, Bangelore; and (v) Institute of Tree Breeding and Genetics, Coimbatore. These Institutes will conduct national level research on a particular facet of forestry while taking care a regional research needs (India, 1987).

## References

*India*, 1987.

Swan, L.W., 1952. Some Environmental conditions influencing life of high Altitudes. *Ecology*, 33 : 109-117.

Singh, O.P., 1989 *Environmental Development; Dimensions of Development Planning*, Vol. I, New Delhi : 133-144.

*Science Monthly*, 1927.

*India*, 1987.

## Chapter 24

# Changing Perception of Wildlife Tourism and Its Impact on Carrying Capacity in the Corbett Tiger Reserve

*Bhagwati Joshi*

During recent years, wildlife tourism has become widely popular in India. Consequently, wild-life has added new dimensions to tourism industry, through exposing untouched and virgin wilderness areas to large numbers of visitors, originating from places of different cultures, economies, attitudes and perceptions of nature. With the emergence of National Parks and sanctuaries, people have become rather crazy to visit nature preserves. Thus while the increasing popularity of wildlife tourism has created vast potentials for the development of tourism and allied activities, at the same time has paved the way to continuous degradation of the fragile and delicate ecosystems of the protected areas by exposing these to unwanted human interference and through the activities which ultimately prove detrimental to the interest of the wildlife in the park areas and also are contrary to the well perceived objectives of wildlife conservation and protection.

The philosophy behind declaring an area as a nature reserve faced the test of time only after the advent of tourism as an industry. This, on one hand provided an economic support to the idea and on the other gave birth to many such problems, which were completely unforeseen. Concentrated public use of National Parks and

Sanctuaries is initiating ecological processes which are far more destructive than are changes due to other processes such as cattle grazing etc. (Darling and Eichhorn, 1967). The transport and other infrastructural support services which have altogether provided increased opportunity for recreation of nature preserves has been causing disastrous degradation of the flora and fauna within the parks and their surrounding areas through clearing forests and allowing concentration of settlements and services.

In India, wildlife tourism has received undersirable publicity through different media during recent years which has highly encouraged the common people to enjoy the wilderness of nature preserves in all possible ways. However, realisation is growing gradually that if not properly managed and directed, people may destroy the very values, the National parks were intended to preserve, and the impact of thousands of visitors on soils, plants and animals can be destructive, far reaching and even irreversible.

Nature preserves are perhaps the only places where no destruction of unknown values takes place or atleast such destructions is theoretically avoided. Thus the fact is gaining ground all over the country that National Parks and Sanctuaries are not only of aesthetic, recreational or conservational values but also are indispensable reserves of biological diversity to be used for reacting man's ecological steps. Such reacting is essential to determine the viability limits of various ecosystems. In the absence of the scientific knowledge of these limits of resilience of natural world, we might disturb the nature, though unknowingly but in such a ways that irrepairable damage is caused.

Thus, the changing concept of tourism has resulted into great degree of conflicts among resource managers, administration, resource users, environmentalists and more particularly between tourists and the parks authorities. The major conflict involves in the reconciliation of short term and often contradictory human interests with the interests of fauna and flora. This conflict of interests is getting more and more acute and visible with the tremendous increase of population in urban centres, popularization of nature preserves for the growth of tourism and with the availability of increased scientific knowledge of ecological processes and system. Due to gaps in policy making and implementation, and also owing to lack of desired co-operation and co-ordination among various agencies and departments entrusted with tasks of tourism development and wildlife conservation, Government is spending crores of rupees on ambitious wildlife conservation projects on the one hand and paradoxically also exposing these protected areas to large population of visitors through publicity and creation of varying range of services, amenities and infrastructure, within the parks and sanctuaries as well as in the surrounding areas. This clearly sounds contrary to the objectives of wildlife conservation and protection for which the National Parks and sanctuaries have been created.

A full and long term resolution of such conflicts, however will require large-scale changes in the wildlife management practices and high level of coordination among different agencies in policy planning and these two within the framework of the present day socio-economic and political situations and pressures. However, it can be argued that the wildlife conservation is a luxury compared to the conservation of other essential bio-physical resources. But wildlife conservation is not incompatible with these resources, rather it enhances them. Wildlife areas are the best protected natural areas and therefore they have greater catchment capabilities.

Natural forests have a greater catchment capability than exploited or plantation forests, or plantation agriculture and horticulture. The wildlife conservation, therefore, may have greater and wider acceptance if these material advantages area stressed instead of highlighting the protection and development of wild animals directly. The advantages of wildlife conservation are not easy to explain to many segments of society. Likewise the potential values of genetic resources, scientific understanding and ecological reference centres are difficult to grasp. The intangible values of culture, heritage, aesthetics, and rights of all creatures to exist cannot be used for people crazy for wilderness and for those who are employed in tourist services.

**The Corbett Tiger Reserve**

The Corbett Tiger Reserve, popularly known as the Corbett National Park, is situated along the Himalaya foothills, in the northern part of the state of Uttar Pradesh. Geographically it is located between the two sub-Himalayan ranges of the Lesser

Himalaya in the north and the Siwaliks in the south, and encompasses the extensive flat floor of south Patti Doon. The preserve has its spatial extension between 29°25′ and 29°45′ North Latitudes and 78°45′ and 79°5′ East Longitudes with a total geographical land surface of 520 $km^2$. Administratively it falls in the Himalayan districts of Pauri Garhwal and Nainital (Fig.1). Ramnagar in district Nainital is the headquarters of the Corbett Tiger Reserve and in an obligatory point to reach the Park. The Park was created back in 1936 as the first National Park of the country simultaneously with the enforcement of United Province National Park Act passed in 1935; and named as Hailey National Park, after Sir Malcolm Hailey, the Governer of the State. In 1956 the preserve was finally renamed as the Corbett National Park as a homage to Jim Corbett, the famous author and wildlife conservationist who spent a large part of his life in this region and made special contribution to the creation of the preserve. Wildlife conservation in India touched a big landmark in April 1973, when the most prestigious and the biggest project, the 'Project Tiger' was launched and the Corbett National Park was declared as the first Tiger Reserve of the country.

The Corbett Tiger Reserve by virtue of its bounded location between two sub-Himalayan ranges, absence of any pocket of human habitation within its boundaries and its comparative insulation from external interferences is very rich in biological diversity. Besides, the geological and geomorphological uniqueness of area has rendered this reserve as an ideal habitat for a variety of fauna and numerous species of trees, grasses, shrubs, herbs, etc. The Reserve consists of some of the finest moist-Bhabar-Doon Sal and moist Siwalik sal forests. The area is interspread with hill ranges and valleys drained by small rivers. The Savannah woodland, bamboo brakes, dry deciduous forests and sissoo-khair forests provide a variety of natural vegetation and as well as the wildlife habitats for different species. It is known as the best protected area of sal forest in Siwalik ranges. The biodiversity of the reserve has however been affected due to the submergence of a large area with numerous species of vegetation into vast multipurpose Kalagarh reservoir created along the river Ramganga within the boundary of the protected area in 1973.

**The Tourist Potential**

The tourist potential in the present context is referred to as the tourist attracting and entertaining capacity of a wildlife preserve.

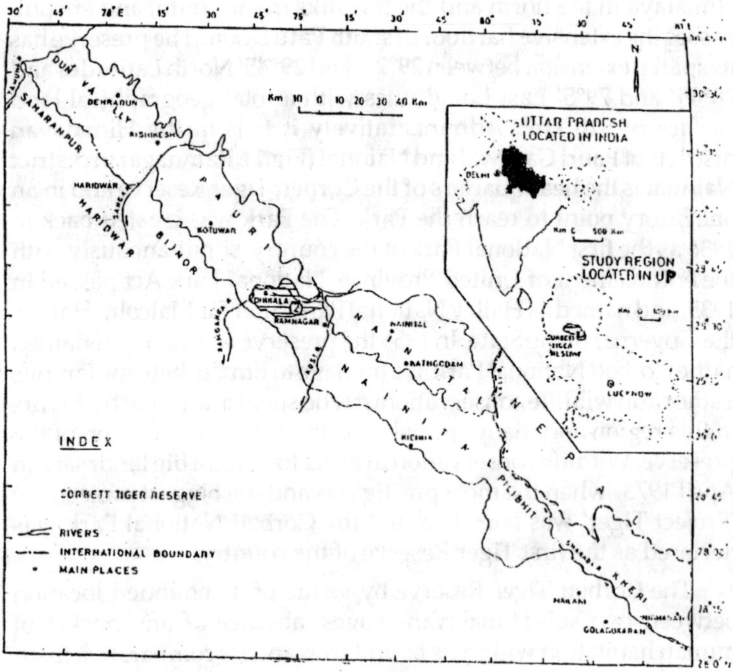


**Fig. 1. : Himalayan Foothills of Uttar Pradesh Index Map**

Tourist potential of a nature preserve can be measured through the analysis and assessment of several parameters such as climatic conditions, scenic beauty and status of wilderness, diversity of fauna and flora, recreation-facilities, accessibility and accommodation, etc.

The Corbett Tiger Reserve owing to its location between the two sub-Himalayan ranges and encompassing the broad flat valley of south Patti Doon enjoys fairly good weather, particularly during winters and summers. During winter, mornings are wonderful but sometime foggy with average temperature varying between 5°C and 30°C. Springs are very pleasant when the temperature ranges from 10°C to 30°C, and the summer temperature normally does not go above 40°C. Besides, nested between the two sub-Himalayan ranges and encompassing the extensive doon floor and grassland with a back drop of stately dense sal forests the Corbett Tiger Reserve is scenically one of the most beautiful wildlife preserves of the country. The river Ramganga (West) meandering through the Reserve adds to to its natural beauty and bio-diversity. Owing to its bounded location and comparative insulation from external interferences, the Reserve enjoys complete wildness.

A wide range of recreational facilities and wilderness amusement provided by a wildlife preserve are the additional attractions for the modern tourists. The Corbett Tiger Reserve owing to its variety landscapes, offers wide ranging recreational facilities, such as, elephant riding fishing, climbing, viewing, swimming, bird watching, etc., to visitors. The Corbett National Park has the largest number of forest rest houses, hutments, tourist cabins and other lodging facilities to accommodate variety of tourists, of all the wildlife preserves of U.P. foothills. Dhikala situated in the heart of the Reserve is one of the biggest tourist complexes located inside the nature preserves of the country. Besides, there are a number of hotels with all modern amenities near the entry gates along the eastern boundary of the Park (Fig. 2). The Corbett Tiger Reserve is easily accessible by all means of transport, particularly road and rail.

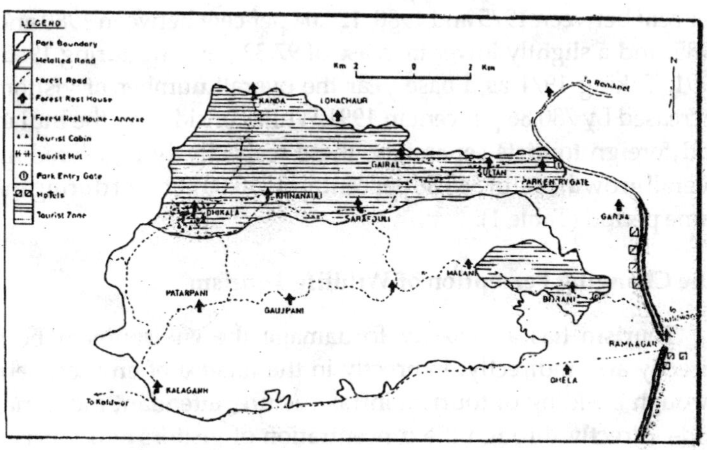


Fig. 2. : Corbett Tiger Reserve Tourist Zone, Transport Network and Accommodation Facility

It is obvious from the foregoing evaluation and assessment of various geographic parameters determing the tourist potential of nature preserves that the Corbett National Park is one of the most ideal wildlife preserves for tourists. The Corbett, therefore, is extremely popular and is the heavily visited park because of its scenic value, good weather, easy access, and for several historical reasons, such as the many adventures of Jim Corbett which find a setting here.

**Pattern of Tourist Growth**

Table 1 reveals that the Corbett Tiger Reserve receives a large number of Indian and foreign tourists and the number of visitors has been increasing progressively. The Reserve was visited by 39791 and 45215 Indian and foreign tourists, respectively in 1987 and 1991. The Corbett Tiger Reserve has registered a phenomenal growth in tourist-population. In 1971 it received 473 tourists of which 10.56 per cent were foreigners; while in 1981 it was visited by 12079 Indian and 1365 foreign tourists. After 10 year in 1991, as mentioned earlier, as many as 45215 tourists including 4400 foreigners visited that Park (Fig. 3). The Reserve maintained a tourists growth rate of 76.15 per cent between 1971 and 1975; 114.13 per cent between 1975 and 1980, 122.04 per cent between 1980 and 1985; and a slightly lower increase of 97.32 per cent during 1985-1991. Taking 1971 as a base year the overall number of visitors increased by 780.86 per cent in 1991.While considering the Indian and foreign tourists separately, these respectively registered an overall growth rate of 865.86 per cent and 780.00 per cent during the same period (Table 1).

**The Changing Perception of Wildlife Tourism**

Tourism has a capacity to damage the environment both directly and indirectly; indirectly in the misuse of environment through building of tourism infrastructure, attendant industries and—directly through the concentration of visitors and tourism support of manpower, equipments and facilities in areas of natural beauty, and in and around the natural reserves. These facilities are put to wrong use and exposed to exploitation without any feeling and regard for their protection and preservation. During recent years the concept of tourism has changed considerably with increasing urban stress and rise in the income level of population, mainly owing to industrialisation. These changes are mainly of two types.

**Table 1 :** Pattern of Tourists Growth in the Corbett Tiger Reserve from 1971 to 1991.

| *Year* | *No. of Tourists Visited* | | | | | | | |
|---|---|---|---|---|---|---|---|---|
| | *Indian* | *Per cent of Total Tourist* | *Per cent change* | *Foreig-ner* | *Per cent of Total Tourist* | *Per cent change* | *Total Tourists* | *Per cent change* |
| 1971 | 4233 | 89.44 | — | 500 | 10.56 | — | 4733 | — |
| 1972 | 5972 | 80.52 | | 1445 | 19.48 | | 7417 | |
| 1973 | 6457 | 82.64 | | 1358 | 17.36 | | 7813 | |
| 1974 | 6527 | 88.73 | | 829 | 11.27 | | 7356 | |
| 1975 | 8070 | 89.25 | 90.64 | 972 | 10.75 | 94.4 | 9042 | 76.15 |
| 1976 | 9732 | 92.31 | | 811 | 7.69 | | 10543 | |
| 1977 | 10629 | 88.01 | | 1449 | 11.99 | | 12078 | |
| 1978 | 8061 | 86.08 | | 1304 | 13.92 | | 9365 | |
| 1979 | | | | | 9707 | 89.56 | | 1132 |
| 10.44 | | | | | | 10839 | | |
| 1980 | 9222 | 89.56 | 14.28 | 1098 | 10.64 | 12.96 | 10320 | 14.13 |
| 1981 | 12079 | 89.85 | | 1365 | 10.15 | | 13444 | |
| 1982 | 16603 | 91.77 | | 1489 | 8.23 | | 18092 | |
| 1983 | 17874 | 91.09 | | 1747 | 8.91 | | 19621 | |
| 1984 | 13458 | 89.63 | | 1557 | 10.37 | | 15015 | |
| 1985 | 20688 | 90.28 | 124.33 | 2227 | 9.72 | 102.82 | 22915 | 122.04 |
| 1986 | 26604 | 91.73 | | 2397 | 8.27 | | 29001 | |
| 1987 | 36881 | 92.69 | | 2902 | 7.31 | | 39791 | |
| 1988 | 36555 | 91.59 | | 3354 | 8.23 | | 39909 | |
| 1989 | 35443 | 91.37 | | 3349 | 8.63 | | 38792 | |
| 1990 | 36161 | 92.54 | | 2917 | 9.72 | | 39080 | |
| 1991 | 40885 | 90.28 | 97.83 | 4400 | 9.72 | 97.50 | 45825 | 9762 |
| Total Change | | | 865.86 | | | 780.00 | | 780.86 |

## Craze to Visit Wildlife Areas

In recent years, the craze to visit National parks and wildlife sanctuaries has increased, particularly among the affluent classes

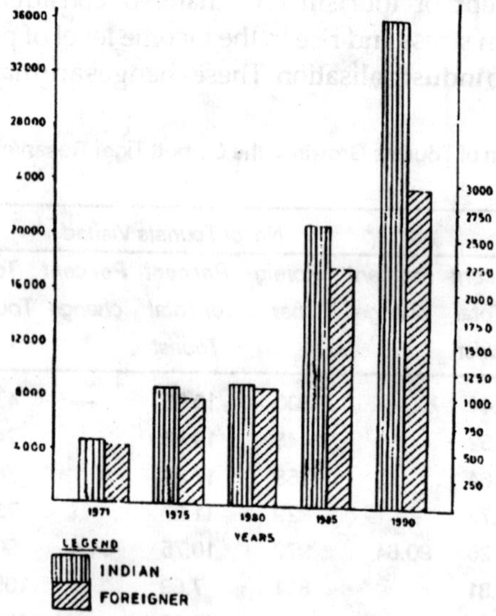


**Fig. 3. (a) Pattern of Tourist Growth in Crobett Tiger Reserve from 1971-1991**

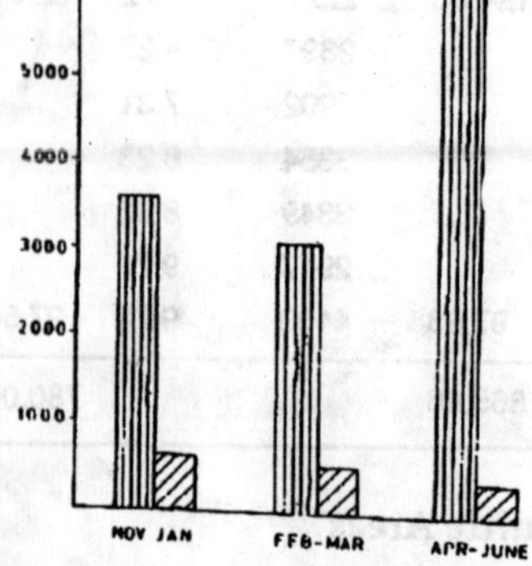


**Fig. 3. (b) Season wise break up of Indian and Foreign Tourists in Corbett Tiger Reserve**

of the society. Consequently these protected areas are receiving a heavy influx of visitors, particularly during holidays and weekend. A considerably high proportion of these visitors belongs to higher income group and affluent classes of the urban population of surrounding and nearby urban and industrial complexes (Table 2). Visitors survey conducted in the Corbett Tiger Reserve revealed that more than 37 per cent tourists visiting the park were from higher income class of society (Table 2). Thus the gaining popularity of wildlife tourism has further been contributing to the depletation of vulnerable ecosystem of our protected areas.

**Table 2 :** Economic Status of Indian Tourists Visiting the Corbett Tiger Reserve

| *Name of the Park* | *Number of Tourists Interview-ed* | *Affluent (per cent of Total visitors interviewed)* | *Middle Income Group (per-cent of Total visitors interviewed)* | *Low Income Group (per cent of Total visitors interviewed)* |
|---|---|---|---|---|
| Corbett Tiger Reserve | 200 | 37.00 | 46.00 | 17.00 |

**Five Star Culture**

Over crowding in some popular national parks is now becoming as frightening as in big cities. As stated earlier a significant proportion of visitors to National Park and Sanctuaries is from affluent society who can afford large amounts and use their own transport (Table 2). These visitors are accustomed to the comforts of the twentieth century and want to enjoy all modern amenities of life available to them, even in wildlife protected areas. Such visitors get initial idea of visiting wildlife areas with all possible modern luxuries and comforts generated from the modern five star culture which has certain essential patameters like luxuries living, expensive travel, specific kind of recreation and amusements.

The Corbett Tiger Reserve which receives the highest number of tourists of all the National parks situated in the U.P. foothills provides classic example of emerging five star culture along the eastern boundary of the reserve, near Ramnagar (Fig 2). A number of high class hotels with all modern facilities of living and recreation have been coming up. These luxurious hotels entertain a large number of tourists all the year round putting heavy stress on the delicate ecosystem of the adjoining reserve, directly as well indirectly.

**New Outlooks and Extreme kind of Law and Order Situation**

As discussed in the preceding section, with the increase in income level, the number and type of visitors to National Parks and Wildlife Sanctuaries has been increasing progressively for the last 10-15 years. Consequently the number of tourists belonging to higher income group and the affluent classes of the society visiting the wildlife area, has increased considerably during recent years. Such visitors have a completely different perception of wildlife and of the ecology of the protected areas and they want to enjoy the city comforts, amusement and recreations amidst the natural surroundings of the wildlife areas. Owing to their high societal status and economic influence they have full political backing. Such type of visitors are often found taking advantage of their position by violating the rules and regulations of National Parks and Wildlife Sanctuaries, creating very serious problems for the management staff and other visitors. Hunting is one of the many inappropriate and illegal activities of such visitors causing direct damage to wildlife population and threatening in basic principles of protected areas. Game in unprotected areas is very rare now while it is much easier in protected areas when such powerful people can taken all advantages.

The visits of such irresponsible tourists is now being highly facilitated by the emergence of high class luxurious hotel complexes around the National Parks. The Corbett Tiger Reserve provides the most ideal example where such culture is developing fast with coming up of a number of modern hotels along its eastern boundary (Fig. 2). The authorities of the Reserve often face situations created by the inappropriate activities of powerful people. Such extreme kind of law and order situations are further compounded by the fact that park managers, answerable for such violation of regulations have no legal power at all to handle the problem effectively, and thus become silent spectators to such destruction. Therefore in order to ensure the effective management, protection and conservation of wildlife in the protected areas the park authorities must be given sufficient legal power to deal with such extreme kind of law and order situations.

### Carrying Capacity

Although the concept of carrying capacity has generated considerable research (Butler, 1972; Stankey and Lime, 1973); investigators have experienced great difficulty in defining the term. Indeed this situation had led, Jackson, Buszywski and Botting (1976, 359) to conclude that there is no generally accepted definition of carrying capacity and no standard approach of how it should be calculated. At the same time, efforts to establish generalizations about carrying capacity have been hampered by the problem specific nature of such studies.

In resource management the concept of carrying capacity received early attention in range and wildlife management (Mitchell, 1979). But carrying capacity is equally relevant to a broad range of problems. Nevertheless, geographers and social scientists have channeled research efforts towards its application in recreational planning. In the early 1960s Clawson and Knetsch (1963, 262) called for more research on carrying capacity related to outdoor recreation. They suggested that carrying capacity was closely related to the concept of crowding, because for everyone there comes a point at which enjoyment decreases because of the presence of too many other people. They argued that crowding could have both physical and psychological components. Lapage (1963, 34) reinforced this interpretation when suggesting that carrying capacity consisted of two components--aesthetic and biotic. Burden and Randerson (1972, 439-40) stressed that the amount of use an area can receive without environmental deterioration depends upon the declared management objectives for an area, as well as upon the degree of naturalness desired. On this basis they defined carrying capacity as the maximum intensity of use an area will continuously support under a management programme without inducing a permanent change in the biotic environment.

### Biophysical Carrying Capacity

The biophysical and ecological carrying capacity relates to changes in the bio-physical environment through both natural processes (erosion, siltation) and human impacts (fire, suppression, recreational activity). Research on biophysical carrying capacity, has covered a range of aspects. Investigators have sought to determine critical thresholds beyond which physical (soils, geomorphology) and biological (vegetation, wildlife) phenomena are altered by human activities. Topically, work has been spilt among inquiries about the trampling effect of hiking on trials or sking on mountain slopes, users impact on natural parks and studies concerning the effect of foot traffic and industrial and transportation activities on the fragile ecosystem of protected and park areas (Mitchell, 1979).

**Behavioural Carrying Capacity**

Behavioural or sociological carrying capacity draws attention to the way in which use of a resource affects satisfaction. Geographers have had more active role in studies of behavioural carrying capacity than in those focussed on bio-physical carrying capacity. The main objective of such studies has been to discover what variables contribute to recreationists having 'satisfying' experiences the extent to which people were happy and unhappy, satisfied or dissatisfied, with a recreational experience. This type of investigation in veiw of Priddle, Clark and Douglas (1973) focussed attention upon the need to understand the behavioural carrying capacity of given environments, especially the number of users who can be accommodated in a particular place—National Park etc. and still be provided with a satisfying experience. Attention has been directed towards an array of biophysical, economic, social, and behavioural variables, with special emphasis given to the role of crowding and congestion as determinants of satisfaction.

La Page (1963) drew attention to the need for studies of behavioural carrying capacity for recreational planning, but it was Lucas (1964) who provided one of the first empirical studies on this topic. This study by a geographer, is generally acknowledged as a benchmark by investigators in many disciplines. Lucas was interested in determining whether boundaries of officially designated wilderness areas corresponded to the wilderness image held by recreationists using the Boundary Waters Canoc Area in Minnesota in U. S. A. He examined three elements of environmental perception for both resource managers and visitors; (i) the importance of wilderness qualities relative to other potential uses, (ii) the area considered to be wilderness, and (iii) the essential characteristics of wilderness and acceptable type of use. The third element came to grip with such issues as what degree of crowding and types of activity had positive or negative effects on recreation experience. Several successful attempts have also been made by geographers to combine together both the bio-physical and behavioural aspects of carrying capacity, using similar strategies. The major objective of such studies has been to test the hypothesis that the quality of recreation experiences is related to conditions of crowding and that different intensities of use cause proportional degree of ecological damage (Burton, 1974).

**Assessment of Carrying Capacity**

It is obvious from the foregoing discussion that the concept of carrying capacity aims at determining the limits of maximum use, within resource and ecological constraints. The basic elements, therefore are two—the user and the resource. The first requires the determination of the attitude of users and their distribution in space and time, whereas the second poses physical and natural constraints to restrict the first. These two parameters will dictate the upper limit to the total amount of use that can or should be derived from the park, without the area undergoing deterioration to a subsequent loss of satisfaction to the present or future users. The limits so defined will need modifications on psychological grounds. Because increasing encounter with other users might start diminishing the users satisfaction before the actual limits posed by physical considerations are touched. Thus the recognition of the problem of carrying capacity requires the acceptance and then determination of basic standards of environmental quality against which to judge the impact of human visitation (Bardwell, 1973).

The difficulty in determining the limit at which benefits to recreationists or environmental quality of an area starts deteriorating arise mainly because decline in environmental quality does not become evident in a shorter period. Whereas in other resource use processes such as agriculture, forestry, dairying etc. production per unit area can easily be measured, and therefore carrying capacity may easily be estimated by declining productivity. Also in case of National Parks, too great a variation in intrinsic requirements, make it more complicated. For instance, there may be a major difference between a park where wildlife is the main attraction and a park where a geological wonder is the main attraction. In one case presence of human being is disturbing and in the other their presence is comforting. Still it is essential and attempts have been made to assess the carrying capacity of natural reserves, protected areas, national parks and Sanctuaries (Singh, 1976).

According to Hogg (1973) the carrying capacity, can be determined by parameters like erosion of soil, exhaustion of natural resources, contamination of water and interference with other landuses. Depending upon the nature of most limiting factor, determination of the carrying capacity could be carried out. In case it is possible to evaluate the limiting factor quantitatively, it will be easier to assess the capacity. For example if physical overcrowding is the limiting factor, a knowledge of space required per tourist can provide measure for the estimation of carrying capacity. But where direct evaluation of limiting factors, such as soil erosion, water pollution of interference with ecological processes of park-environment etc. is not possible, the evaluation of carrying capacity itself can be indirect. On such method could be a close monitoring of the impact of user's activities and consequent assessment of the upper limits, beyond which impact and activities should be treated irreversible. Ovington et. al. (1972) observed that the carrying capacity of recreational areas can be assessed in terms of the degree of environmental degradation suffered per unit of tourist activity. With sufficient data, relating to tourists numbers, tourist activities and related environmental changes, it becomes possible to set realistic limits on tourists activities.

## The Dhikala Tourist Complex : A Case Illustration

It was attempted in the present investigation to assess both the bio-physical and bahavioural carrying capacity of Dhikala tourist complex of the Corbett Tiger Reserve. Dhikala in one of the largest tourist complexes situated inside any National Park of the country and consequently entertains comparatively a much higher number of tourists. Owing to several practical reasons it could not be possible to bring out quantitative statement of biophysical carrying capacity of the area. Hence some indirect and simple techniques of assessing the biophysical carrying capacity have been employed. Various tourists activities and their direct and indirect impact on the ecosystem of the Dhikala area were studied and analysed comprehensively, through intensive field surveys, mapping and photographs. The existing ecological status of vegetation soils, water and animals assessed through this exercise revealed clearly that the pressure of visitors has already reached the threshold limit of the carrying capacity of the area (Table 3).

**Table 3 :** Status of Biophysical carrying capacity of Dhikala Area of the Carbett Tiger Reserve

| *Tourism Parameters* | *Impact observed* |
|---|---|
| 1. Crowding Congregation | Change in Habits and Movement Pattern of Animals Wilderness Affected, Noise Pollution, Accumulation of Litter |
| 2. Trampling | Loss of Vegetation, Plant Communities, Reduced Tree Seedling Destroyed<br>Soil Compaction Occurred, Soil Moisture, Declined, Grassland Damaged |
| 3. Feeding of Animals | Change in Habits and Movement Pattern |
| 4. Cooking, Room Heating, Smoking | Loss of Vegetation and Fires |
| 5. Tourist Vehicles | Noise and Water Pollution, Change in Animals Habits and Movement Pattern. |
| 6. Transport | Loss of Vegetation, Soil Erosion, Hydrological Disturbances |
| 7. Accommodation | Increased Influx of Tourist and Consequent Pressure on Carrying Capacity. |
| 8. Catering | Accumulation of Litter and Waste |

Congregation of considerably large number of tourists at many places causing changes in animal population and affect their movement pattern. A number of plant and plant communities particularly the grassland of the area are highly susceptible to trampling and consequently get reduced in number and their growth is very slow at places. Tree seedling were also found destroyed due to trampling, and soil compaction was resulting into hampered plant growth. Besides, trampling of soil can also be associated with reduced soil moisture and reduced soil aeration at many places where visitors usually concentrate. Noise and pollution generated by a large number of tourist vehicles on the

Park roads seemed to be responsible for noise and water pollution and changes in animals habits (Fig. 4).

Virtually all the visitors who visited the Corbett Tiger Reserve during the 50 years of its existence have visited or stayed at Dhikala, the main attraction and focal point of tourism activity in the reserve. Dhikala is still the destination for everyone who comes to Corbett and it is the point of departure for wildlife viewing and elephant rides by the great majority of them. The scenic beauty of Dhikala is undeniable, the sweeping view of densely forested Kumaon Hills, the meandering Ramaganga river and the wild elephants feeding on the grassy flats of the river flood plain makes it one of the most unique and distinct reserve of the country.

Although Dhikala has attracted a large number of visitors for many years, it has, only recently developed into a problem spot, mainly because of the increase in number and type of visitors. Previously, only the wildlife enthusiasts and outdoors men ready

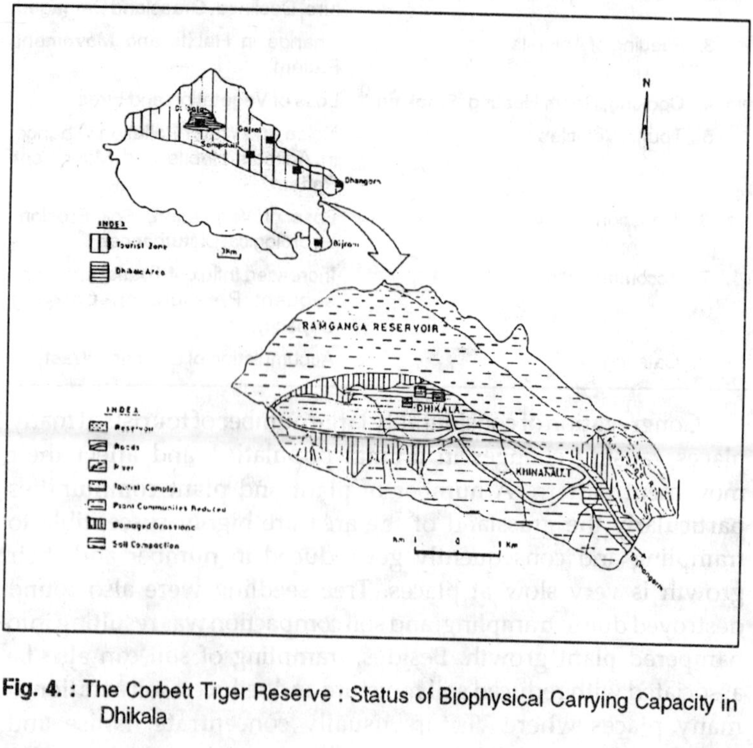


**Fig. 4. :** The Corbett Tiger Reserve : Status of Biophysical Carrying Capacity in Dhikala

to rough it, were attracted to the reserve. Arduous travel is no longer required and now the average visitor expects city comfort. Consequently, the behavioural carrying is being approached at Dhikala. Many of the people who come to observe wildlife on quiet surroundings, no long enjoy their stay because of the coming and going, the noise and the litter. In order to assess the behavioural carrying capacity visitor- interview method was adopted. Six hundred visitors, entering the Dhikala area 200 each on two week-end holidays in December, March and May 1991 were interviewed regarding their wilderness experience in the area. This experience revealed some interesting facts pertaining to the wilderness image of Dhikala. Out of 600 visitors interviewed, about 30 per cent had come to the area to observe the wildlife in the real sense (Table 4). As many as 62.10 per cent of the wildlife enthusiasts were not satisfied with the wilderness status of the area owing to one or the other reason. More than 30 per cent wildlife lovers were completely dissatisfied with overcrowding of visitors, 16.80 per cent complained of noise problem, 9.60 per cent could not enjoy their visit due to litter accumulation at too many places of attraction and more than 18 per cent of wilderness lovers were found most unhappy with the noise, dust and pollution generated by tourist vehicles in the area. Whereas, 24.90 per cent wildlife watchers expressed their dissatisfaction on account of the combined effect of all these factors which owe their origin to heavy influx of visitors (Table 5). The wilderness zones of Dhikala area as emerged from the exercise on behavioural carrying capacity have been shown in Fig. 5.

**Table 4 :** Factors of Motivation Associated with Indian Tourists Visiting the Corbett Tiger Reserve in 1991

| *Name of the Park* | *Number of Visitors Interviewed* | *Urban Stress (per cent of Total Visitors Interviewed)* | *Lover of Wildlife (per cent of Total Visitors Interviewed)* | *Other reasons (Per cent of Total Visitors Interviewed)* |
|---|---|---|---|---|
| Corbett Tiger Reserve | 200 | 22.00 | 40.00 | 39.00 |

It is quite obvious from the preceding discussion on the two different but important aspects of carrying capacity in the Dhikala area of the Corbett Tiger Reserve that the heavy influx of tourists in

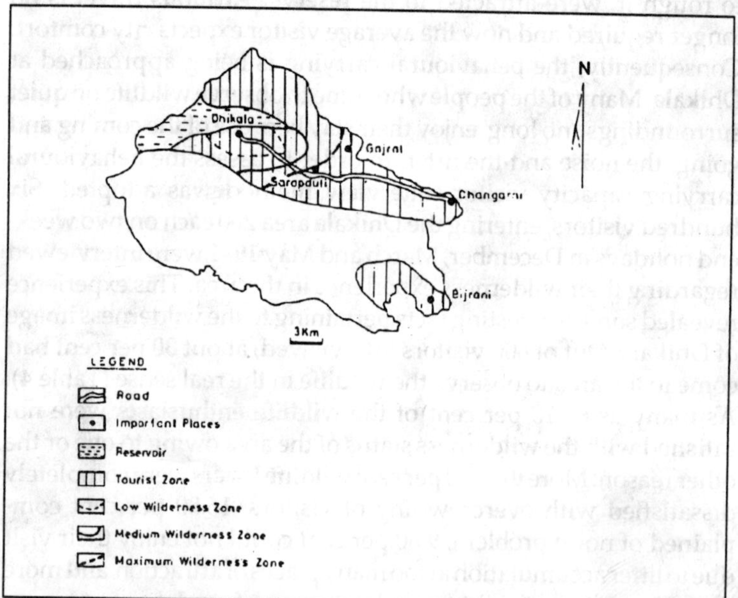


**Fig. 5.:** The Corbett Tiger Reserve : Wilderness Zonation Around Dhikala

the area is causing considerable pressure upon the biophysical as well as behavioural environments of the region. Evidence suggests that most of the impact on biophysical environment occurs within the first few years after opining of an area of a reserve. After that point, the deterioration continues, but at a greatly reduced rate. The results show that while any use has any impact, there are critical stages in the progressive change in the environment (Mitchell, 1979). This information will be useful to managers in deciding what levels of use are appropriate, and in choosing among alternatives. The behavioural investigation indicates that satisfaction is affected by congestion- one of many considerations demanding the attention of the park managers.

## Strategy for the Regulation and Management of Tourism

Tourism has now become one of the important activities of National Parks and Sanctuaries in the country and despite its far reaching ecological consequences, the tourism will be going to continue as a major phenomenon of our wildlife landscape. Many

**Table 5 :** Status of Behavioural Carrying Capacity in Dhikala Area of the Corbett Tiger Reserve

| *Month* | *Number of visitors interviewed* | *Wildlife lovers (per cent of total visitors interviewed)* | *Dissatisfied* | *Satisfied* | *Dissatisfied due to crowding (per cent of wildlife lovers)* | *Dissatisfied due to noise (per cent of wildlife lovers)* | *Dissatisfied due to litter accumulation (per cent of wildlife lovers)* | *Dissatisfied due to vehicles noise & pollution (per cent of wildlife lovers)* | *Dissatisfied due to All Factors (per cent of wildlife lovers)* |
|---|---|---|---|---|---|---|---|---|---|
| December | 200 | 26.00 | 65.00 | 35.00 | 32.00 | 18.00 | 8.00 | 12.00 | 30.00 |
| March | 200 | 36.00 | 57.00 | 43.00 | 29.00 | 15.00 | 7.00 | 24.00 | 25.00 |
| May | 200 | 40.00 | 63.00 | 37.00 | 31.00 | 18.00 | 12.00 | 19.00 | 20.00 |
| Total | 600 | 34.00 | 62.00 | 38.00 | 30.15 | 16.00 | 9.60 | 18.55 | 24.90 |

of our popular National Parks, like the Corbett Tiger Reserve, have started to receive very heavy influx of different categories of tourists, beyond their carrying capacity. Obviously, the ever increasing number of visitors to wildlife protected areas is causing great damage to the fragile ecological system and the bio-diversity of these preserves. Planning and management of wildlife tourism, therefore need to be considered as one of the most important and critical aspects of the wildlife conservation.

As discussed earlier, the Corbett Tiger Reserve, being the extremely popular National Park is visited by a considerably large number of tourists every year. The sprawling tourist complex at Dhikala, in the heart of the core area, is not only a major source of disturbance but also creates several ecological problems. The following measures may be taken for the regulation and management of tourism in the Reserve :

(i) An alternate site should be developed to relocate Dhikala tourists complex. Dhangarhi situated near the entry point could be an ideal such spot. In case it is not practicable to close down Dhikala due to several practical reasons, a possible solution could be to close down atleast 50 per cent of the available accommodation and also reduce the staff complexes.

(ii) Attempts should be made to discourage tourists access to Dhikala. As a first step, management can identify other potential wildlife viewing areas along the road to Dhikala, in the surroundings of existing forest rest houses and in the southern tourist zone, at Bijrani. Ideally each of these viewing areas could be distinct units with their own access and specific attraction. Four such spots could possibily be developed at Bijrani, Sarpaduli, Gairal and Dhangarhi (Fig. 2). Once these areas are developed, the tourist - pressure on Dhikala will reduce considerably.

(iii) The Bijrani area, which is now being developed for day visitors should have a nature interpretation center, and the museum at Dhangarhi should also be developed into a proper interpretation center. Both these proposed nature interpretation centres, and the overall management of tourism within the Park should be under the charge of separate eduction officer. He should also be responsible for training and supervising the guides and naturalists who

**Table 6 :** Control Measure to Increase Carrying Capacity of National Parks (Modified from Gilbert et. al. 1972)

| *Type of Control* | *Method* | *Specific control policies* |
|---|---|---|
| Regulatory | Increased Policy Enforcement | Impose Fines, Increase Surveillance |
| | Zoning Regulations | Spatial Zoning of Uses, Temporal Zoning of Users |
| | Restrictions on Use Intensity | Open or close Access Points, Require Reservation Assign Travel Routes to tourists Vehicles, Limit Size of Parties, Limit Total Park Population |
| Manipulative | Physical Alterations Information Dispersal | Open or Close Access Roads, Improve Access Road, Advertise Specific Park Attributes to Attract Certain Type of Visitors, Educate Visitors Regarding Care of Park Ecology, Develop Alternative Sites and Advertise Underused Areas of Park. |
| | Eligibility Requirements | Charge Constant Entrance Fee<br>Charge Marginal Cost Free<br>Require Demonstration of Ecological Knowledge. |

would be required to provide education to visitors. It should be made compulsory for every visitors to spend at least some time in such interpretation centres, where the tourists of varying perceptions and attitudes could be motivated and educated, in an interesting manner, for the protection and conservation of wildlife and its habitat, before they enter the wildlife zone.

(iv) Acquiring group vehicles preferably mini-buses for the movement of tourists inside the Park should be given high priority. This will not only minimise the ecological impact resulting from large number of private vehicles entering the park but also make transportation available for tourists which is presently lacking.

(v) Appropriate measures should be taken, using electronic and other media, to encourage a specific class of visitors who has some appreciation and acceptance for nature and nature conservation. For this , National Parks and Sanctuaries should not be taken as source of income to government and means of employment for the people.

(vi) In order to increase the carrying capacity of the reserve the park managers may choose from a variety of measures such as zoning of the area, channeling movement, education, regulations, law enforcement, interpretation, management of biological communities, encouraging the use of alternate sites. For practical purposes, the control measures can be classified as regulatory and manipulative (Gilbert et. al. 1972). A regulatory control defines, where, when and how people may travel and visit the Reserve. Manipulative controls are less obtrusive, rather than regulating intensity of use or behaviour, they influence behaviour through manipulating factors used by the visitors in deciding where to travel and how long to stay in a wildlife preserve (Table 6).

## References

Bardwell, S., 1973, *Wilderness Use in South-West Tasmania*, A Pilot Study, Department of Geography, Monash University Monograh. 1.

Burton, R. C. J., 1974. *The Recreational Carrying Capacity of the Countryside : A Research Report presenting the Methodology and results of Ecological and Psychological Survey of Cannock Chase Staaffordshire*, University of Keel Library, Keel.

Burden, R. F. and P. F. Randerson, 1972. Quantitative studies of the Effect of Human Trampling on Vegetation as an Aid to the Management of Semi Natural Areas; *Journal Applied Ecology*, 9 : 439-57.

Butler, K. S., 1972. *Resource carrying capacity : A Literature Index and General Biliolography*, Working Paper No. 9, Institute of Environmental Studies, University of Wisconsin, Madison.

Glawson, M and J. Kenetsch, 1963. Outdoor Recreation Research : Some Concepts and Suggested Areas of Study ; *Natural Resources Journal*, 3 : 250-75.

Gilbert, C. G., G.L. Peterson and D.W. Lime, 1972. Towards a Model of Travel Behaviour in the Boundary Waters Canon Area, *Environment and Behaviour*, 4 : 131-58.

Hogg, D., 1973. *Planning for recreation in the Victorian Alps : The Planning Approach*. Monograph, No. 1 Currg, Australia.

Jaakson, R. M. Buszywski and D. Botting, 1976. Carrying capacity and Lake Recreation Planning; A case study from North Central Saskatchewan, Canada, *Town Planning, Review*, 47 : 359-73.

La Page, W.F., 1963. Some Sociological Aspects of Forest Recreation, *Journal of Forestry*, 61 : 32-66.

Lucas, R.C., 1964. Wilderness Perception and Use. The Example of the Boundary Water Canoe Area, *Natural Resource Journal*, 3 : 394-411.

Mitchell B., 1979. *Geography and Resource Analysis*, Longman, London.

Ovington, J. D., 1972. *Impact of Tourisum on Mt. Olga National Park*, Department of Forestry, Australian National University.

Priddle, G.B., C.D., Clark and L.R. Douglas, 1973. The Behavioural Carrying Capacity of Primitive Areas for Wilderness Travel Paper Presented in the Annual Meeting of the Association of American Geographers, Atlanta Ga.

Singh, R.L., 1976. *Planning and Management of National Parks,* U.P. Forest Department.

Stankey, G. H., and D. W. Lime, 1973. *Rereational Carrying Capacity : An Annotated Bibiliography.* U.S.D.A. Forest Service General Technical Report INT- 3, Intermountain Forest and Range Experiment Station, Ogden Utah.

# Chapter 25

# Integrated Area Development in Himalaya : A Conceptual Approach

*R.K. Pande*

*P. C. Pande*

*G.B. Pant*

Integrated area development a watch word of modern national planning, had opened a new vista for contributive research, basically it seeks its origin within geo-economic and commercial thought. It shows a organizational structure of a space sector which is a unit of an area on the earth surface unified by some particular characteristics. It helps in grading or scaling of ratings of different problems. It is one of the significant concepts serving as a tool of economic development, rational resource use and equitable distribution of resources and amenities in the region. A local and lower order level plan, essentially an integration of various considerations, depends on the rational use of resources and purposeful protection as well as transformation of space environment for improvement of living conditions of the people along-with balanced space organization. The scale and content of planning decisions should be determined in time frame, by the short and long term prospects of the area concerned.

In first and second five years plans, emphasis was given on various problems of regional devélopment, as the resultant later on in hill regions of Uttar Pradesh, which led to a new hill development organization under the control of Hill Development Board.

But in most cases of planning postulates the local, and regional aspects of norms and values have not been considered properly.

Today when the world is passing through a period of scientific revolution, the hill areas have still remained shrouded in the mists of confusion myths, romance and misinformations. Mankind has yet to make a deeper probe into its own resources, socio-cultural mileau and to construct an alternative strategy of regional planning rather than hastily advocating a stereotype and conventional management of resources.

The hill areas constitute a sizeable proportion of country's population. These areas suffer from continuous exploitation and have remained cutoff from the main stream of national life. Inspite of constitutional guarantees given to the tribal and hill people vide article 46 of the constitution and post-independence efforts for development of depressed classes. Tribals and hill people have not been in a position to capitalise the gains of socio-economic development. This has partly due to physical and socio-cultural barriers, but more so on account of lack of involvement of people themselves.

The reality pertaining to the integrated rural area development specially in hill environment has revealed that inspite of several efforts of modernization, westernization and concentration of investment in different sectors- the vast hill area, remained in a condition of stagnation, even also the situation is similar in the case of economic development too. In most part of the hill areas several kinds of spatial gaps, poverty trans can be observed, may be called as the 'kingdom of poverty'. It is advocated that the basic, social, economic and environmental goals, integrated area development, will be adequately achieved only with the change in spatial organisation of the space with the introduction of four basic considerations : (1) maximization of contacts, (2) minimization of efforts, (3) optimization of space and (4) building a quality environment.

Certainly, the process of integration has been widely accepted to promote the idea of development, which involves spatio-functional system in terms of transformation and modification of the rural landscape with a view to maximizing the developmental activities in a particular region or space. The basic idea of rural upliftment is the outcome of a series of quantitative and qualitative changes occurring within a given rural population and whose conversing effects in time span indicates, the rise in the standard of living. It does not refer tothe isolated or the sectoral development of various norms and values such as 'Community Development', 'Rural Amenitification', 'Agricultural Extenstion', or any programme of such type; but rather to comprehensive development of spatial inequalities and imbalances with the organizational system of human dimensions. The structural development system with such changes and transformation may be called as 'GEO-CHOROSYSTEM' which analyse the system of regulation, control and automation with its own techniques so called as 'GEO-GYBERNATICS'.

In formulating the rational development, the impetus need be given to the development from below (local), while activity and guidance from above (the government/planning/research body). Among the more significant factors in spatial organization and its regulation, the following sets are noteworthy; geo-economic, socio-cultural, techno-institutional, political, commercial and ecological. It helps in involvement of qualitative and quantitative informations based on the appraisal of natural and cultural resources and their potentialities for a balanced, optimum and integrated development. It is one of the most significant vehicle which carries the basic as well as new idea of developmental proposals-Co-operative planning rather than monopolostic system of planning.

In the beginning of the Five Year Plans many problems of regional development were covered but no substantial could be achieved mainly due to lack of proper operational structure. In most cases of planning postulates the local aspect of function or activity has been neglected and this avoiding tendency has created mess in developmental organisations. Since development potentials are to be developed within the expanding economy, development of natural resources of the village territory-soil, water, agriculture, health, education, etc. as well as training the people to improve their skill are most important and essential aspects of planning leading to developmental justice; and it will be an optative orbit of the proposed treatise.

The assessment of agricultural situation and development potentials- the Himalaya records about 85 per cent of its population as rural, thus the agricultural activities become the main profession. The basic agricultural problems of the area includes lack of credit facilities, inadequate water supply, lack of chemical fertilizers, poor circularity parts, soil erosion, etc. Owing to purely traditional system of agriculture, the yield of crops is far from satisfactory. It should be advocated that how the pelisse from the problematic face of the village can be removed. For increasing the output of agricultural production it is necessary to provide facilities of agriculture (seeds, chemical, fertilizers, insecticides, etc.) with the proper knowledge of soil characteristics. Well defined arrangement for extending credit facilities for purchasing agricultural equipments and other necessities, with such provision it is expected that the double cropped land will certainly increase and the aim would be achieved.

The basic infrastructure like circularity part and the basic level of education, drinking and irrigational water, electric etc. have to be developed in novice manner. Planned introduction of these amenities would certainly be helpful in developing the rural network to perform a variety of economic functions such as spread of technical know-how and scientific knowledge, procurement of storage and marketing facilities, etc.

In overall scheme of rural development and reconstruction, housing aspect should be given an important place for a 'house is not only a place for protection from the vagaries of weather but also for an enlargement of one's personality; it is an institution in itself' (NBO report). The problem of rural housing has remained a constant problem, neither the government nor the private sector under the community development programmes have properly tackled the housing problem. The major problem of rural housing in the Himalaya (which falls under the intensive earthquake zone) is concerned with introducing the improvements in traditional building methods along with changes in building materials (maximum use of woods in houses with scientific roofing, i.e., angle of roofs and beams). Most of the Himalayan houses vary from one to four rooms unit without conspicuous planning (multifunctional rooms with common shelter place of different age-groups) buildup of locally available materials-stone, slate, mud and wood constructed in close proximity to each other. Modifications and improvements in housing conditions is considered essential to provide a neat, clean and healthy life in order to avoid health hazards caused due to dirty and dingy situation to restrain the life. In the proposed study area the houses are inan dicator of the influence of several factors like physical and cultural; this being more pronounced in case of rural Kumaun.

In order to give healthy environment to the inhabitants of area, the vectors of health should be taken not only in *pathogens* manner but also under the etiology of disease within environmental perspective, so called as *geogens* because the *geogens* are responsible for *pathogenesis* of disease. The factor groups converging into spatio temporal frame and causing the growth of disease, are (1) stimuli from the envrioirment (2) response from a host and (3) conglomeration of thoughts and traits (culture). Undoubtedly, the Himalaya differs from (above mentioned norms, i.e., environmental conditions, response level and cultural aspects) the other parts of the country. Hence, it has own problem of health and disease. In order to know the number of disease and its causing factors, an analysis of different relating aspect would be essential.

If the process of development is taking place under the aegis of geo-economic, socio-cultural and ecological offerings, then it is necessary to improve animal husbandry, as well as the second profession to crops, i.e., livestock.

In order to have a scientific, systematic and rational study following approaches may be made for a consolidated Integrated Rural Area Development:

1. Historical perspective of rural development and planning in different local and regional spaces; topical and case studies, by using the model of genetic approach which deals with the dynamic pattern of human organization and its cultural processes in the frame of space-time continum, which passes through the successive steps of search, occupancy, transformation and after all integration. For this purpose, three basic approaches of morphogenesis, i.e., retrogressive, restrospective and prospective, may be applied, and finally an attempt should be made to study philosophical thesis of cultural cycle.
2. Rural planning is a multi-faced problem in relation to Organization in different local and regional spaces; topical and case studies. Landuse models, organization of rural people, rural labour increasing the fate of participation of both the people and resources in development and production.

3. Policy for developing farmers-Improving the social frame of agriculture-farmers, organization, multipurpose farmers, cooperatives their local and regional integration lying physical (road, electricity, irrigation system etc.) and financial and institutional (banking, credit, marketing storage, school etc.) infrastructures; progressive elimination of gaps in new system of agricultural development. Attempts for spatial integration and functional coordination of settlements -agro-distributional schemes.
4. Integration of farming and industries, rural consumer market-reducing rurai-urban dichotomy; introduction of appropriate technology.
5. Modernizing the rural society and culture-policy and planning for transfer of loyalty and values from traditional technology to modernizing relationship between education and social planning and group responsibility. Evolving training-knowledge-diffusion and communication systems, rural health improvement schemes together with family welfare schemes.

# Chapter 26

# Ecophysiological Models for Amelioration of the Cold Desert in Himachal Pradesh - Spiti

*R. Bawa*
*R.N. Sehgal*

Man for his very livelihood depends upon plants, small wonder that his impact on vegetation is evident particularly everywhere.

Climate dictates the overall pattern of vegetation. Within the framework set by climate, other factors exercise their influence, and the more detailed a study of vegetation of a given area is, the more intimately it will have to inquire into the factors active in this area.

Throughout the Himalayas, landuse is the mainstay of the population; this means that wherever there are people, vegetation displays the results. In the northwest Himalayas, the impact of population on vegetation under the dry climatic condition is tremendous and has led to wholescale destruction of the vegetative cover, beating back the limit of forest growth altogether. Devastration is to the level that slopes have been rendered barren and replaced by *Artemisia* and/or *Ephedra* steppe.

In Spiti, the climate is so dry that the tree growth is restricted to the immediate vicinity of more substantial water courses and irrigated area. The "semi-desert" or alpine steppe vegetation is extremely sensitive to overgrazing and fuel gathering. "Natural conditions" are almost non-existent. As particularly large flocks of sheep and goat are kept in these areas, the vegetation is over large areas uniformly composed of "pasture-weeds". The sparce growth of trees is too severely restricted.

With the further increase in human and domestic animal population, more pressure has been put on the already highly denuded area. Thus the importance of providing a vegetational cover cannot be compromised at any cost. Considerable efforts, however, are now being made in Spiti through various projects sponsored by the State government and Desert Development Project to increase the high-altitude forested areas through reforestation. The DDP and the Forest departments are doing a good job by planting various tree species (Poplars, Willows, etc.) and also developing pasture areas by introducing clovers, but these are mainly restricted to the water courses. But at many places, even if we provide water during the drought periods, tree survival is restricted as it can be seen in case of Gue where Willows planted near water course did not sprout. Thus showing that water alone is not the prime cause behind poor tree and vegetation growth.

Thus, this momentous task of providing vegetation to the bajadas, boulderfields and sandzones, if it is to be successful, must be planned on a sound foundation. An important pre-requisite is the assembly of scientific facts concerning the physical environment and the course of various life processes of species occurring there or in the surrounding areas.

The main problems which encounter the process of providing vegetative cover included environmental hinderances, ecophysiological restriction, erosion, etc.

The present paper deals with the description of the above mentioned hinderances in detail and mechanical methods for providing vegetation to the deserted lands. In treating the subject matter, particular emphasis has been placed on describing as completely as possible by comparing studies carried out in other cold regions of the world, keeping in view that the basic problems and approach are similar.

**Environmental Restrictions**

The environment of Spriti is unique. The winters are long and chilling with mercury even dropping to-40° C in many cases. But a minimum temperature of -20°C is usually experienced. The summer and rainy seasons are nearly totally dry with few rare showers. Temperatures above 25°C are experienced during summer months. The area is lashed with strong winds (40 kmph average) causing dust storms during dry periods. On the whole, the climate is hostile for vegetational growth.

## Ecophysiological Restrictions

The upper limit to the distribution of vegetation is general and forests separates as a fairly sharp discontinuity from the treeless alpine heath-grassland. In principal, there can be three major reasons for the upper survival limit of trees on mountains with seasonal fluctuating climates (Tranquillini, 1979):

1. Negative carbon balance
2. Arrested phenological cycle
3. Inadequate resistance to deleterious factors

## Negative Carbon Balance

The sum total net photosynthesis of leaves during the growing season at high altitudes is promoted by some factors and repressed by others (Tranquillini, 1979). The factors promoting carbon gain include abundant light, high air and soil moisture, acclimation, genetic adaption; while factors limiting are low atmospheric carbon dioxide, low air and soil temperatures, growing season frost, strong winds, and short growing season (Fig. 1).

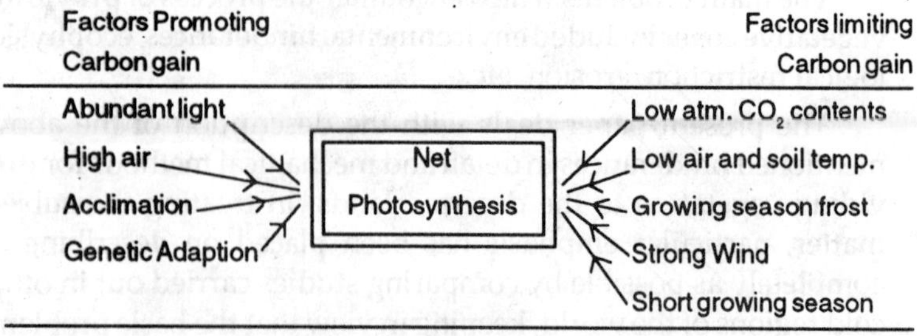


Modified after Tranquillini, 1979

**Fig. 1 :** Schematic diagram showing effects of various external factors on net photosynthesis

Without doubt, factors tending to limit dry matter production predominate timberline, as it is usually noted by the decline in dry matter production with increasing altitude. This reduction in dry matter production, however, lead to a deficiency in organic matter. Dry matter increment diminishes and this has a number of consequences as represented in Fig 2.

Low total annual $CO_2$ uptake

↓

Deficiency in organic matter

↓

Reduced growth increment of

| | Resultant Effect |
|---|---|
| Shoot length and leaves | Reduction in Photosynthetic productivity of the individual (+ve feed-back) |
| Stem diameter | Smaller storage capacity for water and organic reserves, poor transport of water and assimilates |
| Roots | Deficient mineral nutrition and water supply |
| Seeds | Restricted dispersal of regeneration |

**Fig. 2 :** Physiological and ecological consequences from low growth increment as a result of weak photosynthetic productivity.

*Source*: Tranquillini (1979)

## Arrested Phenological Cycle

Shoot extension is of special importance in the annual cycle of growth process. Tranquillini (1979) suggested that the shoot growth and its individual phases are on the one hand promoted and on the other impeded by various factors at high altitudes as described in Fig. 3.

Incomplete cuticular layers and leaf abscission scars increase transpiration and winter desiccation. In such cases, the phenological cycle and the associated acquisition of resistance are no longer synchronised with the climatic rhyth. This incomplete inadequate formation of shoot primardia in terminal buds results in limited shoot increment the following season. Studies of Tranquillini (1974), Baig *et. al.* (1974), Baig and Tranquillini (1976) and Plotter (1976) have shown that thinner cuticles and incomplete development of shoot primordia in terminal buds are widespread phenomena of tree species growing in the transition zone above timberline.

Seasonal course of shoot extension

| Beneficial factors | Stage of development | Restrictive factors |
|---|---|---|
| | 1. Bud Break | — Late shoot activity (Genetically Det.) (E = 1-7) |
| — Early shoot activity | 2. Extension of shoot axis | — Low growth potential of high altitude provenances (E = 2,4,7) |
| | 3. Tissue differentiation (lignification) | — Organic matter deficiency (E = 2,3,5,7) |
| | 4. Expansion of leaf Primardia | |
| | 5. Completion of cuticular layers | — Low temperature (E = 1,2,4,5,6,7) |
| | 6. Terminal bud formation | — Frost during growing season (E = 1,2,3,4,5,7) |
| | 7. Differentiation of shoot primordia in terminal buds. | — Small increment Incomplete maturation |

E = Effect On, *Source*: Tranquillini 1979.

**Fig. 3 :** Interrelation of internal and external factors influencing the annual course of shoot extension.

## Inadequate Resistance to Deleterious Factors

There are many local deleterious factors at high altitudes, but they are of secondary importance. Tranquillini (1979) suggested that the main two factors are forest and disiccation. The major portion of damage occurring to trees in the timberline ecotone is caused by excessive water loss from terminal buds. The realization of this damage depends upon a chain of cause and effect relationships as represented in Fig. 4.

Thus we see that the above mentioned factors acting in unison, they amlify the influence of each other and thus together ensure that above a certain altitudinal zone, trees can no lonġer withstand the incident winter desiccation.

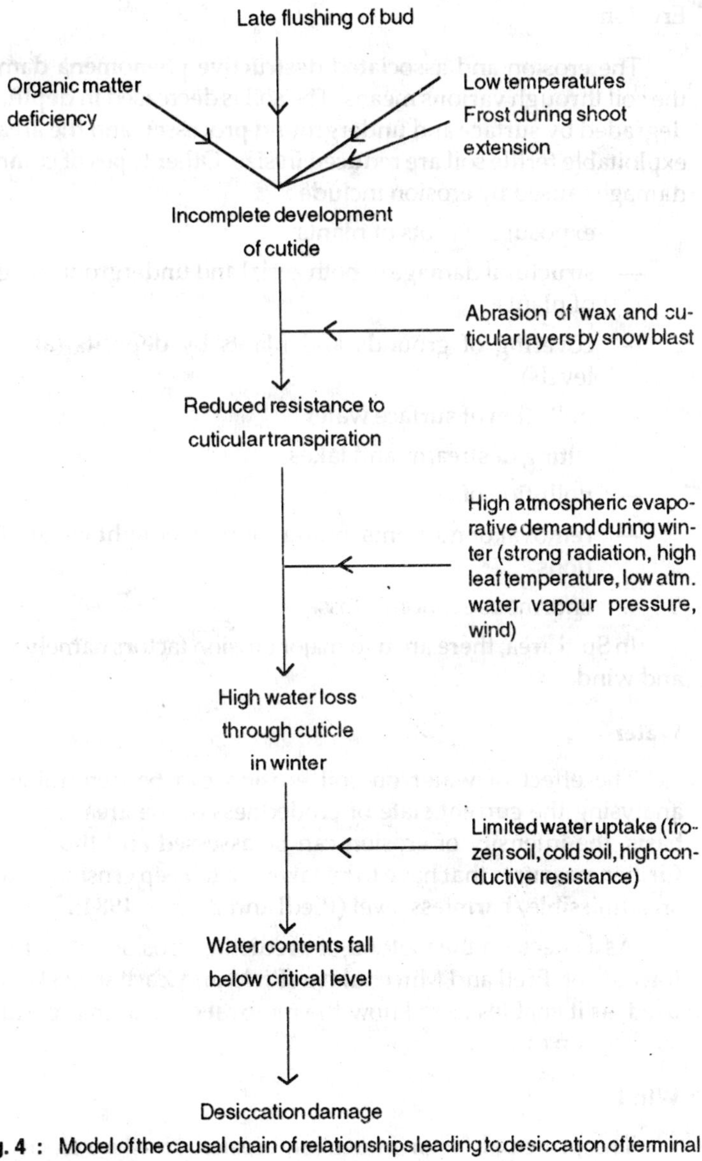


**Fig. 4 :** Model of the causal chain of relationships leading to desiccation of terminal shoots during winters

## Erosion

The erosion and associated destructive phenomena damage the soil through various means. The soil is decreased in depth, it is degraded by surface and underground processer, and the areas of exploitable fertile soil are reduced in size. Other types of common damage caused by erosion include :

— exposure of roots of plants
— structural damage to both aerial and underground organs of plants
— covering of grounds and plants by deposits (at lower levels)
— pollution of surface water
— silting of streams and lakes
— pollution of air
— removal of nutrients, humus and other light weight fractions
— enormous economic losses

In Spiti area, there are two major erosion factors namely water and wind.

## Water

The effect of water on soil erosion can be determined by analysing the current state of erodedness of the area; and on its basis, the intensity of erosion can be assessed and thus assume further measures that have to be taken up to keep erosion down to an admissible/harmless level (Riedl and Zachar, 1984).

As a check on the method of predicting erosion intensity, the formula of Pretl and Mircchulava (Rield and Zachar 1984) can be used, as it enables us to know the removal of solid matter during surface run off.

## Wind

The principal factor in aeolian erosion are the velocity and direction of the wind. A change in the pressure of the wind on rigid surfaces will alter its carrying capacity.

This relationship between wind velocity and the pressure

exterted on the surface can be calculated following Riedl and Zachar (1984). Relationship between wind velocity and the susceptibility of soils of different granularites to aeolian erosion are illustrated in Table 1.

**Table 1 :** Wind velocity (m/s) at which soil grains start moving.

| Gran diameter | 0.01 | 0.10 | 0.25 | 1.00 | 1.50 | 2.00 |
|---|---|---|---|---|---|---|
| Displacement along soil surface | 3.65 | 3.83 | 4.57 | 6.62 | 7.65 | 8.57 |
| Lifting into the air | 3.72 | 5.41 | 6.60 | 10.71 | 13.41 | 16.25 |

Source : Riedl and Zachar 1984.

## Revegetation of Barren Areas

All barren lands are notable for their unfavourable ecological environment in which only a poor cover of vegetation develops, providing the soil with insufficient protection against destructive agents. In Spiti, the conditions are still the worse as at maximum places, even the beadrock is exposed. Such devasted areas usually cannot be afforested by ordinary methods, and special methods have to be selected according to the particular problems of each area.

## Fenced Terracing

On very steep slopes where the soil is unstable and the fine-grained material has been lost fenced terracing (Fig. 5-7) can be employed (Riedl and Zachar 1984). The fences have to be sufficiently strong and durable to hold back soil until the vegetation stand is able to take over this function. The posts are usually 0.8-1.3 m long and 30-50 cm apart, with strong horizontal members interweaving them. These fences terraces may be continuous or interrupted (Fig. 8).

## Palisade Method

As well as fenced terraces of the type mentioned there are many other methods of consolidating unstable soils, one such being the palisade method (Zachar 1982). Where the construction of uninterrupted palisade rows poses a problem, non-continuous palisade terraces or plalisade steps should be built (Fig. 9). Sometimes,

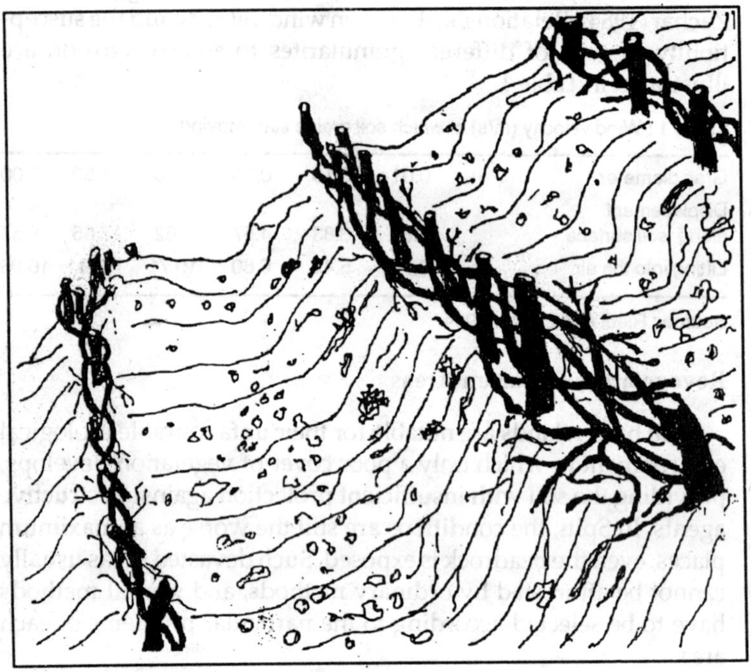

**Fig. 5. : Detail of a fenced terrace used to stabilize barren soil on steep lope**

simple palisade units are constructed in which closely set posts are arranged in a semi-circle and earth is heaped against them so as to form a half-rounded terrace. The single seedling thus planted is protected from the possibility of falling stones.

**Banguette Terraces**

Afforestation on so-called banguette terraces has proved adequate where the surface layer of rubble is shallow and consists of coarse material upto a size of 20-30 cm (Zachar, 1982). A soil terrace is constructed, the seedlings are planted along the mid-line of the terrace, and a small fence is errected to prevent the seedlings from being covered with rubble from above. Here the main function of the terrace is to protect the seedling rather than stabilize the terrace (Fig. 10).

**Fig. 6. : Detail of fenced terrace with earth fill and plantation**

**Step Terraces**

Step terraces are an aid to afforestation, especially on periglacial colluvial deposits, steep-sided dejection cones, and other slopes of moderate to steep gradient. On stony or grass-covered ground, the terraces may be shaped as trenches or as level strips, or they may have an elliptical or circular profile, the size being adjusted to suit the general landscape (Riedl and Zachar 1984). Different types of terraces are shown in fig. 11 (Zachar 1982).

Fig. 7. : Detail of fenced terrace after few years

## Stony Terraces

Other methods for stony areas where stone movement is there, stony terraces as represented in Fig. 12 can be done to prevent falling of stones-erosion, and covering of seedlings etc.

## Soil Working Techniques

Trench-cum-pit type soil working is recommended by Ghosh (1977), in the sloping areas with stony pavements. In between the trenches, pits 60 cm x 60 cm and 1 m deep should be dug at 2m x 2m spacing. Trenches as well as pits should be filled up with a 50 : 50 mixture of the dug up soil and sit brought from the nallahs.

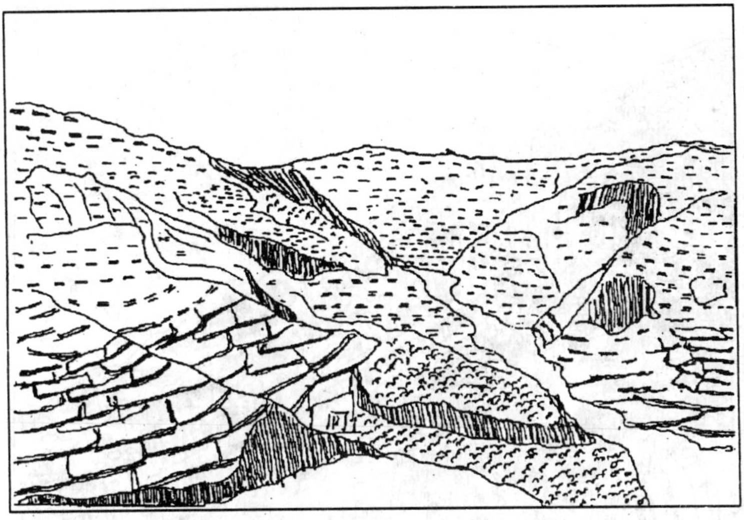

**Fig. 8.** : Stabilizing barren slopes with discontinuous fencing before afforestation

However, Riedl and Zachar (1984), suggested that even cleft and mound plantings too give satisfactory results.

**Irrigation**

Another major problem which will restrict the growth of plantations is of irrigation during the dry summer months. Of course the plantations are carried out near the permanent water channels or at an elevation below them, then irrigation can easily be provided through small water channels (Khuls) or with the help of pipes. But if the plantations are to be carried out above the water courses, then of course the problem can only be solved through water lifting devices such as high drums and by use of water lifting pumps working on solar or wind energy; and the water stored in reserviors (made of polymers or dug up and provided with plastic sheets).

Of course this water lifting project will cost a lot, so in order to maximise the use of lifted water, irrigation should either be through pipes operated manually or through drip irrigation techniques.

**Fig. 9. : Stabilizing of debris with palisade fences**

## Fertilization

As maximum of the area in Spiti is devoid of top soil, excepting on few fan shaped alluvial deposits, proper fertilization is also a requirement of the area for survival and growth of the plantations. Thus fertilizer can be added either during the earth working period

**Fig. 10.** : Detailes of afforestation and banquette terrace

(FYM) and / or it (NPK) can be supplied after the plantations when irrigation is done. An easy way of fertilization can be by additing the fertilizer (NPK) in the irrigation system itself.

**Protection Against Graziers**

Above all, if the plantations are not provided with protection against domestic and wild graziers, especially during periods when natural alpine pastures are covered by snow, till then there is no use undertaking such projects.

Source : Zachar

**Fig. 11. :** Detailes of method of afforestation using circular-arc terrace (above), surrounding the seedlings with boulders (below)

## Conclusion

By knowing the major constraints to be encountered during revegetating the cold desert areas, we can conclude that our

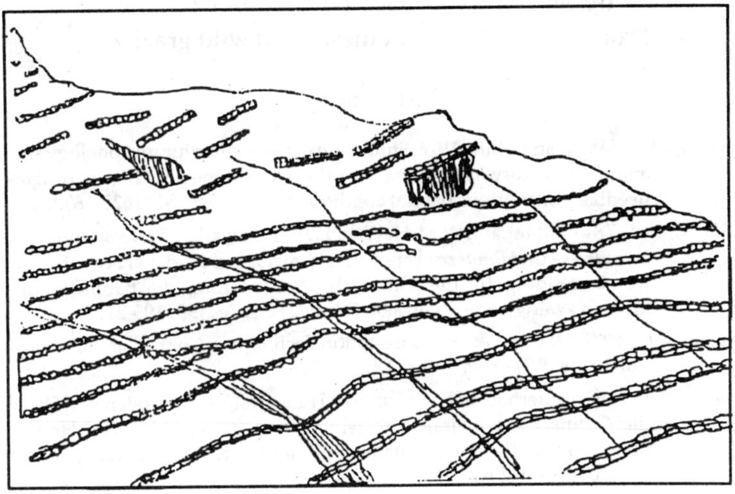

Source : Aachar

**Fig. 12.** : Planting of stony ground using low stony terraces

program should be based on sound scientific observations and logical thoughts which can be summed up as :

— Recording the metereological observations of the area to be planted in detail.

— Note the species presently growing in the area,

— Collect information from old records and other local sources of the species previously growing,

— Screen out exotic species which can withstand the recorded environment.

— Seedlings should be healthy and grown in cold climate only from seeds collected from local or surrounding areas.

— Mixed plantations should be preferred over single species.

— Calculate the amount and type of erosion by various means,

— Prepare the area for plantation depending upon soil and topography to provide stability to the area against erosion and protection of seedlings.

— Perform the needed soil working conditions,

— Provide proper irrigation,

— Requirement of fertilization to be met with,

— Protect the area from domestic and wild graziers.

## References

Baig, M.N. and W. Tranquillini, 1976. Studies on upper timberline morphology and anatomy of Norway spruce (*Picea abies*) and stone pine (*Pinus cembra*) needles from various habitat conditions. *Can. J. Bot.* 54 : 1622-1632.

Baig, MN. W., Tranquillini, and W.M. Havranek, 1974. Cuticulare Transpiration von *Picea abies* - und *Pinus cembra* - Zweigen aus verschiendener Seehohe und ihre Bedeutung fur die winterliche Austrocknung der Baume an der alpinen Waldgrenze. Zentralbl. *Gesamte Forstwes. 91* : 195-211.

Ghosh, R.C., 1977. Handbook on Afforestation Techniques. Controller of Publication Delhi, FRI Press, Dehra Dun : 411.

Platter, W., 1976. Wasserhaushalt, cuticulares Transpiratonsver mogen und Dicke der Cutinschincten einiger Nadelholzarten in verschiedened Hohenlange und nach experimentaller Verkurzung der Vegetationsperiode. Diss. Univ. Innsbruck.

Riedl O. and D. Zachare, 1984. *Forest Amelioration.* Elsevier, Amsterdam : 623.

Tranquillini, W., 1974. Der EinfluB van Seehohe und Lange der Vegetationsceit auf das cuticulare Transpirations vermogen von Fichtensamlingen in Winter. *Ber. Dtsch. Bot. Ges. 87* : 175-184.

Tranquillini, W., 1979. Physiological Ecology of the Alpine Timberline. Ecological Studies 31, Springer Verlag: 137.

Zachar D., 1982. *Soil erosion.* Elsevier, Amsterdam.

# Chapter-27

# Environmental Impact Assessment (EIA)

*Y.C. Pande*

*R.K. Pande*

In Sai basin, a combination of factors lead to a situation where both mass wasting and soil erosion are very active. A hostile natural environment combines active mountain building with a monsoonal climate. High population puts extreme pressure on the land and leads to poor management practices, overgrazing and deforestation for fuel, timber and fodder.

The Sai basin is situated in the Lesser Himalaya, which are one of the world's most active young folded mountain belts and are still being formed as the Indian crustal plate continues to move against and under the Tibetan crustal plate to the north. Eathquakes provide the evidence for this continuing activity, representing sudden releases in stress within the crustal layers.

A combination of geologically rapid uplift and a monsoonal climate, combining high temperatures with high and intense seasonal rainfall, results in very active landscape forming processes dominated by rock weathering and erosion.

Mountain building movements are characterized by alternating periods of high activity when the land rises relatively quickly, and low activity when relatively little movement takes place. In active phases rivers struggle to cut down to new base levels and erosion dominates. In quieter phases rock weathering becomes equally important causing mineralogical changes and soil formation.

Mass movements are the principle method by which slope evolve. As the mountains rise the rivers cut down forming deep-V-shaped valleys (Fig. 1A). The sides of the valleys are undercut by river action, the rocks are being weathered which weakens them and eventually turns them to soil. These weaker material can no longer stand at the same steep slope angles as the original rocks. These slopes then fall when triggered by excessive rainfall or earthquakes (Fig. 1B).

If river levels remain static for a long period (many thousands of years) the slope angles will be reduced to gentler grades and the whole landscape will become more stable. Even when the debris eventually stabilises, part of it continues to be delivered into stream and river channels so increasing their erosive energy and consequently their ability to erode laterally (Fig. 1C) and deposit alluvium.

Soil erosion is also active. This is influenced by rainfall intensity, soil type, soil properties, soil erodibility, slope angle and vegetation cover. In most parts of Sai Basin vegetated soil slopes of up to about 30° tend to be stable against further slope instability. However, removal of vegetation from these slopes can contribute significantly to soil erosion.

The history of slope process in Sai Basin is one of a continuous succession of periods of rapid downcutting, alternating with static periods when a certain amount of adjustment towards a more stable state is achieved. Thus the process described above may be repeated (Fig. 1D).

## Regional Factors controlling Natural Hazards

Geology, climate, weathering, landuse and erosion are found to be vital regional factors controlling natural hazards. Sai Basin, on an average, falls under the Almora crystalline. Sai river flows along the North Almora Thrust (NAT), which is segmented due to the presence of transverse faults. The lithology of the arca is characterized by seven rock formations, viz, garnetiferous mica-schist with interbedded micaceous quartzite, micaceous quartzite with pelitic intercalations (phyllite-phyllonite), granitic gneisses with amphi-

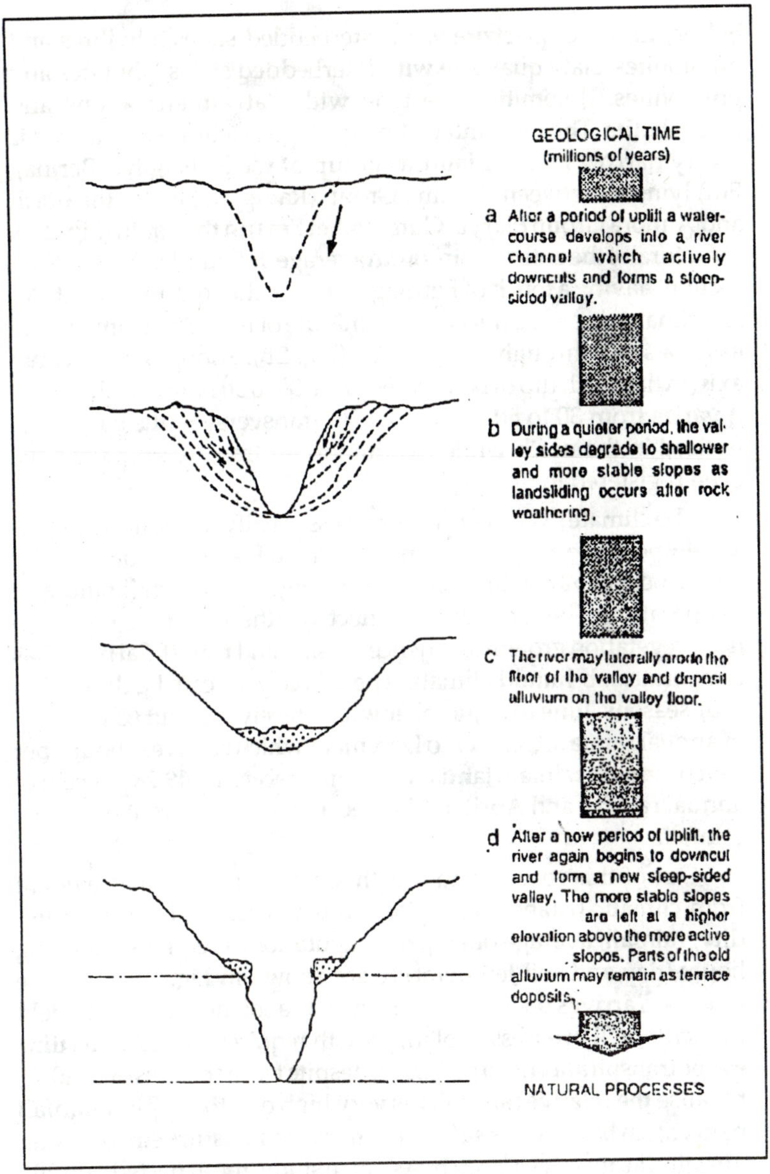


**Fig. 1. :** Cyclic delopment of a river valley during mountain buildings.

bolitic, massive quartzite with interbedded slates, phyllites and phyllonites, slaty quartzite with interbedded slates, phyllites and phyllonites, dolomitic limestone with slaty intercalations and amphibolite. The most interesting geological feature in this area is a very narrow layer of Jaunsar group of rocks (Nagthat-Berinag Fm) lying in between Damtha Group (Rautgara Fm) in the north and Almora Group (Saryu-Gumalikhet Fms) in the South (Fig. 2A). Structurally the entire basin, on an average, falls under Autochthon section, having a touch of Berinag Nappe in the south and west. An anticlinal axis can be traced in the middle of the basin from west to east, passing through Someshwar (Fig. 2B). Along the anticlinal axis, in the north dip of beds varies from 35 to 50 N. and in the south it varies from 30 to 50 S. Four sets of transverse faults have been noticed in the north while two important transverse faults have been registered in the south.

The climate of an area is defined essentially by its mean annual air temperature and mean annual rainfall. Elevation, topography and aspect combine to affect local changes in rainfall, wind and temperature. These conditions affect weathering and soil formation, vegetation growth and hence erosion and runoff. Sai basin has temperate sub-humid climate. The whole year can be divided in four seasons, June to September which receives about 65 per cent of annual rainfall, October to December which receives about 6 per cent of annual rainfall, January to March receiving 19.28 per cent of annual rainfall and April to May receiving 9.27 per cent of annual rainfall.

Out of this second and fourth are deficit periods as evident from climatic balance. Variations in temperature, relative humidity, rainfall, average open pan evaporation, average sunshining hours/day and soil temperature are shown in Fig. 3. When the monsoon arrives the water quantity in the soil quickly rise to field capacity, well in excess of plant growth requirements. At this time evapotranspiration is suppressed, despite temperatures over 20°C, because the relative humidity is very high due to a typical rainfall pattern, Sai basin never suffers from severe moisture stress of long duration at any given stretch. As a case study, the expected rainfall, at 80 per cent probability, is estimated on fortnighty bases for each fortnight, starting from 24th Sept to 7th May. Fortnightly periods are chosen because of low temperature during the period when

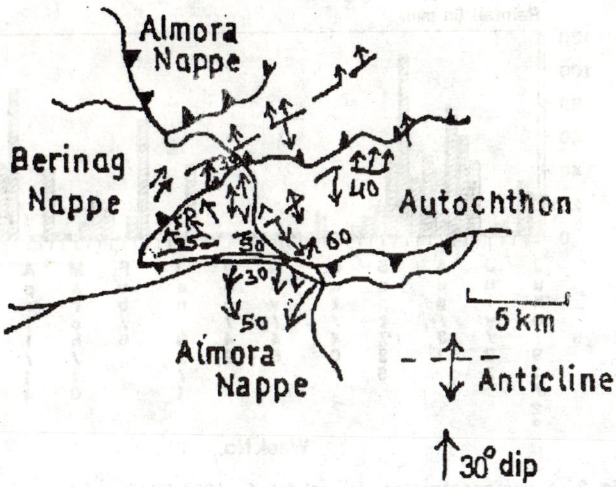


**Fig. 2 (a) :** Structural map in and around Sai basin.

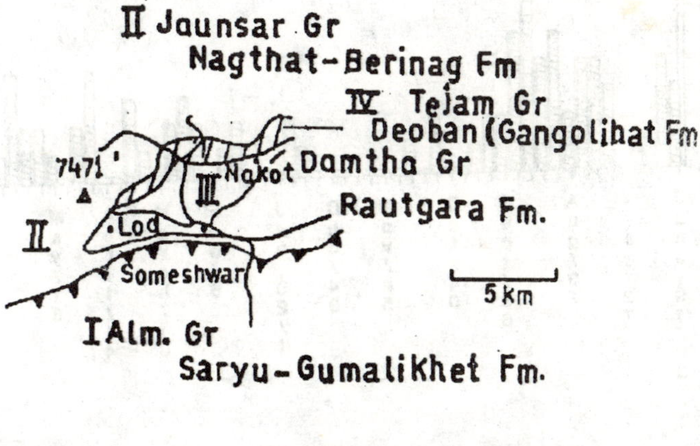


**Fig. 2 (b) :** Lethotectonic zones in and around Sai basin.

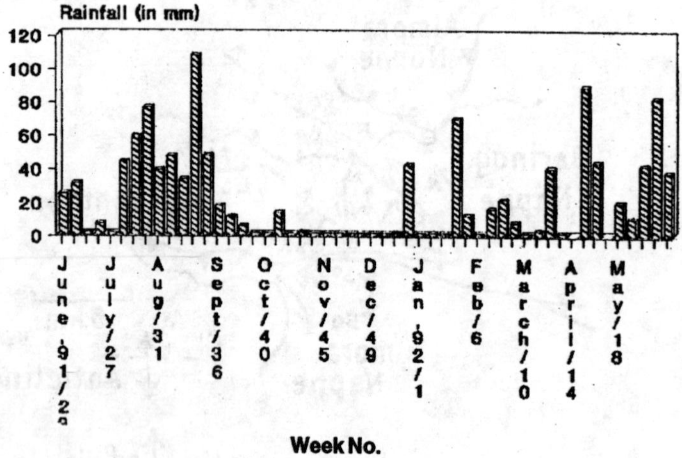


Fig. 3. : Summary of meteorological data for 1991-92 (Someshwar Station)

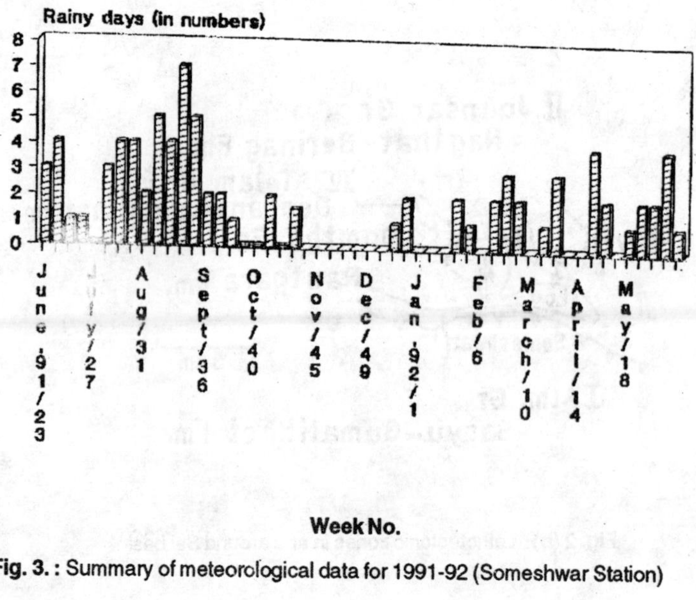


Fig. 3. : Summary of meteorological data for 1991-92 (Someshwar Station)

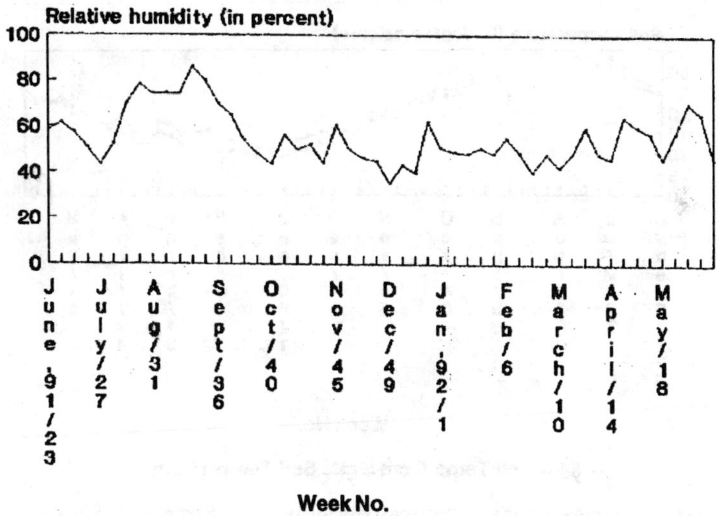


**Fig. 3.** : Summary of meteorological data for 1991-92 (Someshwar Station)

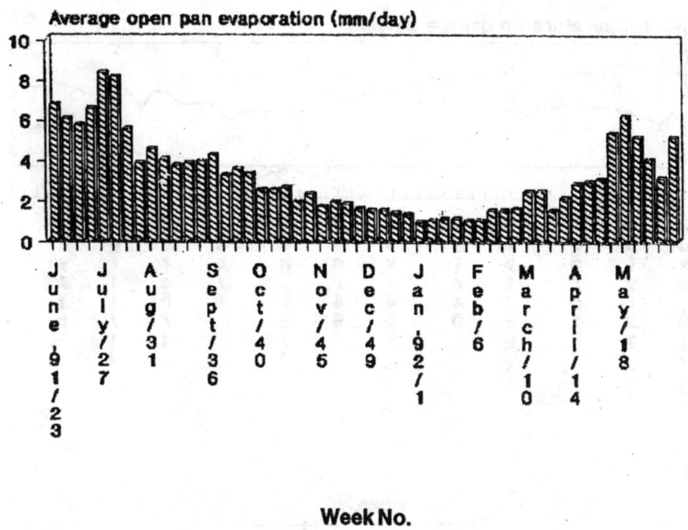


**Fig. 3.** : Summary of meteorological data for 1991-92 (Someshwar Station)

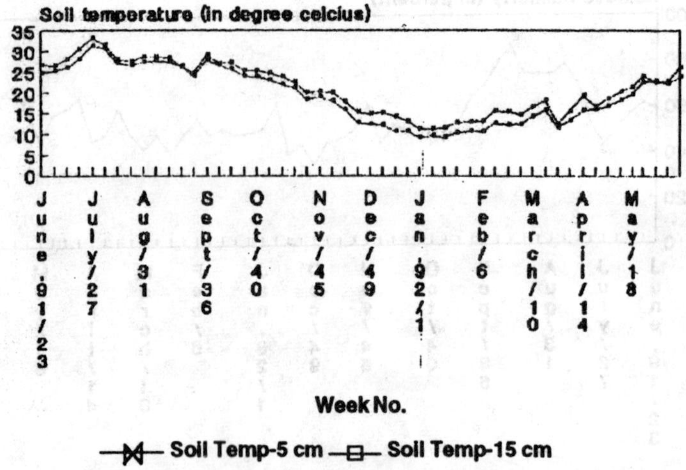


Fig. 3. : Summary of meteorological data for 1991-92 (Someshwar Station)

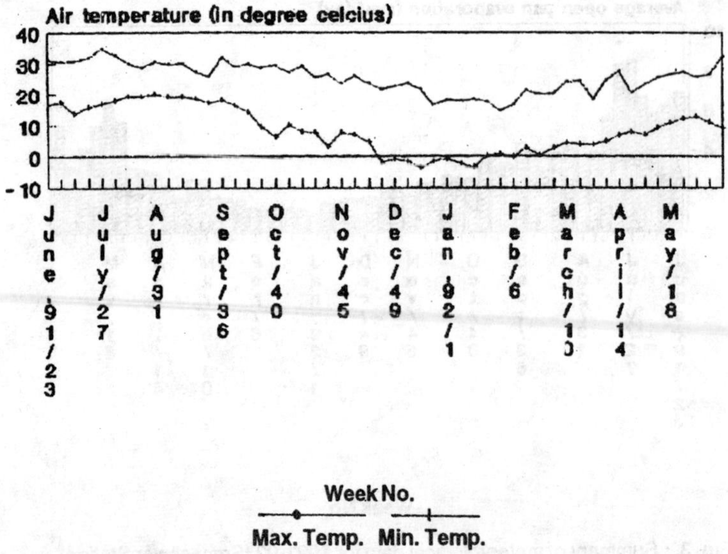


Fig. 3. : Summary of meteorological data for 1991-92 (Someshwar Station)

rainfall is expected. No significant stress is likely to occur due to long dry spell, as the vegetation remains in a slow growth stage. Since every fortnight the expected rainfall is only at 80 per cent probability, the distribution of rainfall among the occasions is estimated by Markov-Chain Model (Saxena and Surrendran, 1973). As only in the first to fifth fortnight rainfall is expected at 80 per cent probability, the distribution is calculated for these fortnights with reference to the 26th fortnight, having no rainfall. Since the rainfall of the 20th fortnight is related to monsoon and it is at for end, so for this fortnight, only expected value of rainfall at 80 per cent probability has been accounted. For the assessment of the soil moisture status, the field capacity and permanent wilting point are taken 10 per cent and 6 per cent, respectively, for the most prevalent sandy loam soil. Apparent specific gravity of 7.5 and soil depth of 60 cm have also been considered. Thus, moisture status has been estimated by-available soil moisture at the end of 'nth' fortnight (Stn)- available soil moisture at the end of (n-1th) fortnight (STn-1) + rainfall in the 'nth' fortnight (Rn)- consumptive use in 'nth' fortnight (an). Calculated at 80 per cent and 50 per cent probability (Fig. 4), rainfall is expected only between 1st fortnight to 5th fortnight. It is noticed that with normal rainfall there is no moisture stress all along the growth period.

Although it can rain continuously for days at a time, the monsoon is often characterised by periods of rain lasting for only a few hours, broken from dry spells of similar length. If the sky clears between showers the sun becomes extremely hot and evaporates surface water very rapidly, sufficient to bake a soft crust on exposed soil surfaces. Another characteristic of monsoon rain is that it is often very intense. Peak intensities of 100 minutes duration at most. Rain of this intensity is very erosive, especially if it follows a period of normal rain during which the oil has become well wetted. The bust of rainfall saturates the upper part of the soil profile, which can liquify and slide downhill in a destructive earth or mud flow.

Climate directly affects the speed at which rocks weather. Rocks in the area undergo decomposition at least twice as fast as those in the cooler areas. The area consists of metamorphic and very soft sedimentary rocks. All are highly folded which has produced numerous joints and cracks throughout the rock mass. The monsoon rains, coming at the hottest time of the year, create a

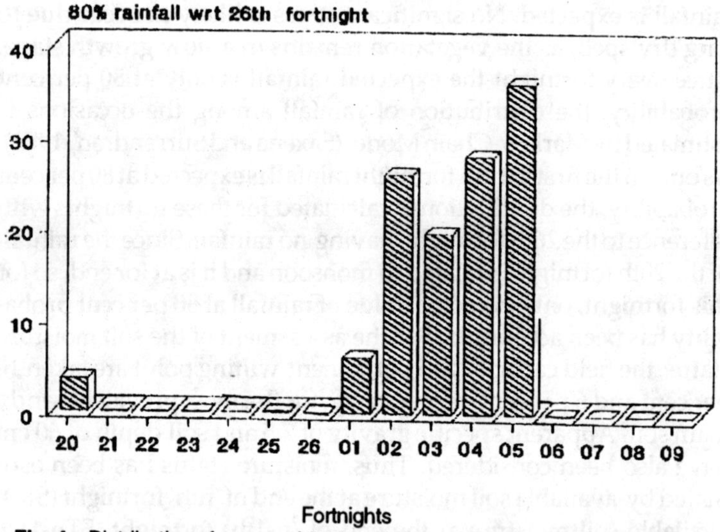


**Fig. 4. : Distribution of seasonal 80% probable rainfall w.r.t. 26th fortnight.**

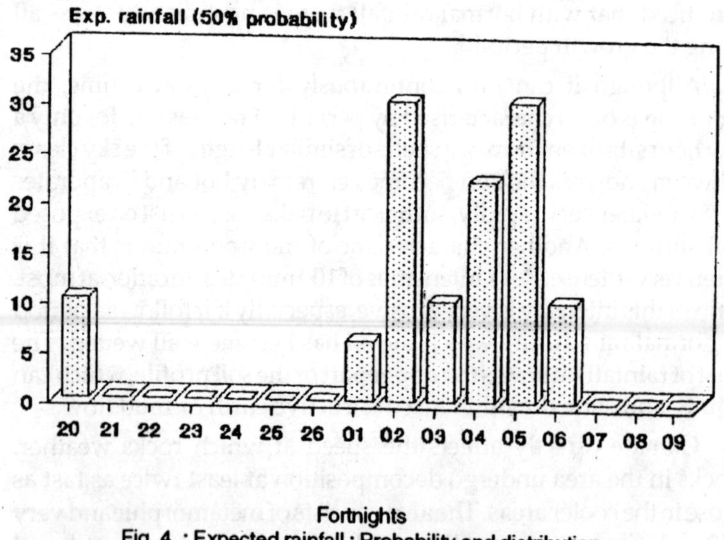


**Fig. 4. : Expected rainfall : Probability and distribution (September 1991-September 1992)**

highly active chemical weathering environment which vigorously attacks the rock fabric to a depth of several metres. Water penetrates all the joints and micro-fractures in the rock.

The minerals, which in many cases have a low resistance to weathering are reduced to clay, silt, mica and quartz. Initially the surface zone decomposes, together, with these zones adjacent to joints and fissures. As weathering continues the fresh strong rock changes to weak rock and eventually to a residual soil. Between the parent rock and the soil are transitional layers of increasing weathered material of decreasing strength.

The strength of the rock is drastically reduced by weathering and the weathered layer shears when part of the slope is over-steepened. It is the strength of the transitional weathered layers which often controls the depth of landslides, particularly debris slides on steep slopes (Fig. 5).

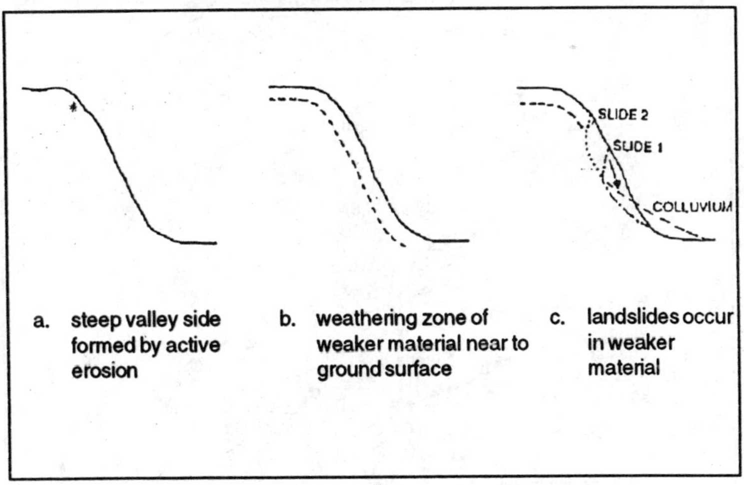


Fig. 5.: Weathering control on debris slides on steep slopes.

Vegetation can provide a protective cover or boundary between the atmosphere and the soil, the major effects being hydrological and mechanical (Fig. 6). Vegetation influences the way in which water is transferred from the atmosphere to the soil, ground water and surface drainage systems. In affecting the volume and rate of flow along different routes vegetation influences the process

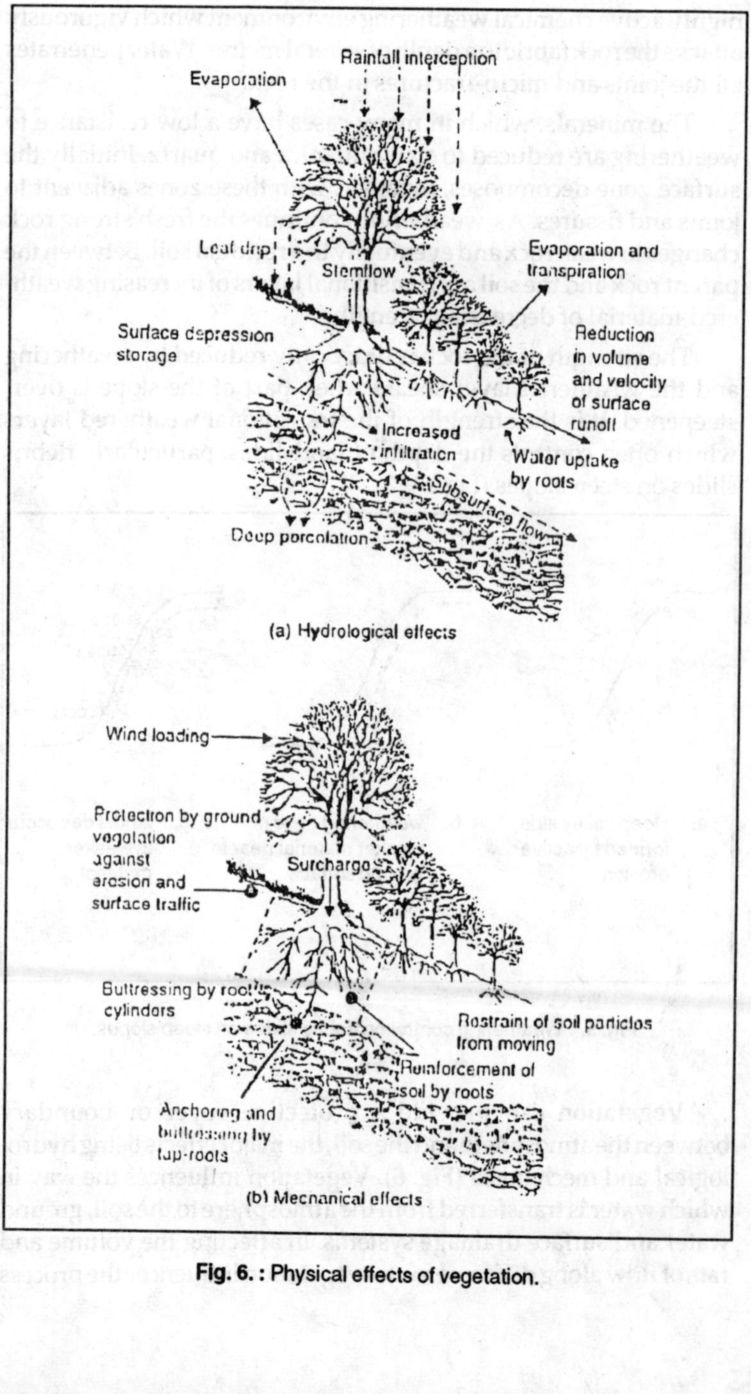


**Fig. 6. : Physical effects of vegetation.**

and extent of soil erosion. It also modifies the moisture content of the soil and thus its shear strength. Mechanically, vegetation increases the strength and competence of the soil in which it is growing and therefore contributes to its stability.

The importance of vegetative cover is demonstrated by the effect of its removal. In the year following clearance of a forest cover for fuel, timber or agriculture, soil erosion and mass-movements increases dramatically because of the removal of the protective layer from the ground surface. Regrowth of the vegetative cover causes erosion rates to decline in subsequent years, but this trend is frequently offset some five to seven years later by a rapid rise in soil loss due to mass failure of the soil as the old roots rot away. Examples such as this illustrate the importance of recognising that existing vegetation may clearly be playing a major engineering role. Failure to appreciate this can result in unforeseen and often disastrous consequences if the vegetation is removed, or replaced by inappropriate species and varieties. The engineering effects of vegetation must therefore by considered not only when proposals are being made for reforestation of unprotected slopes and unstable landslides, but also when policies for the management of vegetated area or for the clearance of vegetation are being assessed.

Vegetation is found extremely effecting in preventing rainsplash erosion by maintaining soil infiltration rates and preventing surface casting. The proportion of vegetation cover determines the amount of ground protected from direct rainfall impact. Maximum protection is given by covers of 70 per cent or more. As a result of a combination of a surface roughness, infiltration and interception, volumes of surface water runoff from vegetated areas are much less than from bare soil. Changing landuse from forest of dense grass to a more cover results in higher runoff volumes. Vegetation also reduces the velocity of runoff because of roughness presented by the stems and leaves of the plants. Tall undergrowths retard overland flow more than shorter ones. Laterally spreading surface root systems are more effective in reducing surface soil erosion than vertically structured ones with top roots. Trees, large bushes and shrubs are able to prop unstable or loose boulders and stones, preventing them from rolling downslope. Superficial vegetative litter and upper layers of soil containing a dense mat of roots running parallel to the surface provide improved drainage on slopping ground through subsurface water flow. Vegetation also protects the soil from trampling by directly absorbing the impact of human and animal feet. Shrubby species disappear more rapidly than do grasses under impact from trampling. The mechanical effect of the roots of vegetation is to enhance the shear strength and resistance to sliding and increase the strength of the soil / root mass through the binding action of roots in the soil. The top roots of many tree species penetrate in to the deeper soil and rock layers and anchor them against downslope movement. The surcharge weight of trees planted at the bottom of an unstable hillslope increases the stability of that slope with time. The result of a decreasing soil moisture content, as a result of evapotranspiration, is to reduce the pore water pressure in saturated soils and to increase soil suction in soils below saturation, so increasing the overall slope stability. The effect of evapotranspiration in vegetation varies with species and rooting depth but the effect can also be seasonal.

However, the vegetation also adheres some adverse effects. Leaf drip, comprises raindrops that coalesce on the surfaces of leaves and stems and fall to the ground as larger drops at greater erosive velocities than rainsplash. If these drops fall from canopies less then 0.5 mt in height there is no significant in soil detachment as the drops do not gain sufficient velocity in descent. Grasses and small shrubs therefore result in lesser impact velocities than trees. Broad-leaved trees produce larger drips than needle-leaved trees. Irregular and tussocky vegetation covers locally increase the erosive capacity of overland flow through vortexing, eddying and turbulence. Vegetation on the top of an unstable slope provides additional surcharge weight on the slope so increasing the downslope forces and the tendency for sliding. Wind loading is usually only significant when the wind is strong and gusty. The forces induced in trees by wind are sufficient to disturb upper soil layers and thus directly initiate mass-movements. Root wedging has a potentially destabilising effect whereby fissures and joints in rocks are opened up by the advance and growth of roots in the quest of water.

## Specific Factors Controlling slope Instability

The more the rock structure dips into the valley the greater the potential instability (Fig. 7). Valley sides where the structure dips into the valley have more serious landslide problems. The steeper the slope angle, the greater the potential instability and the greater

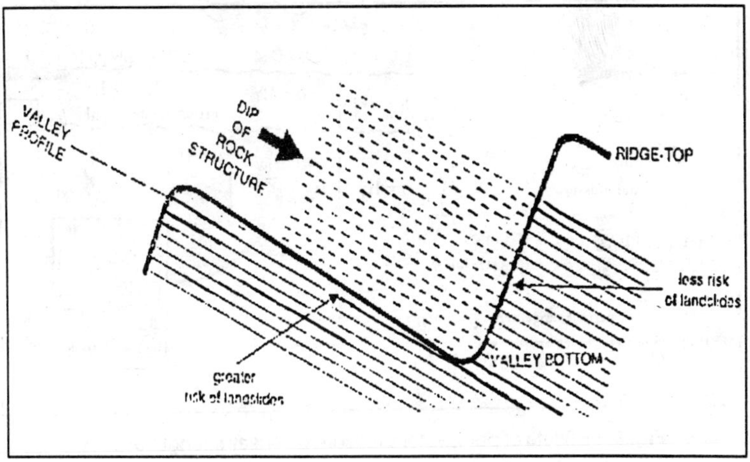


**Fig. 7. : Influence of rock structure on valley profile.**

the slope height, the greater the potential instability. Often the cut slope to man-made structures like road, canal etc. are cut too steeply, even vertically. These soon fail and cause much larger problems. Similarly, the lower the shear strength of the soil, the greater the potential instability. The shear strength of a soil depends on its mineralogy, its grain size distribution and its density. In general, clay soils have a lower shear strength than sandy soils.

The higher the groundwater level, the higher the pore water pressure, the greater the potential instability (Fig. 8). When groundwater is present in a slope it can be seen as a spring or a seepage. Groundwater rises in the monsoon season and so that the water pressure rises. Water pressure reduces the shear strength and causes failure of the slope.

A major cause of slope failure is inappropriate design and construction. The most common cause of canal bank failure is deforestation along canal alignment and the oversteeping of the canal bank during construction. A canal bank failure blocks the channel which then causes a washout and a major landslide movement is then initiated.

## Slope Instability

In general geological reactions go on in a slow motion and their

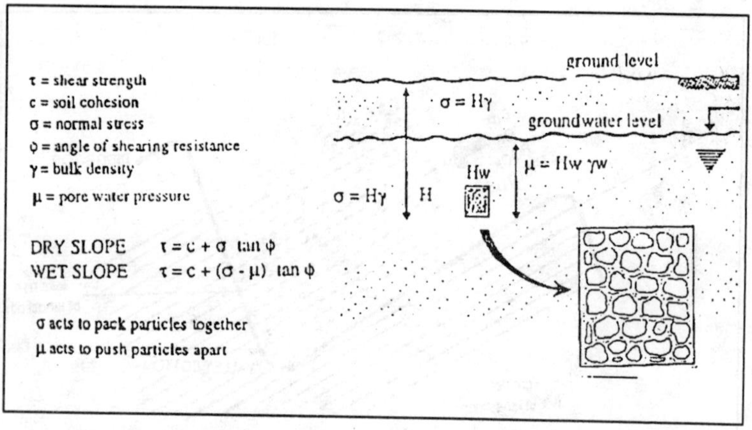


**Fig. 8.** : Effects of pore water pressure on shear strength of soils.

results are obtained after a long period which may be disastrous too. However, following are the major geological disturbances recorded in the study area (Fig. 9).

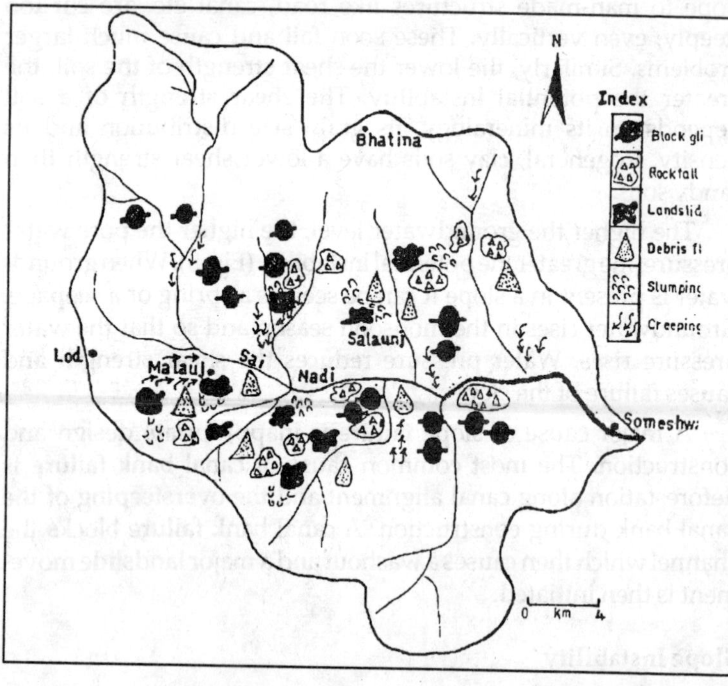


**Fig. 9.** : Mass movement types of Sai basin.

**Flowage :** Registered mainly in Dungal and Harnwal areas. It means downward movement of unconsolidated or loosely packed mass along no definite surface of failure. In Dungal mass-movement is not very visible and down-slope movement is only 0.7 cm per year. In monsoon period, mass-movement in Harnwal is easily visible and is about 1.46 mt a day on an average.

**Sliding** : Recorded in Dungari, Nag Joshi, Lod and Billori areas. In this case the mass fails by moving along a definite surface or plane of failure. The mass above this surface is unstable resting on a stable base. In Dungari slumping is the major form of sliding, while in Nag Joshi rock slides and falls are the main movements. In Lod debris slide is registered and in Billori debris fall is the main geologic disturbance.

**Landslides :** Generally landslides are caused by unfavourable geological sequence of formations, dips and joints, insufficient toe and lateral support, hydrological (pressure of excessive sub-surface water) conditions and topographical (steep slope) features. Five types (Fig. 10) of Landslides are recorded in Bhatina, Kwerali, Baragaon, Bayala Guth and Naikana. In Bhatina circular failure is noticed in this case rotational failure has involved fine grained soils and is the movement of material downslope along a concave-upward slip surface. The failure has breaked up into a number of smaller slips and in very wet conditions the material moves in a slump or earthflow. Since its start it has continued to work back up the hill-side and the whole valley side from ridge-top to channel floor is affected.

In Kwerali planar failure has occurred along a week surface parallel to the ground surface and has involved sliding of the weathered rock debris over the strong or less weathered rock.

Wedge failure has been noticed in Baragaon area. In this case of mass-movement, the failure planes are controlled principally by two intersecting fractures in rock and the debris in chiefly consisted of rock fragments.

In Bayala Guth toppling failure is the geologic disturbance. Here, the failure planes of mass-movement are controlled by three parallel and steeply dipping fractures in rock and the debris is mainly consisted of rock fragments.

Sapalling is the main landslide type recorded in Naikana area, is the fall of individual fragments of rock from a rock cliff as they have become loosened by weathering and human activities.

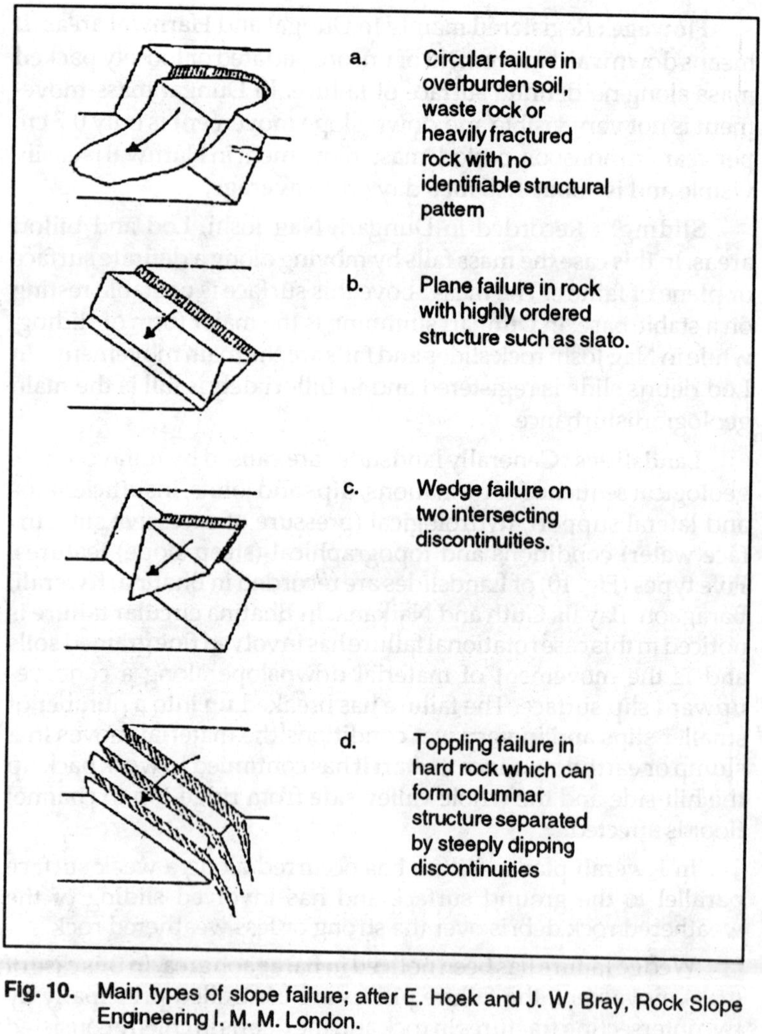


**Fig. 10.** Main types of slope failure; after E. Hoek and J. W. Bray, Rock Slope Engineering I. M. M. London.

## Soil Erosion

Soil erosion is a two-phase process consisting of the detachment of individual particles from the soil mass and their transport by erosive agents. When sufficient energy is no longer available to transport the particles a third phase, deposition occurs.

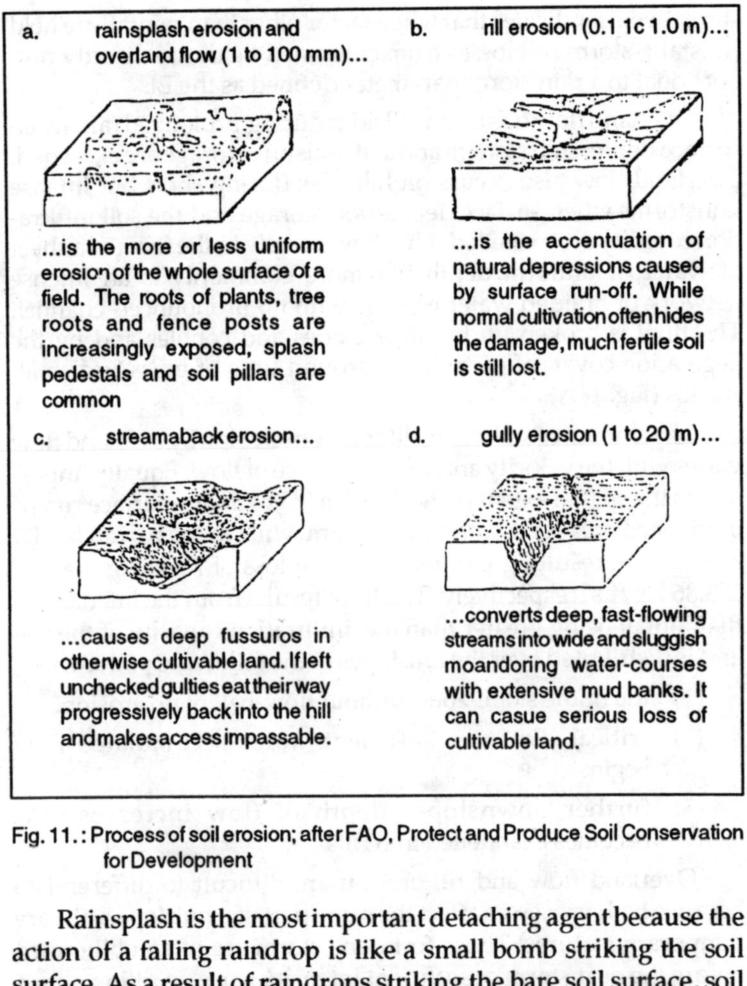


Fig. 11.: Process of soil erosion; after FAO, Protect and Produce Soil Conservation for Development

Rainsplash is the most important detaching agent because the action of a falling raindrop is like a small bomb striking the soil surface. As a result of raindrops striking the bare soil surface, soil particles are thrown through the air over distances of several centimetres. The transference of soil particles also has the effect of compacting the soil into a surface crust which in turn reduces infiltration capacity and thereby promotes greater surface runoff. Since rainsplash erosion acts uniformly over the land surface its effects are seen only where stones or tree roots selectively protect the underlying soil and splash pedestals or soil pillars are formed (Fig. 11A). Such features frequently indicate the severity of erosion. In Sai Basin the total rainfall depth and $EI_{30}$ values for individual storm events indicates that when factor other than rainfall are held constant, storm soil losses from cultivated fields are directly proportional to a rain storm parameter defined as the EI.

Overland flow occurs on hillsides during prolonged rain when surface depression storage and soil moisture storage are exceeded. Overland flow also occurs on hillsides during short but intense rainstorms when surface depression storage and the soil infiltration capacity are exceeded. The flow is rarely in the form of a sheet of water of uniform depth but more commonly is an intense network of braided water courses with no pronounced channel. The flow is broken up by large stones and cobbles and by the vegetation cover, often swirling around tufts of grass and small shrubs (Fig. 11A).

The amount of soil loss resulting from erosion by overland flow varies with the velocity and the turbulence of flow. Equally important is the spatial extent of the flow. In Sai basin the surface runoff in forested area is recorded as 6.57 mm while in the barren land it is 17.8 mm, resulting into the sediment loss of 71.56 Kg/ha and 153.86 Kg/ha, respectively. The flow results from the intensity of the rainfall being greater than the infiltration capacity of the soil and is distributed over the hillslopes in the following pattern, viz.,

(1) top of the slope-zone without flow-belt of no erosion.

(2) critical distance- sufficient water accumulated-flow begins.

(3) further downslope—depth of flow increases-flow becomes channelled into rills.

Overland flow and rill erosion are difficult to differentiate because both processes affect the same part of a hillslope. Rills are temporary features. Those form one storm are often obliterated before the next storm of sufficient intensity to cause rilling, when the channels may form an entirely fresh network. Most rill systems are discontinuous in that they have no connection with the main drainage system. The concentration of flow increases the rill erosive power downslope and the channels deepen and widen and migrate both upslope and downslope (Fig. 11B). Rill flow transports large grains of soil and accounts for the bulk of sediment removal from a hillside. In the crystalline rocks of the Sai Basin, sufficient overland flow from steep slopes provides enough discharge to the rills and, therefore, a little source area flow depth is able to initiate a rill. However, in the case of dolomitic limestone most of the overland flow is absorbed by the local rocks. Consequently, source area have to be enlarged to produce sufficient flow depth to initiate a rill.

Gullies are relatively permanent steep-sided water courses which experience temporary flows during rainstorms. Compared to stable river which has a relatively smooth, concave-upwards profile (Fig. 11C), gullies are characterized by a headcut and various steps or kniek points along their courses. These rapid changes in slope alternate with sections of very gentle gradient, either straight or slightly convex in long profile. Gullies also have relatively greater depth and smaller width, carry larger sediment loads and display very erotic behaviour (Fig. 11D). Gullies are almost always associated with accelerated erosion and therefore with landscape instability. The main cause of gully formation is too much water brought about by deforestation. i.e., burning of the vegetation and overgrazing resulting in greater runoff. Gullies do not develop from large rills. The sequence of gully formation is shown in Fig. 12. Gullies are also initiated where linear landslides have left deep, steep-sided scars and through ground surface subsidence caused as a result of piping erosion.

The formation of continuous pipes or underground channels is most common in soils that when dry take up water very slowly and reluctantly, but once saturated have no cohesion and flow like mud. These soils are usually found where there is some severe imbalance such as excessive sodium and complete defloceulation, e.g., Mowe. Piping has occurred due to surface water infiltration through the soil surface and its downward movement until it comes to a less permeable layer. In case of an outlet the water flows laterally through the soil over the less permeable layer and the fine particles of the more porous soil are being washed out. In such situations lateral flow increases, and the pipes enlarged sometimes leading to roof collapse, and the tunnel becomes a gully.

**Mountain Zone Classification**

When considering the alignment, design, construction, rehabilitation and maintenance of civic structures it is important to recognise where mass-wasting or soil erosion is occurring and distinguish between different types of slope failure and erosion processes.

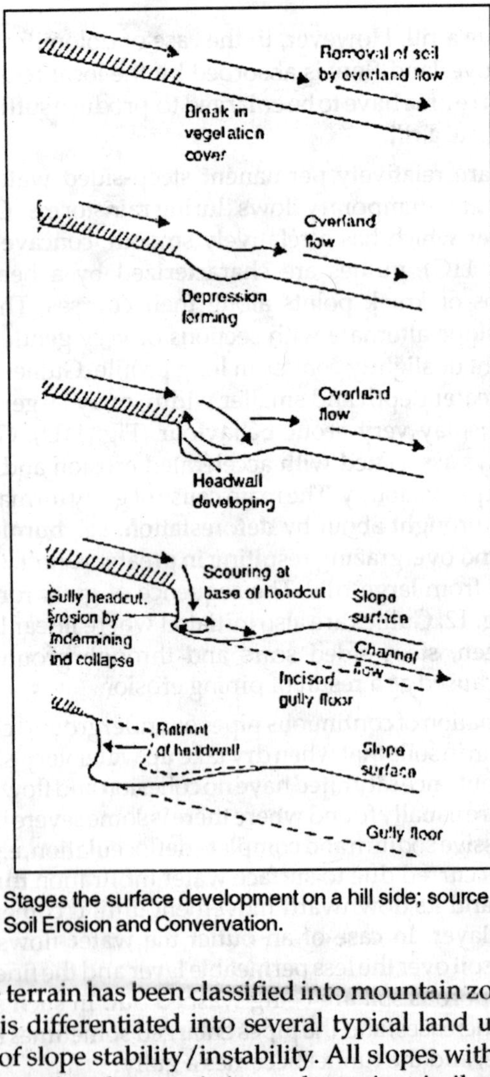


**Fig. 12. :** Stages the surface development on a hill side; source R. P. C. Morgan, Soil Erosion and Convervation.

The terrain has been classified into mountain zones and those in turn is differentiated into several typical land units based on criteria of slope stability/instability. All slopes within a land unit display similar characteristics and require similar conservation (engineering and vegetative design) inputs for environmental protection measures. Therefore, by identifying these land units it is possible to formulate a framework for design, construction rehabilitation and maintenance of civic structures which will reduce the risk of mass-movement and soil erosion. Fig. 13 illustrates a mountain zone classification which is described in Table. 1.

**Table 1:** Mountain Zone Classification

| *Zone* | *Land Unit* | *Description* |
|---|---|---|
| 3 Degraded Middle Slopes < 35° | 3A | Ancient Erosional Terrace. Covered with in situ weathered profile of soil upto 3 m thick. Slope angle generally < 35°. Relatively stable. Often Farmer terraced. Surface water erosion high. |
| | 3B (4D) | Degraded Colluvium. Transported slope debris or land-slide debris comprising gravel, cobbles, boulders bound in silt/clay matrix. Slope angle < 35°. Relatively stable. Often farmer terraced. Variable permeability. |
| 4 Active Lower Slopes > 35° | 4A | Bare Rock Slopes. Steep slope angles often > 60°. Stability dependant on orientation of discontinuities, i.e. bedding/joints. |
| | 4B | Rock Slopes with shallow (<2m) loose debris cover. Slope angle 45° - 60°, Shallow instability and debris slides. Mass instability as 4A. |
| | 4C | Active Colluvium. Thick landslide debris often with toe being eroded seasonally by river. Slope angle > 35°. Actively degrading, highly unstable. |
| | 4D (3B) | Degraded Colluvium. See 3 B but less stable. |
| 5 Recent Side Terraces and Valley Floor <20° | 5A | Higher terraces. Oldest terraces often heavily dissected by drainage gullies. Thick weathering profile up to 3m. Slope angle generally 0°,-20°. Generally stable but can be highly erodable if surface residual soil cover is penetrated. |
| | 5B | Terrace Scrap Faces. Steep, often sub vertical. Actively degrading usually as sudden slumps. Highly erodable forming cones of loose debris. |
| | 5C | Low Terraces. Usually above normal flood levels, but susceptible to periodic high floods. Soil cover generally < 30 cm. Highly permeable and erodable. Slope angle 0° - 10°. |
| | 5D | Flood Plain. Gravel/Cobble banks, highly mobile and covered by annual flood levels. No established soil cover. Highly permeable and erodable. |

Fookes, *et. al.* (1985)

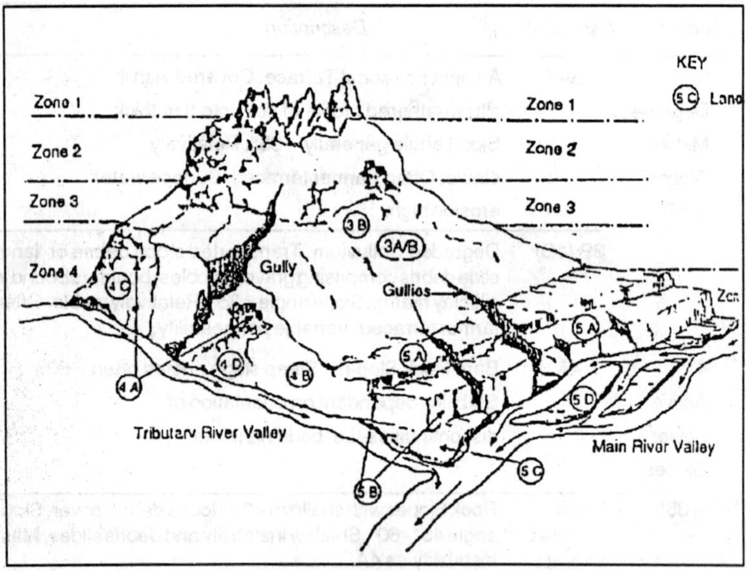


**Fig. 13. : Mountain zone classification; adapted from Fookes *et.al.* 1985.**

The landscape forming process tends to develop in a series of cycles. After a period of earthquake activity and mountain uplift the rivers and watercourses erode the landscape by cutting down and forming narrow incised valleys. The steep sides of these valleys are very unstable and consequently mass-wasting is widespread. These steeper active valley sides are mountain zone 4. In periods of very low tectonic activity the downcutting action of the rivers reduces but mass-wasting continuous until the side slopes to the valleys become shallow and more stable. These are mountain zone 3.

A deposit of alluvium is formed by the river on the valley floor. This is mountain zone 5. When a period of activity begins, the river again cuts down and forms a new narrow incised valley with unstable sides. The deposit of alluvium remains on the hillsides above the new river level and this is called a river terrace deposit. Above the steep sides of the new valley are more gradual slopes formed during the earlier quieter period, these are partly composed of colluvium (old landslide debris) and partly in-situ weathered materials.

**Mountain zone 3 :** Zone 3 comprises ancient degraded slopes now located high above the present river valley. These are relatively stable zones having undergone a long geological period of degradation. The near surface materials are being chemically weathered to red soil at a greater rate than their removal by erosion and consequently the present slopes are generally at angles less than 35° (Land Unit 3A). These areas are usually farmer terraced but they are only locally irrigated from spring sources or from the upper reaches of channels by small farmer canals (gcll). In some areas (Land Unit 3B) colluvium may be present, also at slope angle of less than 35°. This is generally stable but care in drainage design is needed to prevent saturation in wet periods and reinitiation of movement.

**Mountain zone 4 :** Zone 4 comprises the active lower slopes immediately above the existing river and its nearest terraces. The slopes are steep, always greater than 35° and because they have been formed by the most recent phase of river downcutting, are in the process of active degradation. They comprise steep, bare rock faces generally steeper than 60° (Land Unit 4A); steep slopes of between 40° to 60° (Land Unit 4B) in which the rock is covered by weathred rock materials and debris; a range of failed slope forms including mounds of landslide debris at angles steeper then 35° that are progressively unstable (Land Unit 4C); and similar debris that has now degraded to less than 35° and is marginally stable (Land Unit 4D).

**Mountain zone 5 :** Zone 5 comprises the river channel and its flood terraces together with up to several levels of terrace formed in the recent past, the highest of which is up to 86mt above present river level. These terraces are all composed of coarse gravels, cobbles and boulders with sand. The highest and oldest terrace has undergone the longest period of weathering and has been degraded and been uplifted so that the typical flat top fronted by a steep scarp slope is now replaced by a gently sloping surface (20°) (Land Unit 5 A). The upper level of the terrace has been weathered to a red day up to 3 mt in thickness. This becomes thinner on the lower and more recent terraces. These terraces are fronted by steep slopes of partially cemented gravels, cobbles and boulders which are highly erodable and unstable (Land Unit 5B). Very active erosion of these fronted slopes occurs where water courses flow over terrace edge. The more recent river terraces (Land Unit 5C) are two young for much weathering to have occurred and generally consists of very permeable grey-brown sands, gravels, cobbles and boulders. They are a few meters above river level but can be inundated in exceptional floods. Within the river channel, gravel banks occur which are covered each year in times of high flood (Land Unit 5D).

**Appropriate Conservation Techniques**

It is very difficult, if not impossible, to stop the slope instability once the movements start. However, their severity, frequency and magnitude can be minimized by taking appropriate preventive measures. The measures taken for the mitigation of slope instability should be compatible with the local conditions and socio-economic constraints and such that peoples participation can ensured.

**Mountain zone 3** : Zone 3 is characterized by in-sites weathered profile of soil up to 3 mt thick or consolidated landslide debris, shallow slopes (generally less than 35°) and relatively stable with little active landsliding. The problems arises when rapid soil erosion occurs once vegetation is removed or surface water is allowed to concentrate. The preventive measures are as follows :

* Cut soil slope should not be steeper than 45°
* Careful spoil disposal. It is better to distribute spoil in this layers and compact it well; It should be ensured that material rests at its natural angle of repose (35° to 40°) and revegetate quickly.
* A balance in cut and fill should be maintained along the whole construction.
* Measures for revegetation and types and species of plant should be selected according to the farmers experience suitable methods include-for spoil/new fill areas -brush layering or brush matting or live stacking with seeding or live stacking with grass and shrub plantations for newly cut areas-grass and shrub plantations or jute netting with seeding or jute netting with grass and shrub plantations or sodding.

**Mountain zone 4** : Zone 4 land is characterised by steep slopes (generally more than 45°) and is highly unstable with active mass movement and active gully erosion. For conservation purposes zone 4 is subdivided into four land units :

A. In the land unit 4A, the stability of the bare rock slopes, often with slopes angles in excess of 60° is dependent on two factors, one natural and one man made, viz.,

   a. The effects of rock structure, i.e., orientation of discontinuities such as bedding and joints, on the stability of the rock face.

   b. excavation techniques for the construction of roads canals, tracks, rails etc.

   Inappropriate excavation techniques are often cause of slope instability. Hand excavation of open benches with outer masonry walls by chiselling causes little disturbances of the rock face. In the case of rock falls and sliding due to highly joined or foliated nature of the rocks steel and concrete piles in rows can be driven deep to secure anchorage for the unstable bars. On the road sides steel bolts can be fixed to prevent blocks or sheets of rocks from falling or slipping. This measure is to be accompanied by sealing of joints, cleavages and cracks by grouting or other means.

B. In the land unit 4B, there is high risk of mass-wasting as the loose, weak weathered rock layer slides off the underlying sounder rock in a debris slide. The stability of the steeply sloping (45° to 60°) shallow loose debris (less than 3 m depth) of land unit 4B is dependent on good preventive measures. The use of conventional lowland construction technique have caused considerable environmental damage in the area. Although these may work well in the alluvial plains they are harmful when applied to unstable steep slopes. All water courses and channels of the area intercepted and diverted around the crown of the slide or the potentially dangerous site through properly lined deep ditches that have adequate gradient. The springs are likewise entrapped and channelized out through pipes. The deep ditches would have to lined with impervious material such as clay or polyproplyline sheet in order to prevent leakage of water. The open drains should have stilling basins (which serve as shock absorbers) at intervals or their beds made rough and irregular by putting a series of drop structures with adequate aprons. This is done to let the energy of the water current be spent. The ditches may also filled with gravel or graded mixture of gravel and coarse sand to check clogging and choking from debris or slipping and falling of litter in ditches. The slopes should be protected with appropriate vegetative measures, viz:

a. for newly cut slopes in soil :

* brush layering, or
* live stacking with grass and sharp plantation, or
* jute netting with grass seeding, or
* jute netting with grass and shrub plantation, or
* sodding

b. for eroded of degraded natural slopes ;

* live stacking with pioneer plantation, or
* vegetated drystone masonry wall.

C. Unlike units 4A and 4B, land units 4C and 4D are actively degrading and highly unstable and any disturbance in the slope profile will lead to further instability.

Once the reasons for instability have been established the geomorphologist has a choice for several techniques. Fig. 14 shows that the use of rubble drains, both in trench and herringbone form

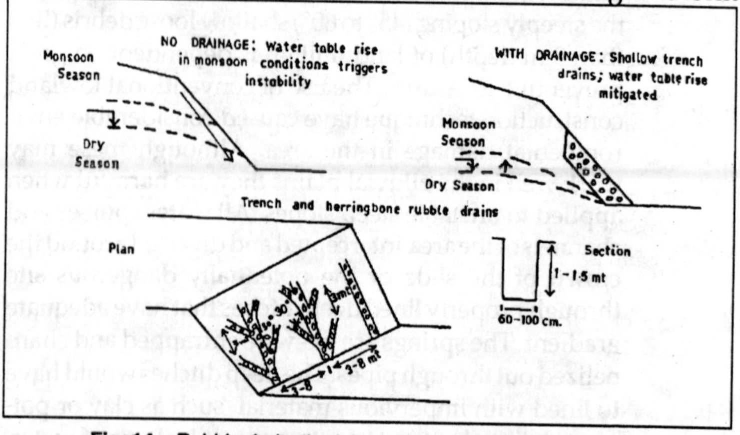


**Fig. 14.: Rubble drains both in trench and herringbone forms.**

and their effect on the drainage of a unstable zone susceptible to high seepage. Rubble drain should be used in combination with the following vegetative measures for improving drainage conditions in the active unstable zone.

* grass and shrub plantation, or
* brush layering, or
* live stacking

River training works include all engineering works constructed on a river in order to control, if possible, lateral shifting, vertical bed level changes and to enable floods and sediments to move effectively. Wherever man utilises the water resources of a river, e.g., for irrigated agriculture, the river is changed from its natural state. These changes often result in the losses of valuable land, property and/or infrastructure which, in turn, justify the construction of river training works to control the river. Scouring is the destructive force in the erosion of the river bank in Sai basin. Scouring is the localised lowering of Sai river bed during times of high flow of around artificial structures that have obstructed the natural river courses. Revetments are constructed to maintain a river bank in its existing position and reduce bank erosion, e.g., toe erosion of landslides. The following measures are the most common form of river training works.

* bamboo fence with sand filled bags,
* rip rap revetment with apron,
* gabion revetment with mattress.

Live wood and aquatic grasses can be planted in the joints of a rip rap and gabion boxes to bind the structure together, thereby turning discrete blocks into a well-anchored mass. This can then become an effective erosion control measure and a productive piece of land, e.g., grass for thatching and mats and green wood for fencing, baskets and furniture, as well as a means of improving the visual appearance of the environment. In using such vegetative measures one should remember to select plant types and species which are able to tolerate periodic flooding and waterlogging, e.g., reeds and willows, viz :

* brush wattles (slope fascines),
* vegetative dry masonry wall/rip rap,
* vegetated gabion

**Mountain zone 5 :** This zone is characterized by river terrace deposits with flat or very gently sloping surfaces and very steep front faces. For prevent on purposes this zone is sub-divided into two groups :

A. The terraces of land units 5A and 5C are generally stable but highly permeable and erodable if surface residual soil cover is penetrated. The younger lower terraces and flood plains are susceptible to seasonal flooding. All bare surfaces resulting from cultivation should be revegetated as quickly as possible to reduce the risk of rainsplash erosion overland flow and rilling. As with mountain zone 3, measures for revegetation and types and species of plant be selected according to the local availability and the farmers experience, viz.,
   * grass seeding, or
   * grass and shrub plantation

B. Like land units 4C and 4D, the terrace scarps of land unit 5B and highly unstable with active instability. However, where construction is required across land unit 5B the use of gabion revetment walls in combination with appropriate vegetative measures is strongly recommended. Once on existing scrap face has been disturbed by either trampling or excavation activities it will only become stable once the angle of repose has decreased to 34° cutting back is generally not possible and so gabion retaining wall should be used as support for the steeper scrap faces (Annex B). Similarly with existing scrap faces which have not been disturbed, it is recommended that slope be protected with gabion revetment walls. This must be carried out if their is a risk of toe erosion.

Gullies occurs throughout mountain zones 3,4 and 5. The gully lead is usually located in zone 3. The most actively eroding part of the gully is usually located in zone 4 and here the gully sides are highly active and should be treated. In zone 5 there is usually a gully accumulation zone where debris is deposited as an outwash fan destroying cultivated areas. Simple and highly effective techniques should be employed to arrest the erosion at the gully head by employing boulders or bamboo planting (Fig. 15). In the

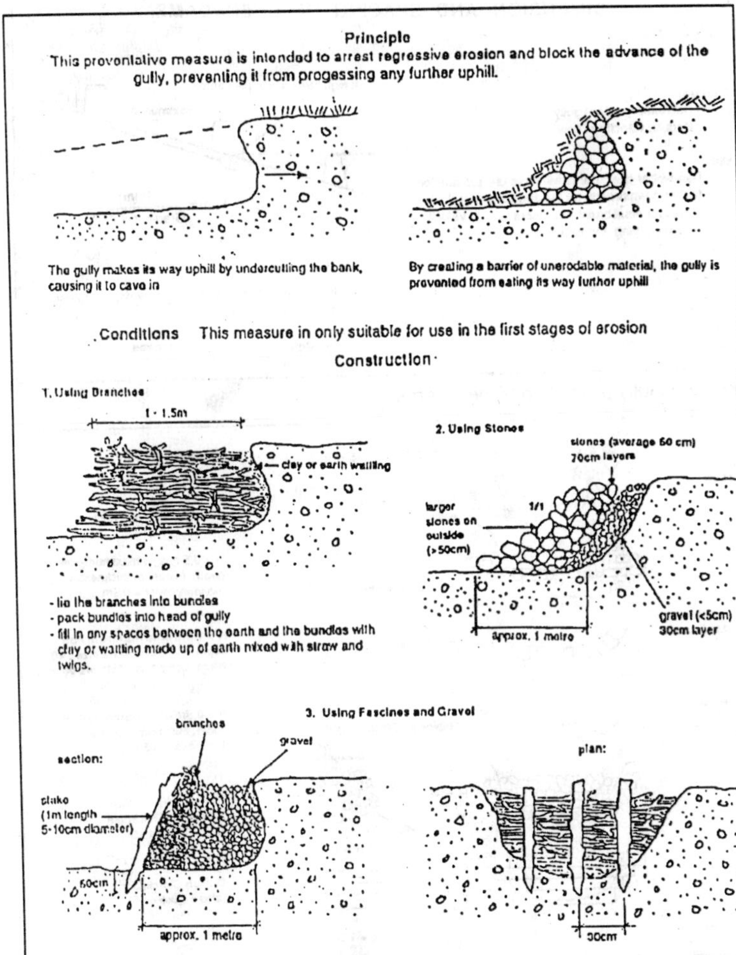


Fig. 15. : Gully protection in the head of a valley; after ILO/SPWP Training Element No. 3 Gully Correction, 1985

more active areas checkdams (Fig. 16) should be used to decrease the gradient of the gully and there by reduce flow velocities. Gully protection structures can include :

(I) Engineering structures

* rubble masonry,
* drystone pitching,

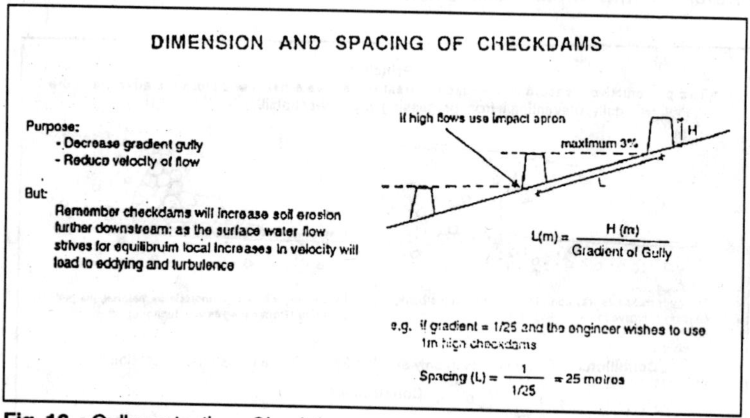


**Fig. 16. :** Gully protection : Checkdams.

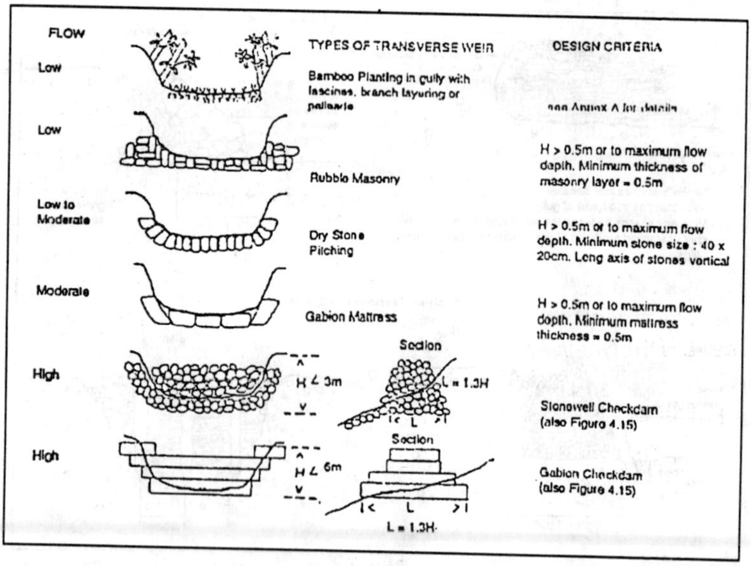


**Fig. 16. :** Structure methods of gully erosion protection; after ILO/SPWP Training Element No. 3, 1985.

- gabion mattress,
- stonewall checkdam,
- gabion checkdam.

(II) Vegetative measures

- bamboo planting,

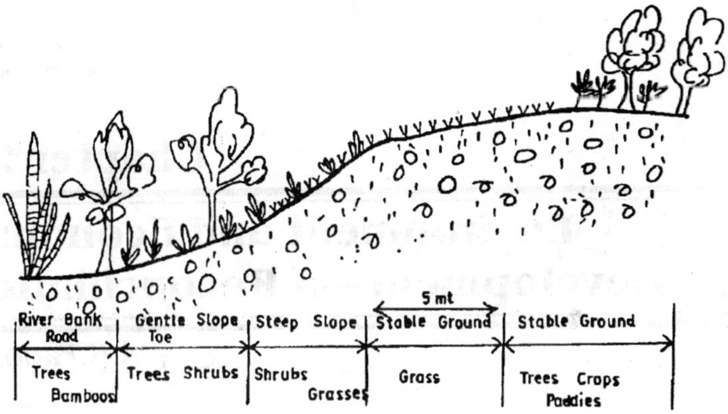


**Fig. 17. :** Afforestation of bare slopes and landslides.

* branch layering,
* vegetative palisade.

The final objective of any slope instability reclamation measure is to establish either a forest or agriculture land. It is necessary to carefully plan which species should be planted in which areas of suitable vegetation cover is to be achieved. An unstable land will always be a very fragile area and it is important that the farmer manage the growth of the vegetation at every stage of development. If the final vegetation is to be forest then certain rules must be followed (*Fig. 17*). This system also applies to road/irrigation cuts and fills or any kind of bare slope.

# Chapter 28

# Environment and Economic Development—A Reconciliation

*A.K.Padoshi*

In recent past, scientists, administrators, social thinkers, and the general public are becoming increasingly conscious of the problem of environmental degradation. There have been numerous suggestions, from many quarters, about minimising the degradation and about achieving economic development without destroying the environment. The present paper offers a few suggestions to reconcile environment and needs of development.

**The Challenge of Development**

The experience of economic development is not particularly happy. In fact the development of the world leaves so much to be desired and so much undone and unfullfilled. During the last 40-50 years or so, the distribution of development world over has been extremely uneven. The benefits of economic development have not been percolating to the developing countries of the world. Most developing countries have been caught in the poverty trap the escape from which appears extremely difficult. According to the World Development Report 1990, around 33 per cent of the population in the developing world is living below the poverty line. In the absolute terms the number of the poor in the world is 1116 million. It is a pity that considerable economic co-operation, mutual assistance and grants to the poor on the humanitarian grounds, the world has not been able to answer the problem of poverty and malnutrition.

In addition to poverty, the developing world is also frequently exposed to the natural calamities like floods, prolonged famines, droughts, epidemics and others. Each time, the calamities take a heavy toll of human life. It is a pity that even after centuries of modern economic development there are many unfortunate pockets in our world today where people still die of hunger and malnutrition. Clearly, the development is still incomplete and it merits serious and urgent attention especially in the Third World.

This, however, does not mean that the developed world is free from the problems of development. The rich countries have to face the problem of development but of a different nature. The rich countries are worried about their peculiar problems like inflation, cyclical unemployment etc. Their anxiety is how to maintain the development that has been achieved by them in the past.

Thus the challenge of development for the rich countries is the question of maintaining the tempo of economic development while the challenge of development in the poor countries is how to escape from the poverty trap. In many parts of the developing world the challenge of development consists in the struggle for just physical survival.

The challenge of development necessitates the exploitation of the environmental resources. But unfortunately, today the mankind is not in a position to go on exploiting the environmental resources indiscriminately for economic development. The rapid expolitation of the environment in the past, has created, what can be called 'The challenge of environment'.

**Challenge of Environment**

Though it is essential for economic development, the exploitation of environment has been posing problems. Development involves changes—qualitative and quantitative—in the economic structure. In its quest for the desired change, the human race has been engaging itself in economic activities that have to go on forever, so long as the human race inhabits the earth. These activities would go on infinitely. The human activities are now capable of altering the bio sphere itself. But unfortunately, the environment on which development depends, is not infinite. It is finite. And this disparity between the infinite economic activities and of mankind and the finite environment creates the 'Challenge of Environment. It consists in echieving economic development without endangering the environment. The mankind has to solve its economic problems within the environmental constraint.

In recent years, environment degradation has become a matter of central concern for both the worlds *viz.* the developed and the developing. The aspects of environmental degradation are nemerous and almost limitless. But the critical ones include global warning, the green house effect, damage to the ozone layer, tropical deforestation, overuse and misuse of pesticides, siltation of dams, soil erosion, desertification of land, exposure of the soil to the dangers of floods and droughts, acid rains, movement and duping of the nuclear waste and so on. Even today we do not know any safe method for the disposal of the radioactive waste. The profligate industrial activities, especially in the developed world, have been mercilessly assailing the environmental resources and disturbing thoughtlessly the delicate ecological balance in the entire world. It has created grave problems before the entire mankind.

Global warming, the possible risks of which are too high to justify complacency of evasion, is no longer a mere theoretical concept but is almost round the corner with stunning certainly. The industrial activities have been emiting greenhouse gases in the atmosphere for a considerably long period. The green house gases mainly consists of Carbon dioxide, methane and chloroflurocarbon (CFC). However, more than half of the greenhouse effect is produced by carbon dioxide produced by burning fossil fuels like coal, oil, natural gas and by deforestation. Fossil fuel combustion produceds over 500 million tonnes of carbon emissions per year. Since 1960, the global concentration of carbon has increased from 316 parts per million to almost 350 parts per million. Most climatologists expect rapid climatic changes which are going to be harmful to the mankind, to take place over the next two decades. The atmosphere in which we live and our future generations would be living, is going to be uncomfortably warm, dangerously carbonised and unsafe for life.

Deforestation in general and the tropical deforestation in particular, which injects 1000 to 2600 million tonnes of carbon into atmosphere spells disastrous consequences for the human race. Deforestation which is indiscriminate is responsible for the loss of forest wealth, destruction of plant and animal life, soil desertification and many others. It is a pity that the rate at which forests are being displaced is much higher than the rate at which they are replaced. The ratio of forest displacement to forest replacement is 5 to 1 in Asia, 10 to 1 in South America and as much as 20 to 1 in Africa. We are literally eating the forest wealth. In many countries, the percentage of land under forests is much lower than the internationally accepted minimum necessary percentage which is around 25 to 33 which is essential for maintaining the ecological balance. (In India the percentage of area under forest is only about 20 per cent of the total geographical area). If forests will be destroyed at this rate, very soon, the earth would be a massive heap of barren, open and desertified soil incapable of growing sufficient food. There would be too many human beings and too few trees to purify the air that human beings will be breathing. Animal life will be running short of vital oxygen. One can easily imagine the consequences of such a scenario.

The environmental degradation, which we have been discussing, is general in character. The effects of such environmental degradation affect the entire world. They transcend the economic, social and also the geographical boundaries. Greenhouse effect, ozone destruction, disposal of industrial waste in water and the resultant impurification of water affect the entire mankind without discrimination. In general, environment degradation exposes the life on earth to grave consequences by making the earth inhospitable, atmosphere hostile and living unsafe.

However, in this context, we must draw a distinction between the environmental degradation caused by the developed, rich and the industrial world and that causes by the developing, poor and the agrarian world. These two degradations differ in their causes as well as their consequences.

It must be remembered that the assault on the environment by the rich world is caused by their desire for future development and to accumulate more and more wealth. The environmental degradation for which the developed world is mainly responsible, includes greenhouse effect, ozone destruction, nuclear waste and general impurification of air and water. It affects the entire mankind without discrimination. The assault on the environment by the poor developing world, on the other hand, is caused by their grinding poverty and by the aspiration of their poor, half clad and undernourished population. It must be noted, in this, context, that for many countries of the developing world, development is not in the real sense. But development for them is just mere survival without any betterment whatsoever in their already pitiably low living standard. These countries causes much of their environmental degradation out of sheer need to survive. And the degradation caused by the poor affects only their local people. Unlike the degradation caused by the rich, the one caused by the poor is incapable of harming the mankind at large.

The difference between the cause and consequences between the two degradations is of vital importance for adopting a suitable strategy for the future economic development in the world and for a reconciliation between economic development on the one hand and conservation of the environment and the ecological balance on the other hand.

Thus, the challenge of development nesessitates the use and exploitation of environment while the challenge of environment makes such exploitation at least undesirable. The solution lies in the reconciliation of the two.

**The Indian Scenario**

The pervasive nature of the environmental degradation has created problems for all countries of the world and India cannot be an exception. The problems created by the environment destruction are quite serious for India and they call for proper understanding, serious consideration and effective tackling. As has been noted earlier the problem for the countries like India is a doubly serious because they have to face the twin problem of achieving development and conserving environment at one and the same time. Many times, due to the conflicting nature of needs of development and the needs of environment India had to face the tough task of making choice out of difficult alternatives. We are forced to decide how much of our future we have to forgo to ensure our today's survival.

When India began her modern economic development around 1950, the environmental resources were generally thought to be plentiful and industrialization meant prosperity and almost unmixed development. By 1990, India has already completed nearly 40 years of planned economic development. What have we achieved in the last 40 years in the field of economic development?

It is a matter of great pride that India has achieved considerable economic transformation during the last 40 years. The country has witnessed industrial expansion and diversification to a considerable extent. India has successfully completed the green revolution in agriculture which has enabled us to achieve food self sufficiency and face successfully the famines that have occurred in the recent past. Similarly, there has also been a remarkable development of the modern tertiary sector of our economy. Educational facilities of all types have expanded considerably. India today, can boast as the third largest country in the world as regards the number of technicians and scientists who are indispensible for modern economic development. As a result of our success in the economic field the life standard of an average Indian has improved remarkably. The rise in the standard of living has happily, reflected itself in the improvement of longevity in India from around 30 years in the 1930s to about 56 years at present.

However, the success story should not make us complacent. We will have to see the other side of our economic performance also. Unfortunately the other side is far from being bright and impressive. In fact, all is not well with Indian economy. Despite planned development we have not yet solved the problems of poverty, unemployment, ignorance, socio-economic disparities and others. Even by our conservative definition of poverty (of about 2400 calories per person per day) Indian had as much as 28 per cent of her population below the poverty line in 1987-88. It must be remembered that those who are just above the poverty line are, by no means, economically comfortable placed. They are just in a position to lead a very poor standard of living. If we relax the definition of poverty a little more and include the bare minimum comforts of life in the definition the percentage of poor people in India is bound to increase several fold. It is unfortunate that even after 40 years of economic planning we have not been able to provide just two square meals to every Indian. In addition to proverty the problems of unemployment, undernourishment, inequalities, ignorance and the life will be staring at us even by the end of the present century. In addition, as has been pointed out by several eminent economists, the economic growth achieved by India, so far, has been slow, expensive, and has aggravated the economic disparities. So, we have the depressing picture of 'poverty amidst plenty'. Clearly, what has been achieved is not enough for us. The condition merits urgent attention.

On the environmental front the condition in India is no less serious. The problems of environmental degradation which are especially serious for India are the problems of air and water pollution and deforestation.

India's urban industrial areas have become as polluted today as those of much, more industrialised countries in the world. Unfortunately our approach about pollution due to industrialization is not that of 'anticipate and prevent' but that of 'React and cure'. As a result of this faulty approach the air and water pollution in our industrial centres like Bombay, Delhi, Calcutta, Madras, Hyderabad and Kanpur has already become dangerous to human life. There has been a serious increase in (i) suspended particulated matter (ii) sulphur di oxide and (iii) dust fall in the atmosphere of these cities.

As regards the deforestation in India, we have to note, sadly, that the agroeconomic system and waste land accounts for nearly 77 per cent of out total land and only 11 per cent of the land is under natural system having protective values. Similarly, out of the total 266 million hectares of land which has some productive use in India as much as 175 million hectares (*i.e.* about 66 per cent) has been suffering from degradations like erosion, shifting cultivation, salinity and alkalinity etc. The area under forest in India has shrunk from 67 million hectares in 1952 to 28 million hectares in 1987. It is only 20 per cent of the total geographical area and much below the world average of 29.5 per cent. The consequences of environment degradation have already started appearing in India in the form of loss of soil fertility, reduction in rainfall, warm winters, exposure of the country to floods, impure air and water and so on. Clearly, we have to face the stiffer and twofold challenge of development and environment in the near future. In these circumstances of dilemma of development vs. environment what is essential is a reconciliation between the two.

**Need of Reconciliation**

The challenges of environment and that of development have posed serious problems and difficult choices before mankind. To meet the challenge of development effectively it is essential to exploit the environment especially in the developing world where the vast humanity is somehow managing to survive with no hope of a better tomorrow for the millions. On the other hand, exploitation of environment threatens to upset the delicate ecological balance with grave consequences for the entire mankind. Clearly, there is an urgent need of meeting both the challenges effectively and successfully. Overemphasis on either development or environment is bound to jeopardise the other. The need of reconciliation between development and environment calls for handling the twin problem of achieving economic development without unduly degrading the environment. This necessitates sound policies about economic development which require a better understanding of the cause of the environmental degradation as well as the socio economic significance of it.

The pervasive nature of environmental problems dictates a new approach and that is the integration of environment management into the economic policy making at all levels of government, designing economic incentives in such a way that they induce environmentally sound practices. Development and environment should be made mutually supportive. Development nations, in particular, will have to find a path of growth different from that traversed in the past by the developed world. Because the developing world can neither ignore development nor environment. Similarly, the rich countries will have to modify their economic behavior, avoid excessive use of resources and minimise the waste. There is a growing view that economic development strategies may need to be rethought and perhaps substantially revised if further degradation of global environment is to be avoided.

The need of the reconciliation between development and environment and the manifold problems that it gives rise to, poses a great challenge to the very relevance of the conventional economic thinking about development which consists in assessing the efficiency of the investment or resource use by the only yardstick of the increase in the national production. It does not consider or taken into account the destruction of the environment caused by the investment. It must be noted that due to the degradation of environment, the effective, real growth of the national product is often less than the one measured by the GDP statistics. Also, many times, the investment which destroys ecology to a great extent, endangers future economic development, and may, perhaps, leave us poorer in the long run. Thus the future is sacrificed for the sake of the present. But the conventional economic theory has no place for such as analysis.

It is necessary to emphasis the macro economic importance of environmental problem. Because, the shortcomings of the national income accounts are fairly well known. There is no accounting for the drawing down of natural resources and the loss of national output due to this drawing down. It is possible that growth based on such a process (which does not consider the loss of resources) may not be sustainable and the impressive national income figures will be relevant only for a short period.

However, if the reconciliation is to provide guidelines for the future public policy it is essential to quantify the costs of the environmental losses. Unfortunately, it is not always easy to quantify the costs of the environmental degradation. For example, the adverse effects of pollution on human health and human life cannot be quantified.

Again, mere assessment of the loss of environment and future development is not sufficent. Because, environmental and economic problems often involve conflicts of interests. Many times strong vested interests may entirely thwart any attempt of reconciliation between environmental and development if it goes against their interests. In most cases of conflict between various interests it is the strong who damage the weak. Therefore, a strong political will is required to overcome the vested interests what may come in the way of the desired reconciliation.

## The Way to Reconciliation in India

In the context of the reconciliation between economic development and environmental degradation, the concept of 'Sustainable Development' has been advocated. Sustainable development depends on managing the resources and not exhausting them. The unilateral or monolythic emphasis on development has been shifted and due attention is paid (in sustainable development) to the preservation of environment as well. Thus sustainable development consists in achieving development without indiscriminately degrading the environmental resources. Thus it stands for the integration of development and environmental policies. Any investment programme, under sustainable development will have to satisfy not merely economic criteria but also environment conservation criteria. However, if there arises a conflict between development needs and environment preservation and if conservation of environment is of the paramount importance, development will have to respect environment. But this is difficult in India where development for million of people is just survival.

In India, a reconciliation between development and environment in the field of industrial development necessitates that industries be dispersed as much as possible. Concentration of industries in only a few localities be avoided to the extent possible. It may be argued and objected that dispersal of industries will result in cost escalation and the industries will not be cost effective. However it must be remembered that the cost which the society would have to bear due to the loss of environment resulting from overconcentration of industries would be too heavy. Moreover, if we examine the benefits of an industrial dispersal from the view point of social, benefit cost it would surely be more beneficial.

As regards agriculture, Indian agriculture would have to play a special role in the development of Indian economy in future. As is well known, despite 40 years of planned economic development, India remains predominently an agricultural economy with agriculture contributing around 30 per cent to the national output and absorbing as much as 65 per cent of the labour force. It must now be recognised that the western model of development in the field of employment which consists in developing the non-agricultural sector for withdrawing labour from agriculture, is not at all likely to succeed in India due to the rapid rise in the already massive size of the labour force (300 million at present) and the decline in the labour intensity in the industrial employment. In India, agriculture would have to bear the burden of creating more jobs in future and thus help mitigate the problem of unemployment. Similarly, Indian agriculture, in future, will have also to help save the country from environmental degradation and also endeavour to restore the loss of environment. Clearly, we need that model of agricultural development which would meet the challenge effectively. The stragegy of sustainable agriculture is suggested for this.

The responsibility of sustainable agriculture, which consists in designing farming practices in such a way as to maintain an optimal yield indefinitely, will be to produce more and meet the requirements of the country without degrading the environment or harming the long term productivity of the agricultural resources. This has become necessary in view of the fact that there are many indications that the gains from the green revolution are rapidly tapering off and they cannot be maintained. There is an urgent need to restore the lost fertility of the soil, produce more for the growing population, spread green revolution to other regions and other crops, generate more employment, remove poverty, take care of the environment and above all, economise the use of land in view of its acute scarcity in the very near future.

With a view of achieving the above manifold objectives, apart from the often suggested remedies like land reforms, irrigation, credit, extension etc. which are extremely necessary and useful, and must be pursued with greater vigour in future, a new technique of farming is strongly suggested. And the new technique is *'Bio intensive mini farming'*. The technique of biointensive minifarming is especially suitable for the labour surplus and land scarce economies and it consists in planting plants closer in small areas. It results in the saving of substantial amount of land, energy, water and money. The technique uses one fourth of the amount of water and one hundredth amount of energy used in the conventional techniques of farming. It does not require costly pesticides and herbicides. The vital difference with the technique of 'Biointensive minifarming' is that the land is cultivated to twice the depth of the conventional cultivation. So, in the biointensive minifarming, the roots of the plants so vertically down (and not horizontally as in the conventional farming) in search of water and food. It, thus, accommodated more plants in a given area of land and thus greatly economises the use of land. It can make even a tiny farm economically viable. It is capable of saving resources, producing more output, generating more employment, help alleviate poverty and stop further degradation of environment. Thus it would be an effective way of reconciliating the development requirements and the environmental requirements. What is needed on our part is an unambiguous policy and sincere action.

## References


Chandler William, U., 1990. Development and Environmental Change. *Economic Impact* 2.

Conable Barber, V. 1989. Development and the Environment. *Finance and Development.*

Financial Express, 16-4-1990.

Krishna Raj, 1988. Ideology and Economic Growth. *Indian Economic Review.*

Mellor John, W., 1988. The Intertwining of Environmental Problems and Poverty. *Environment.*

Oram, P.A., 1988. Building the Agro-ecological Framework. *Environment.*

Singh, S.P., 1989. Present Issues of Environment and Development in Particular Reference to Nehru's Plans for India, *Jawaharlal Nehru Legacy and Legend,* Kumaun University Publication, Nainital.

Warford Jeremy and Partow Zeinab, 1989. Evolution of the World Bank's Environmental Policy. '*Finance and Development.*

# Index

41